CONTESTED WATERS

April R. Summitt

CONTESTED WATERS

AN ENVIRONMENTAL HISTORY OF THE COLORADO RIVER

UNIVERSITY PRESS OF COLORADO
Louisville

© 2013 by University Press of Colorado

Published by University Press of Colorado
245 Century Circle, Suite 202
Louisville, Colorado 80027

First paperback edition 2019
Printed in the United States of America

 The University Press of Colorado is a proud
member of the Association of University Presses.

The University Press of Colorado is a cooperative publishing enterprise supported, in part, by Adams
State University, Colorado State University, Fort Lewis College, Metropolitan State University of
Denver, Regis University, University of Colorado, University of Northern Colorado, Utah State
University, and Western State Colorado University.

∞ This paper meets the requirements of the ANSI/NISO Z39.48-1992 (Permanence of Paper).

Library of Congress Cataloging-in-Publication Data

Summitt, April R.
 Contested waters : an environmental history of the Colorado River / April R. Summitt.
    pages cm
 Includes bibliographical references and index.
  ISBN 978-1-60732-201-6 (cloth : alk. paper) — ISBN 978-1-60732-211-5 (ebook) —
ISBN 978-1-60732-908-4 (paper)
 1. Colorado River (Colo.-Mexico)—History. 2. Colorado River Valley (Colo.-Mexico)—History. 3.
Colorado River (Colo.-Mexico)—Environmental conditions. 4. Water supply—Colorado River Valley
(Colo.-Mexico)—History. 5. Water rights—Colorado River Valley (Colo.-Mexico)—History. I. Title.
 F788.S89 2013
 979.1'3—dc23

                                    2012048830

Design by Daniel Pratt

To my mentors,
Harry Leonard and Gary Land,
For gently guiding me toward excellence

To Harry Leonard,
For instilling in me a love of history
and teaching me that knowledge is only the beginning of understanding.
For demonstrating that kindness could go hand in hand with discipline
and that following a dream is more important than others' expectations.
For teaching me that being a history professor is about so much more than
books, papers, and classrooms.

To Gary Land,
For showing me how to be a teacher and scholar
and how to balance the many demands of academic life with integrity.
For helping me find my own voice in my research and teaching
and for being my mentor and a leader I was always proud to follow.
For teaching me how to persevere in the face of adversity
and steadfastly remaining my friend.

Contents

STANDING ASTRIDE THE COLORADO RIVER is a mass of concrete stretching 660 feet across the deep, sandstone canyon and reaching up its steep walls to a height of more than 726 feet. When it was completed in 1936, the Hoover Dam was the largest concrete structure ever built. Costing $49 million and 112 human lives, this massive triumph of human engineering still inspires awe in the nearly 1 million visitors who view it every year. It symbolizes the human ability to control nature, to harness a river. Its seventeen giant turbines turn the river's power into 4.2 billion kilowatt-hours of electricity each year for California, Nevada, and Arizona.

Below Hoover Dam, the Colorado flows at a gentle pace for almost 300 miles before it is almost completely halted at Morelos Dam. Once the river reaches that point just across the Mexico border, its flow is diverted at a right angle to the west to irrigate Mexicali farms. Only a small trickle makes it through the dam to the delta, reaching the sea on rare occasions. Partially sustained by salty agricultural runoff, the Ciénega de Santa Clara wetlands provide vital habitat for at least six endangered species and migrating birds on the Pacific flyway. Eco-tourists and environmental nonprofits seek restoration of the delta region through bi-national cooperative efforts. Nearby, the Cucapá people live in dwindling homelands and struggle to hang on to their way of life. The Colorado's delta waters are as important to them as the blood in their veins.

At the opposite end, approximately 1,450 miles north of the delta, are the river's headwaters, where snowmelt drains off the Rocky Mountain slopes and gathers in narrow valleys. Tipped westward by the Continental Divide and temporarily pooled in Lake Granby, the Colorado begins its long descent through seven states and into Mexico. Beginning at 9,000 feet above sea level, the river drops all the way to 100 feet at the border. Draining a watershed of more than 246,000 square miles, it provides water to about 30 million people in the arid West.

The story of the Colorado River is filled with tales of conquest, control, division, and depletion. It is, according to scholars, the most litigated and regulated river in the United States, perhaps even the world. While much bigger rivers exist, no other river is more divided and overused. Seven major dams bridge its main channel, with dozens more on its tributaries. Some have called it a disappearing river; a waterway that once flowed as a wild, canyon-carving power and now only trickles at the most southerly end; a river running dry. When one looks at the river today, one thinks that perhaps human settlement was never meant to flourish in the desert.

Today, the Colorado still flows, although harnessed, altered, and reduced. The civilization built on its life-giving water remains while we ponder the implications of lowering reservoir levels. The river's story has always been one of power relationships between peoples and nature. For centuries the river reigned supreme, flowing according to its own dictates regardless of human efforts. Eventually, American ingenuity harnessed the wild river, stored its waters, and divided the spoils. The cities of Los Angeles, San Diego, Las Vegas, Phoenix, Tucson, Denver, Tijuana, and Mexicali all depend on water from the Colorado. Much of US alfalfa and other feed grain crops are grown in the watershed, especially in the upper basin region. A large amount of the table fruits and vegetables sold across the country year-round comes from the lower basin states, and an orange in a Christmas stocking is no longer a special treat.

As the twentieth century drew to a close, power over the river shifted away from traditional centers like local water districts and state governments toward indigenous communities, individual farmers, and the river itself. The concept of sustainability has only recently entered negotiations and policy-making relating to the Colorado River. At present, the pressures of population growth and climate change make real and sustainable management of the river a serious and as yet unmet challenge. Yet our lives are bound to each other. If the river flourishes, so do humans living in the Southwest. If the river dies, so does our oasis-like civilization planted amid rocks and desert where perhaps we were never meant to be.

Only a few previous works are exclusively about the Colorado River. The earliest was Frederick Samuel Dellenbaugh's *The Romance of the Colorado River*. Published in 1902, it is a narrative of the Spanish discovery in 1540 through John Wesley Powell's explorations. Only eighteen years old in 1871, Dellenbaugh accompanied Powell on his river expedition and included his experience toward the end of the book's 386 pages. Other books about Powell's exploration and Hoover Dam include Michael Hiltzik's recent *Colossus: Hoover Dam and the Making of the American Century* (2010) and older classics such as Norris Hundley Jr.'s *Water and the West* (1975). The only truly comprehensive examination of the river's history is Philip Fradkin's *A River No More,* published in 1968 and updated in 1995. In this work, Fradkin examines the long story of human interaction with the river and attempts to raise public awareness of water issues. His lament that the river truly is "no more" represents a tradition of declensionist environmental history that chronicled the decline of nature from pristine to damaged beyond repair.

One of the most important works that set the standard for comprehending western water issues in general is Donald Worster's *Rivers of Empire*, published in 1985. In this groundbreaking work, Worster uses an adaptation of Karl Wittfogel's world systems model to explain the development of the arid West. According to Worster, the West became a "hydraulic society" in which wealth and power were concentrated around the control of water. The ownership of water created a new western culture dominated by powerful water elites. While much more than a simple narrative of the decline of nature, his work nevertheless fits into the category of scholarship that sees a few powerful people determining the history of the western environment, both natural and built.

First published in 1986 and updated in 1993, Marc Reisner's *Cadillac Desert* is another comprehensive examination of the arid West and its water, with two chapters focusing on the Colorado River, the "American Nile." A fascinating journalist's account of personalities and power conflicts, Reisner's social history of western water raised public awareness of limited resources. Also lamenting human damage to nature, Reisner argued that governmental policies of development and reclamation created long-term effects on its natural environment through power structures similar to the ones Worster identifies.

Fradkin, Worster, and Reisner collectively introduced a new kind of western water history that radically changed the way we examine the region today. Previous histories mostly chronicled adventurous exploration or individual heroism in the interaction with a wild frontier landscape. In these works, the individual struggles against the challenges of the West while taming it through much effort and innovation. While parts of these stories are often accurate representations of western history, they fail to communicate what Fradkin, Worster, and Reisner highlighted: the darker power struggles that left the individual behind and the environment irrevocably changed.

Other seminal works that examine western water issues include two edited collections: Gary D. Weatherford and F. Lee Brown's *New Courses for the Colorado River,* published in 1986, and Joseph Finkhouse and Mark Crawford's *A River Too Far*, published in 1991. More recent studies include Donald J. Pisani's *Water and American Government* (2002), Evan R. Ward's *Border Oasis: Water and the Political Ecology of the Colorado River Delta, 1940–1975* (2003), and William M. Lewis Jr.'s *Water and Climate in the Western United States* (2003). Many other studies of individuals or specific issues and western state government approaches to western water continue to add to the overall literature on water and the Colorado River. All of these works give scholars valuable insight into the western story and, taken together, constitute an impressive examination of western water issues.

Yet within the scholarship on the Colorado River, three issues need more attention: (1) the role of western water law in its over-allocation, (2) the complexity of power relationships surrounding the river and western water in general, and (3) the concept of sustainable use and how it has been either ignored or applied in recent times. In this book I argue that while prior appropriation law drove the over-allocation of the Colorado River, power relationships shifted over time from one group to another and have democratized what was once a river of empire. The story is much more complex than a struggle between the common water user and government entities or water elites. Although an understandable evolution of water law in a region where riparian rights were not practical, prior appropriation guaranteed a rush to over-allocate a scarce resource. Because western states wanted as much water as possible for agriculture, urban expansion, and industry, they competed for larger and larger shares of the Colorado River to avoid its appropriation by others. Even though the river basin states did reach sharing agreements at various times and awareness of shortages slowly dawned, basin states continued to push for allocations that overtaxed the river. The compelling question is, why do people press on with unsustainable policies in the face of impending disaster? How does a population change its attitudes, perceptions, and laws relating to a scarce resource before catastrophe strikes?

Although incorrect estimations of the river's flow created serious problems from the start, prior appropriation law meant that no agreement on sharing would control development. Yet as problems with the quality and quantity of water eventually became clear, interpretations of the law's "beneficial use" principle began to change.

In recent decades, environmental activists have successfully argued that maintaining stream flows for habitat was indeed a beneficial use. This new interpretation made the law itself more adaptable, shifting influence over water to other players. Although previously ignored, the riparian habitat and wildlife that depend on it have recently gained power. Once sidelined in the water struggle, Native American communities now hold some influence in the human-to-river relationship. Finally, the river is itself a power with the ability to sideline all other players in the competition for its water. Instead of a neutral setting in which power struggles, heroic triumphs, and environmental damage occur, the Colorado River is a dynamic force all its own, shaping the human societies that have tried to control it.

This book is not intended to be the only authority on the Colorado River but instead a comprehensive synthesis of scholarship and research on what is a very large topic. I believe that to solve present-day environmental challenges to rivers like the Colorado, we must know the history and understand how conversations about rivers have changed over time. To move beyond political rhetoric and historical theories, I believe we should view Colorado River history as a whole within the framework provided by environmental humanities education. As Mitchell Thomashow argues in *Bringing the Biosphere Home*, a place (or a river or watershed) can serve as a "microcosm of global environmental change."[1] In *Earth in Mind,* David Orr asserts that today's environmental challenges begin "with the inability to think about ecological patterns, systems of causation, and the long-term effects of human actions."[2] The history of the Colorado River can, I believe, provide this microcosm through which to study global environmental issues.

To grapple with this complex topic, the book's chapters are organized in two parts. Part I is a chronological narrative of the river's history up to the twenty-first century, with analysis of some long-term issues. Chapter 1 is an introduction to the Colorado River's past, tracing human interaction up through the mid-1940s and illustrating the growth of human power over nature. In chapter 2 I examine the specific tensions among southwestern agriculture, the Bureau of Reclamation, and the disconnect between regional planning ideals and realities. In chapter 3 I discuss the environmental movement's impact on water policy and the river and the fatal flaws in western water law. Chapter 4 presents a discussion of the recent recognition of these flaws and problems of water quality and quantity.

In part II I examine in depth four specific issues related to the river. Chapter 5 presents a discussion of metropolitan perceptions of water and the eight major urban centers that depend on the Colorado. In chapter 6 I focus on the American Indians' struggle for rights and access to the river's water. Chapter 7 is an examination of US-Mexico relations regarding the river and the delta ecosystem the Cucapá people call home. In chapter 8 I look at recent water marketing issues and the tension between viewing water as a social right and considering it a commodity. In the conclusion I highlight recent efforts to address the major challenges of quantity and quality that have always dominated the Colorado River's history. While human actions affect both of these issues, the river itself will ultimately have the final word.

NOTES

1. Mitchell Thomashow, *Bringing the Biosphere Home: Learning to Perceive Global Environmental Change* (Cambridge: MIT Press, 2002), 8.

2. David W. Orr, *Earth in Mind: On Education, Environment, and the Human Prospect* (Washington, DC: Island, 2004), 2.

Acknowledgments

MANY PEOPLE SUPPORTED ME in the process of writing this book. First, I would like to thank the staff at the University Press of Colorado—Darrin Pratt, Jessica d'Arbonne, Laura Furney, Beth Svinarich, and Daniel Pratt—as well as copy editor Cheryl Carnahan, who worked with me and were patient as they waited for drafts. I am especially thankful for the excellent critiques I received from the external reviewers of this manuscript, who so generously took the time to evaluate earlier drafts. I am grateful for the support of Arizona State University and my colleagues at the Polytechnic Campus, especially the faculty head of interdisciplinary humanities, Ian Moulton. I thank the research archive staff who assisted me in this process, especially Director Linda Vida of the Water Resources Collections and Archives at the University of California, Berkeley (now located in Riverside); and the staffs at Penrose Library, University of Denver; the University of Arizona Special Collections Library in Tucson; the Bureau of Reclamation Library and National Archives, Rocky

Mountain Region, Denver; and the Hayden Library at Arizona State University, Tempe. I am also thankful for all the encouragement and support of my wonderful friends and family scattered across the country.

Finally, I dedicate this book to two of my most important mentors: Harry Leonard, my undergraduate professor at Newbold College, England, and Gary Land, my teacher, friend, and first department chair at Andrews University in Berrien Springs, Michigan. Without the help and encouragement of both of these scholars, I would not be where I am today. I am forever grateful.

CONTESTED WATERS

Part I

A River through Time

1

Conquering the Wild Colorado

The River before 1945

OF THE IMAGES THAT COME TO MIND WHEN ONE thinks of the arid American West, one of the most prominent is the Hoover Dam. Constructed from 1931 to 1936 during the most painful years of the Great Depression, this colossal structure symbolized multiple ideals for struggling Americans: the power of humans over the environment, the successful joining of federal power and individual ingenuity, and the validation of American capitalism and democracy in the midst of crisis and doubt. By far the tallest and largest dam on earth when it was constructed, this concrete structure is still one of the most impressive. Although today at least thirty-two dams worldwide are taller and even more have larger volumes, the Hoover Dam remains one of the most inspiring structures in the United States.

While seven large dams and around a dozen smaller ones straddle its banks, the Colorado River itself is less impressive. The Colorado River ranks seventh in length and watershed size in the United States, but its somewhat meager flow places it far below at least twenty other American rivers. The importance of this long and unpredictable ribbon of water with an erratic cycle of flood and drought, however, cannot be overstated. At least 30 million people depend on the river's waters.

As the River Flows

Before people dammed and harnessed it, the Colorado was a wild and unpredictable river, prone to cyclical floods and drought according to the seasons. The main source for its water comes from snowpack in the Rocky Mountains that melts in the spring and summer and pours down into valleys, going wherever gravity and landscape take it. Flowing through seven southwestern US states and two Mexican states, the river and its many tributaries drain approximately 246,000 square miles. With an average flow of 15 million acre-feet (MAF) annually, it is the lifeline of the entire region.[1]

Along the 1,500 miles of river are many canyons, the deepest of which is the Grand Canyon, one of the only natural landmarks on earth that is visible from space. There are many other canyons, gorges, and a variety of landscapes along this long river. Close to seventy tributaries feed the main stem, but the four primary ones are the Green River in Wyoming, the Gunnison in Colorado, the San Juan that passes through New Mexico into Utah, and the Gila in Arizona. The Colorado's waters once contained one of the world's largest numbers of fish species native only to its ecosystem. Biologists assert that at least sixteen unique species once lived exclusively in the Colorado River.[2]

Water comes to the Colorado River watershed primarily from two sources: precipitation and accumulation of snowpack in the Rockies, and the summer North American monsoon. These two annual events create interesting seasonal and inter-annual hydrological variability in the basin. When snowpack melts in spring, the water seeps into the ground, recharging the aquifers, while the remainder flows down into streams that feed rivers—including the Colorado. As this melting snowpack increases water flow, the Colorado has experienced dramatic spring floods, at least during the years before human alteration of the river. In the summer, the monsoon is a change of wind pattern that brings moisture from the subtropical Pacific and the Gulf of Mexico up into the lower basin region. Most of the river's annual flow comes during the months of April to July from melting snowpack. Monsoon storms add to the flow of the main stem and tributaries in the lower part of the basin in July and August.[3]

In addition to the seasonal changes in flow, there are inter-annual variations caused by atmospheric circulation patterns, including El Niño or La Niña events, as well as the Pacific Decadal Oscillation (PDO). An El Niño event is caused by trade winds

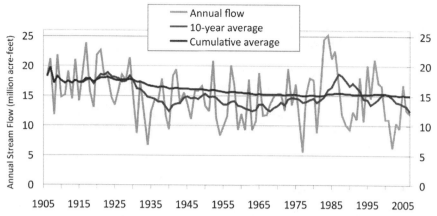

FIGURE 1.1 *Natural flows of the Colorado River at Lees Ferry, Arizona, from 1905 to 2005. Courtesy, Jeff Lukas, Western Water Assessment; data from US Bureau of Reclamation, http:// www.usbr.gov/lc/region/g4000/NaturalFlow/current.html.*

that blow west across the tropical Pacific Ocean, creating a warming effect. Lasting generally between six and eighteen months, El Niño usually creates increased precipitation and flooding, especially in the lower Colorado basin. La Niña events are periods when trade winds increase over the eastern and central Pacific, generally causing colder ocean temperatures. This event often results in a much lower amount of precipitation, with less snowpack in the mountains and low river flows or drought. The less familiar PDO event creates periods of warmer or cooler temperatures in the Pacific for much longer cycles of 30 to 50 years. The water will be either predominantly cool or warm for stretches of 15 to 25 years during one or the other variation.[4]

These details mean that the Colorado River watershed has a natural pattern of variations in climate and precipitation both within years and within long stretches of time, creating cyclical wet periods alternating with long and sometimes very intense drought. During the early twentieth century, when records of the river's flow were kept, a lot of variation occurred. Between 1905 and 1922 the river's flow was high, averaging 16.5 MAF. More recently, US Geological Survey records show that the highest flow occurred in 1984 at 22.2 MAF and the lowest was in 2002 at only 3.8 MAF (see figure 1.1).[5]

Currently, the flow is affected by climate change as well as other factors such as dust. Recent studies have shown that when dust covers snow because of high levels of agriculture and overgrazing, the snow melts faster, causing a quicker rate of evaporation from vegetation and resulting in less runoff. Climate change studies predict that increasing global temperatures could affect the Colorado River by lowering flow levels between 7 and 20 percent.[6]

Although current figures are disturbing, understanding the river's past is important for finding solutions in the present. Beyond expected climate change, other actions of humans specific to the river have changed its nature, altered its flow, and impacted the entire bioregion. What follows in this chapter is a tracing of the early history of the Colorado River up to 1945. Within this story, one can see two distinct phases in the human relationship with the river. In the first, explorers and settlers wrestled with the wild river, finding themselves at the mercy of its whims. In the second phase, engineers and politicians took control of the Colorado during a great dam-building era and bent it to human will. While this apparent victory of humanity over nature symbolized for westerners the triumphant march of progress, conflict over shares of the Colorado's life-giving water soon dominated the human partnership with the river. The relationship became unsustainable.

Early Exploration

Scholars assume that the first non-indigenous people to see the Colorado River were the Spanish conquistadors Coronado sent north to search for the fabled "seven cities of gold."[7] The first written record of the Colorado River comes from Francisco de Ulloa, who explored part of the river's mouth at the Gulf of California in 1539. Sent by Hernán Cortés to explore up the Pacific Coast, Ulloa is credited with drawings of the Baja Peninsula that made cartographers assume California was an island. The next record comes from another Spanish explorer, Hernando de Alarcón, who worked with Coronado's exploration of western North America. In 1540 Alarcón explored the Colorado River up to the present site of Yuma, Arizona.[8]

A few other Spaniards explored the river right after Alarcón's expedition, including Melchior Díaz and García López de Cárdenas. The latter was the first non-indigenous person on record to view the Grand Canyon. After the 1540s, no other exploration of the area entered the records, but cartographers illustrated the mystical region and gave the river several names. The most common name from the Spanish maps and carried on by other European mapmakers was *Rio del Tizon*, meaning "River of Embers." This name was presumably given by Díaz, who named the river for the way he saw local native peoples staying warm in cold weather.[9] Eventually, the name *Colorado* or *Colorade* began to show up on various maps of the region, sometimes referring to the Gila River or other tributaries and sometimes to what we know as the Colorado today. By the 1740s, some maps started to replace the name *Tizon* with Colorado, Spanish for "red river."[10]

In the years that followed, the Colorado River basin region remained largely unexplored and unmapped. Most maps of North America showed a large blank in the Southwest, with only a few notations and incorrect assumptions that a river led to an enormous inland lake or that California was an island. Some fur trappers and explorers went down part of the Colorado River around 1812, not long after the explorations of the Missouri and Columbia Rivers by Meriwether Lewis and William Clark. Knowledge of what became known as the "South Pass" through the Continental Divide was shared among trappers during these early years but had been largely forgotten a little more than a decade later when Jedediah Smith began his travels through the region. Smith rediscovered the pass in 1824, and other mountain men and traders started using this crossing, which took them in and out of the Colorado River basin on their trek to California and the Pacific Coast.[11]

The South Pass would eventually provide a major migration route for more explorers, gold prospectors, Mormons, and many others who went west in the years after Smith's travels. Passage along that route took many across the Green River, the major northerly tributary of the Colorado River. Eventually, maps began to show more and more of the Colorado's headwaters and tributaries but still left blank a great deal of space in the river's middle. Once the Mormon migration began in 1847 and the California Gold Rush started in 1849, caravans of people moved across the upper basin of the Colorado River through present-day Wyoming. In later years, Mormon migration south into Arizona would bring them across the river at other places far below the Grand Canyon.[12]

In the decades that followed, more than 300,000 settlers migrated to California, passing through the Colorado River basin region. No one stayed there; the goal was to get through the dry and desolate country as quickly as possible to reach California, the Promised Land. At the end of the Mexican-American War in 1848, this desirable paradise became part of the United States by the Treaty of Guadalupe Hidalgo. Forced to the peace table, Mexicans agreed to cede their claim to the North American Southwest—including all of present-day California, Arizona, New Mexico, Utah, and Nevada. Texas had already gained independence from Mexico a little more than a decade before, and in 1848 it was officially annexed. With the exception of a small piece of land along the present Mexico-Arizona border known as the Gadsden Purchase of 1853, the geographic shape of the United States was complete.[13]

As various motives pushed and pulled Americans to the far West, an interesting kind of organization evolved as a way to find order in wild chaos. Although it would be challenged and fought over, a new water law developed to address the region's specific conditions. In the rest of the country and in much of Europe, riparian water law formed the basis of most water rights. Growing out of English common law traditions, this riparian doctrine stipulated that whoever owned the land adjacent to a water source owned the right to a proportional amount of its water. Such rights also implied that water users would not impede the rights of others who used the same water source. Under this law, water could not be transferred outside of a watershed or be sold separately from adjacent land.[14]

This principle worked adequately in the eastern part of the country but not in the West, where water was often located many miles from where humans lived and farmed. Growing out of ad hoc rules gold miners had settled on to keep the peace, the doctrine of prior appropriation meant that whoever arrived first and mined the gold or land or water had priority rights to it. These miners' codes eventually became legal statutes, and the influence of prior appropriation is evident in the 1862 Homestead Act and the 1872 General Mining Act. As it evolved, whoever had physical control of water had the right to divert it to any "beneficial use." This stipulation was meant to ensure that individuals or companies did not hoard water, a scarce resource in the arid West. One could not simply "own" water and keep others from using it. Beneficial use dictated that whoever diverted or claimed the water had to actually use it for some tangible and beneficial purpose.[15]

One of the assumptions behind prior appropriation law and particularly the beneficial use principle was that water users would seek legal redress if another user took more than his or her share or polluted water used by those downstream. One of the problems with this assumption is the cost of lawsuits. Financial costs of litigation might be higher than the costs of dealing with pollution, for example. As the twentieth century progressed, it became increasingly difficult to identify pollution sources and thus the party to pursue in court.[16] Prior appropriation law encouraged people to use more water, regardless of whether they needed it, before someone else took it without enough incentive to ensure protection of other users' interests. This unsustainable relationship with the river remains the basis of present western water law.

Although western migration and exploration were slowed by the outbreak of the Civil War in 1861, the years that followed saw huge waves of migration west across the Colorado River. Thirty-six-year-old Civil War veteran John Wesley Powell became the river's next and perhaps last great explorer. In 1869 he led an expedition down the Green River to the main body of the Colorado River, determined to explore the length of the Grand Canyon and complete the map of this great southwestern river. Against all odds, he and most of his men made it through alive to tell their stories. Powell led another expedition in 1871 and published detailed records and maps of the river after that journey.[17]

In the years following his expeditions, Powell used his expertise as a geologist to work as director of the US Geological Survey. In this position he was responsible for further exploration, mapping, and planning for irrigation and land use. In 1878 Powell issued a report on the American West and his views of its needs and potential. This landmark study was largely ignored at the time of its publication, but within it are the foundations for reclamation projects in the twentieth century.

In his report, Powell paid specific attention to the Colorado River basin, labeling it the "arid region." He further subdivided the basin into districts he called irrigable lands, forest areas, and pasturelands. These three areas would be valuable for different reasons, but he did argue that low areas near streams—irrigable lands—would prove amenable to agriculture with the right irrigation support.[18] Early in his report, he asserted that agriculture might actually never work well in the Colorado River basin because it experienced regular periodic droughts.[19]

Powell's main argument, however, was that the arid region of the West would never succeed agriculturally without large-scale government planning. He argued firmly that individual farmers would seldom be able to construct or afford the necessary irrigation infrastructure. He used the Mormons as an example of successful irrigation, with a church organization to organize and fund irrigation projects. Powell believed that without this kind of structured planning, most western irrigable lands would never be used successfully.[20] He further argued that large water storage reservoirs would have to be constructed to ensure adequate water supply year-round. In general, however, Powell's vision was one of yeoman farmers settling on homesteads in the West and cooperating with others in creating water districts that would conform to basic government laws for western water. He warned against a lack of control and advocated passing strict laws on land allocations, keeping land grants small enough to properly irrigate, and creating strong oversight of all water issues.

Powell's advice was heard but largely ignored because representatives of western regions and states did not like his suggestions. They wanted government support for large-scale projects, and, unlike Powell, they foresaw transporting water out of one watershed to another when needed. Economist Lisi Krall has argued that while Powell understood the need for cooperation, he had a naive belief in the notion of a kind of "enlightened capitalism where people are sensitive to the limitations of the land."[21] In fact, market capitalism assumes individualism and accumulation of property and, in Krall's words, "alienation from the land rather than sensitivity to it."[22] Such an alienation or disconnection from the land also meant a similar disconnection from the water that made that land usable for agriculture.

In the years following the Civil War, the country turned its attention to the Industrial Revolution and economic expansion. Fed by unlimited immigration from Europe, northern industrial cities became crowded and overstretched, leading many to look west for better conditions and opportunities. The passage of legislation in

1862 authorizing the transcontinental railroad and its eventual completion in 1869 opened the West to further settlement and enterprise. Although gold rushes continued sporadically, those seeking fortunes in the West turned more often to cattle ranching or farming.[23]

At first, settlers moving west passed hastily through the land between the Mississippi River and California, but between 1870 and 1900 they began to stop and stay in the Great Plains. The 1862 Homestead Act made access to public land a basic American right, and many homesteaders headed out to try their hands at farming the West. Free land was there for the taking, and if someone could survive and develop a 160-acre homestead for five years, the land became theirs. Eventually, the best of these grasslands were claimed, and any further pursuit of free land would have to occur in the arid Southwest. Individual farmers, however, could only use these lands if major reclamation projects provided water through irrigation.[24]

Many believed reclaiming the deserts of the Southwest was necessary for the entire nation. Frederick Jackson Turner's 1893 speech on the closing of the American frontier simply confirmed a pervasive fear that there was no more free land.[25] If there were no more frontiers for individuals to claim and tame, the problems that accompanied industrialization and massive immigration in the East could not be alleviated. A quasi "back-to-nature" movement began in which people looked back nostalgically at an earlier time before the Industrial Revolution. Those who observed the unrest of labor unions and the political agitation of sharecropping farmers hoped reclamation in the arid West would release some societal pressure.[26]

To further settle the West and respond to pressure from a growing western presence in the US Senate, Congress passed the Desert Land Act in 1877. In this first federal act to affect western reclamation, the government gave land grants of 640 acres to married couples (320 acres to single males) if they could prove that they would irrigate the land within three years. Congress thought this act would attract the right kind of settlers with some capital to construct irrigation systems for their land grants. From the beginning, however, there were problems with fraudulent claims and simple underestimations of the challenges of irrigation works. Very few land grants were irrigated through this act.[27]

While John Wesley Powell argued for locally controlled, farmer-run irrigation districts in the West, congressional members from the new western states argued for government support. Then came the particularly painful economic depression of 1893–98. Boom and bust in the US economy had long been the norm, but the length and depth of this depression made many assume that Turner was right. Some of the most violent labor strikes in US history occurred during this economic crisis, and the number of farmers who lost their land and became tenants increased exponentially. To make matters worse, a severe drought hit the western states from 1893 to 1895.

Many in the West already felt the region was the country's poor stepchild. During the nineteenth century the US Congress had put large amounts of money and effort into reclamation projects for eastern rivers and streams. While there had also been government focus on the West, there was always a feeling that the region was last in line for federal spending. Mapping and surveying projects in the West, the building of military forts and railroads, and encouraging settlement through the Homestead Act did seem to focus a fair share of government attention and money on the West, however. Some scholars argue that the West had become accustomed to federal money during the crisis in the 1890s and now expected such subsidies.[28]

As ranchers and farmers struggled during western droughts, many fled Plains states like Kansas and Nebraska for either homes back east or new farms further west. Taking advantage of the opportunity, Wyoming and Idaho encouraged farmers to move to their states, but they wanted to find a way to provide irrigation so they could absorb more people. Wyoming engineer Elwood Mead eventually gained the support of Wyoming senator Francis E. Warren, who began sending various irrigation bills to the US Congress. Named for Senator Joseph Maull Carey of Wyoming, who first introduced the bill in 1892, the Carey Act (also called the Federal Desert Land Act) was finally adopted in 1894.[29] Warren did not believe the government would build dams and canals, so he lobbied for the government to grant public lands to the states so the lands could be sold and the profits used by private companies to build them. The provisions of the Carey Act gave western states millions of acres in 160-acre homesteads to sell for small fees.

The Carey Act failed the same way earlier attempts had. Western settlers were often unable to manage the irrigation projects they needed and gave up their claims to move on to better, watered pastures. While some private companies managed to set up irrigation projects in various places, many others failed, convincing even more congressmen that more federal money and action were needed to build dams and canals in the arid West.[30]

Even before the federal government officially began its western reclamation projects, several individuals tried to create irrigation works from the Colorado River. One of the 49ers who came to California looking for gold was Dr. Oliver M. Wozencraft, who believed he could divert the river into what was known as the Colorado Desert, a very hot region near an ancient lake bed. Also called the Salton Sink (Lake Cahuilla by the region's indigenous people), the lake had dried up many centuries before and the surrounding soil was very rich, excellent for agriculture if only it could be irrigated.[31] In 1856, geologist William P. Blake, who worked for the Pacific Railroad

Survey, made a similar observation, asserting that almost anything could grow in the soil. Wozencraft probably read Blake's observations, published in 1856, and persuaded the state of California to approve his plan to develop about 1,600 square miles of the desert. However, in spite of long efforts and numerous bills introduced, Congress never funded his scheme.

The next serious attempt to divert Colorado River water to irrigate the desert occurred in the 1870s when Californian O. P. Calloway developed a plan for what is now known as the Palo Verde Valley. He needed financial help for his plan, however, and finally found an investor named Thomas Henry Blythe. Blythe had purchased land in 1851 that later became the main downtown area of San Francisco, and he became very wealthy from his investment. Although Calloway did not live long enough to see any real progress on his dream, Blythe began filing claims for land and water rights along the Colorado River. By the early 1880s he probably owned around 175,000 acres of land and 400,000 miner's inches of the river's water.[32] For whatever reason, he never cultivated any of it.

On the other side of the Colorado River delta, one of Blythe's former business partners was Mexican businessman Guillermo Andrade. Blythe bought into Andrade's struggling company, and both men continued to purchase land along the river until they owned nearly all the land on both banks from the US-Mexico border to the Gulf of California. In 1883, just when it looked as though the two men had an enormous region ready for development, Blythe died abruptly. After a lengthy legal battle, Andrade obtained legal rights to a large portion of the original landholding.[33]

In the mid-1890s Charles Rockwood, like Wozencraft before him, also thought he could divert the Colorado River into rich soil in Southern California. Eventually, he found a funding partner named George Chaffey, a private irrigation engineer. Together, the two men planned for a canal to divert water from the Colorado River into the flat desert valley. An old flood channel known as the Río Álamo seemed the best route for a canal and by far the cheapest to build, but it ran through Mexican land. Once Andrade's legal claims had been settled, he sold Rockwood and Chaffey the right-of-way for the canal. In 1901 the Mexicali Valley in Mexico and the California desert region became prime agricultural land, thanks to the Colorado River.[34]

Renamed the Imperial Valley, the rich land on the US side began to draw settlers. More than 7,000 came in during the first three years, many from Arizona's Salt River Valley. Developers jumped at the opportunity. Soon, three boomtowns sprang up, and railroad construction began to connect them to markets outside the region. Rockwood, Chaffey, and the other partners of the new Colorado Development Company (CDC) must have felt elated by their almost overnight success.[35]

What happened next surprised nearly everyone. The Colorado River is one of the most sediment-filled rivers in the world, and Rockwood's canal and its irrigation ditches were quickly clogged with silt. It was the CDC's responsibility to clear the channels, but it was a losing battle. As more and more land was cultivated and more water demanded, more silt ran through the channels; by the summer of 1904 the canal was almost completely blocked. The worst area was the spot where the river intake connected to the canal; although several bypass intake channels were cut, each one quickly silted up. Finally, Rockwood decided he had to cut a new and larger bypass intake just below the border. Mexico agreed to the new intake with the understanding that 50 percent of the canal water would go to farmers in the Mexicali Valley. Rockwood had no choice but to agree.[36]

The new, temporary channel's gates were flimsy, meant to function only until the old one was dredged out. Before that could be done, spring floods pushed abnormally high amounts of water down the Colorado. The weak gates gave way, and the entire Colorado River left its channel and flowed into the Imperial Valley. By midsummer 1904 over 90,000 cubic feet of water per second were pouring into the valley, which quickly became a lake. In desperation, Rockwood asked Edward Henry Harriman of the Southern Pacific Railroad for help. Harriman agreed, but only if he was given complete control of the CDC. Reluctantly but with little other choice, Rockwood handed him control.[37]

For the next year and a half, Harriman paid engineer Harry Thomas Cory and a team of Indian workers to stop the river. Over and over, the powerful Colorado washed away all their attempts and continued to pour its full flow into the valley. Finally, in February 1907, Cory's men managed to plug the wall of the riverbank with tons of rock and stop the flow.[38] The river returned to its natural channel, and the crisis had ended. Thousands of acres of farmland had been destroyed by water and silt, and a new lake had been created.[39] By the 1920s the Salton Sea had become a major recreation and tourist location. In recent decades, however, the sea's health has been threatened by increasing salinity levels and nutrient pollution from the irrigation runoff that feeds it. Causing eutrophication, these high levels of nitrogen and phosphorus feed excessive algae growth that depletes the water's oxygen levels.[40] The Imperial Valley is still one of the most important agricultural locations in the country. This first diversion of the river was plagued with disaster, but later canal projects along virtually the same routes annually flood the Imperial Valley with approximately 3.1 MAF of water each year.[41]

To many, Rockwell's disaster in the Imperial Valley demonstrated the need for federal policy and support for reclamation of the desert. The late nineteenth century had been fraught with economic depression, drought, labor unrest, unrelenting immigration, and other side effects of the Industrial Revolution. Some Americans began looking back nostalgically to an older "Agrarian Myth" as a cure for some of these side effects. What the country needed in the new century was more independent farmers and more attention to the good in nature.[42] Some of the movements in the late nineteenth and early twentieth centuries—the Back to the Land Movement, the City Beautiful Crusade, the Parks and Playground Campaign, and others—addressed a loss brought about by urbanization, the loss of a connection to nature.

In some ways, the creation of a federal reclamation policy was an outgrowth of these ideas: that America's salvation might be found in irrigating arid land in the West. There was a very large, unsettled portion of the country to which the Turner thesis could still be applied, if only it had water. Supporters of federal reclamation projects believed reclaiming arid western lands would provide an important safety valve for America's cities. Some individuals, such as George H. Maxwell and William E. Smythe, tried to apply social planning principles in the West and hoped federal reclamation would allow cheap land to be reclaimed for small farmers. They argued for the passage of a bill that would bring federal help for constructing dams and canals and distributing land to help settle poor urbanites on small farms.[43]

People like Maxwell and Smythe made a number of assumptions about the West and its water. First, planners assumed there was enough water for large-scale settlement without worries about shortages. The West was enjoying one of the wettest decades in the entire twentieth century, so no one thought of shortages. There was also no awareness of the role evaporation would play in future scarcity. While the idea of damming and storing water sounded good on paper, collecting water in a lake increases the surface area exposed to heat and wind. It is estimated that Lake Mead loses approximately 800,000 acre-feet per year to evaporation—almost 2.5 times the amount of Colorado River water allocated to Nevada.[44]

By 1900, both the Republican and Democratic Parties were supporting some kind of federal reclamation policy, although no one could agree on the shape it should take. Western senators worked very hard to emphasize the need for federal help for arid lands and used some of the same arguments other social planners did to justify the projects. While many eastern politicians opposed such a bill, the ability of western members of Congress to filibuster and block other legislation finally caused the opposition to crumble. With support from President Theodore Roosevelt, the Reclamation Act of 1902 (sometimes called the Newlands Act) passed with the help of Francis G. Newlands, US representative from Nevada. He and other western politicians were mostly interested in what reclamation projects could do for the West, while Smythe and Maxwell believed they could save the entire country.[45]

In short, idealists who hoped to conduct some social engineering or "home making" envisioned a specific future for the Colorado River. Irrigating the West would allow small families to create homes independent of the fluctuating value of life experienced in crowded eastern cities.[46] Some, such as Frederick H. Newell, who later became the Bureau of Reclamation's chief engineer, believed there was a shortage of good farmland in America and that population pressures made creating new farmlands necessary. Irrigation projects would solve the major societal issues of the day.[47]

Others supported federal reclamation because of political pressure. Even Roosevelt supported reclamation at least in part because he needed western support in Congress. The whole idea of federal reclamation also appealed to him in ways it would not have to earlier presidents. Roosevelt appreciated the attitudes of those who saw independent farmers carving successful enterprises out of the desert. All they needed was a little water, and eventually it would all pay for itself.[48]

Taming the River

Thus, with presidential support and a strong western lobby in the US Senate, Congress passed the Reclamation Act, setting aside federal funds from public land sales in the West for irrigation projects in sixteen western states (seventeen when Texas was added in 1906). Irrigation projects would be spread fairly, and money from land sales would be used for projects within those states. To administer the project, the act created the United States Reclamation Service, housed in the US Geological Survey. A few years later it was renamed the Bureau of Reclamation and housed in the Department of the Interior, where it remains today.[49]

Those involved with the Bureau of Reclamation in its early years had high hopes for its success. Roosevelt, Newlands, Newell, Smythe, and Maxwell all saw the new bureau as the answer to many social challenges of the time. In retrospect, however, the bureau never fulfilled its social planning goals. Smythe, Maxwell, and Newell eventually admitted that their ideas for the bureau had failed. While Smythe and Maxwell blamed the bureau for abandoning their goals, Newell blamed the farmers. The fact that so many gave up and left their project farms seemed to him a failure of spirit. He also admitted that perhaps there had never really been a shortage of farmland in America at the turn of the twentieth century. Land speculation created uneven settlement patterns and soaring costs for farmers, who often defaulted on their loans.[50] As the bureau abandoned social engineering for dam building, Newell viewed the 1902 act as an unnecessary and expensive failure.[51]

In spite of what some saw as failure, the Bureau of Reclamation became one of the most important players in the story of the Colorado River. While the bureau created many other projects on other rivers and in other river basins, its work on the Colorado continues to be its largest and most significant effort. The Gunnison (later the Uncompahgre) River Project in Colorado was the bureau's first diversion of a Colorado River tributary to provide irrigation for area farmers in western Colorado. Efforts were already under way to have private companies and the states build canals to irrigate the Uncompahgre Valley. Federal assistance made it possible to turn the project into a much grander one, not fully completed until 1925.

Although the first major dam construction by the Bureau of Reclamation would be Roosevelt Dam on the Salt River in Arizona, this first diversion of the Colorado happened where it did because of political connections. Congressman James Shafroth of Colorado was a member of the House Committee on Irrigation and had helped Newlands draft the Reclamation Act.[52] The first major phase of the project—a diversion dam across the Gunnison River and a diversion tunnel to the Uncompahgre River—was opened and dedicated by President William Howard Taft in 1909. It was a difficult project from the start, drilling through 2,000 feet of rock filled with water that kept interrupting the workers. At the cost of twenty-six lives and just under $3 million, water from the Gunnison River finally flowed to farmers.[53]

At nearly the opposite end of the Colorado River, the second major diversion of the river was on the Gila, another major tributary. The Gila River empties into the Colorado at the present-day city of Yuma in the southwest corner of Arizona. Although the Hohokam and later Euro-American settlers farmed the valley, periodic flooding and extreme drought made regular success difficult. However, very soon after the Bureau of Reclamation was created, Yuma became a candidate for federal help. Approved in 1904, the Yuma Project soon boasted irrigation canals, and construction on the Laguna Dam began soon thereafter. Built from 1905 to 1909, it was the first federal dam on the Colorado River and diverted water to farmers in the Yuma Valley. Upkeep of the earthen dam proved costly, however, and Imperial Dam, built during the 1930s, eventually replaced it.[54]

Over the next few years, federal reclamation took on projects in almost all the western states to garner support in Congress and prove its real value. Newell hoped that extending water projects as fast as possible would enable small farmers to settle, preventing land speculators from gobbling up large tracts. His actions, however, only spurred land speculation and drove up land costs to exorbitant levels. Many farmers defaulted on their debts and left the Bureau of Reclamation with the total cost of water projects.[55] While grappling with these issues, bureau engineers continued earlier surveys for a possible dam and large reservoir on the lower Colorado River—one that would provide enough water storage for Southern California's rapidly growing population and farming enterprises. Los Angeles needed such a dam to provide electricity, and engineers soon identified two possible sites. Before any construction or even serious planning took place, however, another issue needed resolution: water distribution.

For many decades, Southern California's population had been growing, Los Angeles's in particular. By 1913 the city's leaders feared a water shortage and eventually constructed the Los Angeles Aqueduct to bring water down from the Owens Valley (see chapter 5). Farmers in the Imperial Valley, however, needed a better water supply to maintain their profitable agriculture. Water still flowed from the Colorado River through repaired channels that took water south of the border with Mexico and then back north into the valley. However, in 1910, Mexico began a revolution that turned into a civil war, lasting in varying degrees of severity until the early 1920s.[56] Because of all the chaos south of the border, California politicians began seeking support for a canal to the Imperial Valley—which they named the All-American Canal—that would be completely within US borders. The Imperial Irrigation District (the new name of Rockwood's CDC) sent lawyer Philip David Swing to Washington to lobby for the new canal. Swing and California politicians thought the federal government would help them and that the new Bureau of Reclamation could take over the project. They were very surprised, however, when Arthur Powell Davis, the director of the US Reclamation Service (as it was then called), refused to support their plans.[57]

Davis, a nephew of Major John Wesley Powell, was not really opposed to building the All-American Canal but believed the Colorado River was the key to western reclamation. He thought developing the river would be an enormous and complex project that needed comprehensive planning, of which the All-American Canal would be only one part. So he campaigned and received support for a large-scale study of the entire Colorado River basin. Completed in 1922, the resulting "Fall-Davis Report" laid out extensive plans for developing the Colorado River. Its centerpiece was a large dam in Boulder Canyon with hydroelectric generation and enormous storage capacity in the reservoir. It also recommended the construction of the All-American Canal from Laguna Dam to the Imperial Valley.[58]

The question of utmost importance to western politicians was how the water would be divided. They had to settle this issue before they could support any bill authorizing a large dam project. Accordingly, representatives from each of the seven states affected (Wyoming, Colorado, New Mexico, Arizona, Nevada, Utah, and California) held a series of meetings in 1922 to try to reach an agreement. The main impetus for these meetings came from Colorado attorney Delph Carpenter. Like many others, Carpenter worried that if the states did not actively make major decisions about water sharing, the federal government would end up with too much control over state issues.[59]

As the meetings progressed, it became apparent that reaching an agreement might be impossible. Herbert Hoover, the secretary of commerce, was attending the meetings, representing President Warren G. Harding's office. Hoover had long been an active conservationist. Early-twentieth-century conservationists viewed nature as an important resource that should not be wasted. Conserving meant making the best human use possible of something like a river. Hoover was enthusiastic about cleaning up harbors and streams and managing natural resources. He believed in harnessing rivers to stabilize and encourage the growth of fisheries while simultaneously controlling flooding and improving navigation. Water that flowed to the oceans without being put to good use was a waste. A dam on the Colorado would fulfill this important criterion.[60]

Hoover's experience as an engineer gave him credibility in the eyes of the compact negotiators, but the situation was much more politically complicated than he had first realized. California was intent upon securing as much water as possible from the river and believed Arizona was not entitled to any water from the main stream. Arizona politicians argued that the state deserved a sizable share in spite of the water it used from the Salt and Gila Rivers, two of the Colorado's tributaries. California believed water from the two rivers should count as part of Arizona's allocation of Colorado River water. Neither side was willing to budge on this issue. As a possible solution to the impasse, Hoover suggested that the basin be divided in half. Surely the states could then agree on a division of water between the two halves at least. They could later negotiate specific distributions. Most delegates thought this idea was a perfect solution and signed what became known as the Colorado River Compact on November 24, 1922. The member states then headed home to convince their constituencies to ratify the document.[61]

Although Sonora and Baja California were important users of the river's waters, representatives from those Mexican states were not invited to the meeting. Most of those present believed that while the United States might want to make some kind of agreement with Mexico over Colorado River water, Mexico did not have any legal rights to that water. Since all the tributaries and sources for the river originated inside the United States, Mexico could not demand a right to its water. Most recognized that some kind of agreement would eventually be made, but Hoover worried that any record of the commission mentioning Mexico could be used to prove an assumption of rights. Accordingly, Hoover asked that any mention of Mexico be expunged from the record, and the commissioners agreed.[62]

The Colorado River Compact created two divisions: the upper basin consisted of Wyoming, Colorado, Utah, and New Mexico; the lower basin included California, Arizona, and Nevada. According to rather optimistic hydrological reports, the Colorado River contained an average annual flow of approximately 18 MAF. As was later understood, this estimate was made during a time of unusually high water flows, so later years had much less than 18 MAF to divide. The problem was not seen in 1922, however, and the law governing Colorado River water usage was based on this erroneous assumption.[63] As a consequence, no one asked whether the water would keep flowing or whether the allocation arrangement would be sustainable in the long term. This narrow view was largely the result of a longstanding attitude toward the natural environment. Historian Roderick Frazier Nash has argued that from the time of the earliest explorers of North America, settlers developed an attitude of fear toward the untamed wilderness that might destroy them. Military phrases such as "conquering" or "defeating the wilderness" became commonplace when speaking of early America's relationship to the wilderness. Attitudes toward the mighty and wild Colorado River, from the first Imperial Valley farmers to congressional committee members, considered the river something that must be conquered to bring the benefits of civilization to the arid Southwest.[64]

According to the Colorado River Compact, each basin would receive 7.5 MAF of water each year. The lower basin was also allowed to take an additional 1 MAF if needed, and 2 MAF were set aside in a reserve, part of which would be allocated to Mexico if a treaty were eventually signed. The bill divided the watershed into two basins mostly to reassure states such as Colorado that there would be water for their projects, somehow protected from California farmers. Each of the involved states ratified the compact fairly quickly except for Arizona, which feared that without specific allocations, California would take much more than its fair share. The compact division of water between the two basins set aside what had been the basis of western water law: the law of prior appropriation. The law did not, however, set aside prior appropriation among the states of each basin.[65] Arizona feared California would manage to claim all the lower basin's share of the Colorado, since its water districts and organizations were already functioning well and planning canals.[66]

Finally, in 1928, Colorado suggested that the compact be considered law if ratified by six of the seven states, with a specific stipulation limiting California's total water allocation. If California's consumption is left unchecked, California could easily appropriate the entire allocation for the lower basin, and Arizona might later appropriate its share of water from the upper basin. Since Arizona had not signed the compact, it would not be bound by it and could simply establish prior rights to water Colorado had not yet been able to appropriate. Congress began debating these issues in earnest as it held hearings on the Boulder Canyon Bill.[67]

Other pressing questions had to be answered before the bill could pass, however. One concerned the location of the dam, and Congress authorized studies of two dam sites (Boulder and Black Canyons). Eventually, the resulting study showed that Black Canyon would be the better location, but by the time this decision had been made, the entire country was already referring to the project as the Boulder Dam or Boulder Canyon Project. The name stuck and would remain until the completed dam was rechristened Hoover Dam in 1929. Meanwhile, Congressman Philip David Swing and Senator Hiram Johnson of California drafted a bill, alternately called the Swing-Johnson Bill or the Boulder Canyon Bill, that authorized Boulder Dam as well as canal construction to bring water to the Imperial and Coachella Valleys in Southern California. Not surprisingly, most Arizona residents were adamantly opposed to the bill, firmly believing California was out to steal all the water it could from the Colorado River, leaving none for Arizona's deserts. The state's politicians became almost obsessed with a determination to fight greedy Californians at every turn. In Yuma, Arizona, however, farmers and local business owners supported the bill because they wanted the promised flood control and federally subsidized water.[68]

There were other opponents of the Boulder Canyon Bill. Eastern politicians thought the project was an enormous waste of federal dollars, and electric-power interests did not want the government entering their business, fearing a government-run monopoly. A duel between two large newspapers added to the discussions. Harry Chandler, publisher of the *Los Angeles Times*, did not support the bill because he feared plans for the All-American Canal would take away all the water from farmland he owned in Mexico just south of the border. William Randolph Hearst used his *Los Angeles Examiner* and other newspapers to champion the bill.[69]

To garner support from the Colorado's upper basin states, Congress made important adjustments to the bill. First, it agreed that the Colorado River Compact could become law with only six of the seven states ratifying it. Second, it inserted a limitation of 4.4 MAF of water annually to California. After much wrangling in various congressional committees, the bill (in its fourth version) reached debate on the floor of the US Senate, where the senators from Arizona filibustered. Finally, however, the bill passed both houses on December 21, 1928. The Boulder Canyon Project Act authorized the spending of $165 million for the dam and the All-American Canal.[70]

In many ways, this enormous project was Herbert Hoover's creation. During his tenure as President Warren G. Harding's interior secretary, Hoover was very active and vocal in support of conservationism and reclamation. He, like others, became intrigued by the idea of harnessing the Colorado River both to prevent devastation by its periodic floods and to provide water to a burgeoning population in California and elsewhere in the West. Hoover, once a resident of California, understood the water pressures there and also saw the federal government as the best supporter for such a large-scale project. While he had little to do with its actual shape and construction, Hoover considered the dam that was to bear his name one of his greatest legacies.[71] Without his efforts to obtain the 1922 compact, the dam might never have been possible.

The dam is actually a combination of two types of dams: the "arch" and the "gravity" dam. Made of concrete, Hoover Dam was both the largest concrete structure and the largest electric-power generator in the United States and remained so until the construction of Grand Coulee Dam on the Columbia River in 1945. Built from 1931 to 1936, Hoover Dam "contains enough concrete to pave a strip 16 feet wide and eight inches thick from San Francisco to New York City."[72] While statistics are conflicting, approximately 112 people died during the dam's construction. Its building is an epic unto itself, filled with stories of success and failure, trial and error. When it was finished, however, the dam generated approximately 4 billion kilowatt-hours of electricity every year and stored up to 28.5 MAF of water in Lake Mead when it was full.[73]

The Hoover Dam has changed very little over the years and remains one of the largest and most identifiable dams in the United States, becoming a major tourist destination after World War II.[74] Water stored by Lake Mead helps to irrigate more than 1 million acres of land in California and Arizona and an additional 400,000 acres in Mexico. In fact, Hoover Dam makes possible much of the produce grown in the United States today. Construction on the All-American Canal south of the Imperial Valley started in 1936 as soon as the dam was completed. Built eighteen miles north of Yuma, Arizona, Imperial Dam diverted water into the All-American Canal and carried it eighty miles to the Imperial Valley. The Coachella Canal was completed in the early 1950s and sent water another 124 miles north of the Imperial Valley. In 2000, the Imperial and Coachella Valleys alone produced more than $1.6 billion worth of farm products.[75]

To distribute the water California would receive from Hoover Dam and its reservoir, various water users in Southern California met to hammer out an agreement. This second major "law of the river," called the Seven Party Agreement, apportions the majority of California's water share to agriculture. In 1929, California agreed that its share of the Colorado should be 4.4 MAF annually and that it would try not to exceed that amount. Arizona refused to sign any agreements on apportionment, so California was actually free to exceed its allocation whenever it chose. California did, however, sign the 1931 agreement that listed priorities and water amounts granted to seven major water users: Palo Verde Irrigation District, Imperial Valley Water District, Coachella Valley Water District, Metropolitan Water District (Los Angeles), City of Los Angeles, City of San Diego, and the County of San Diego, listed in order of priority. One of the problems with the Seven Party Agreement was that it apportioned far more than 4.4 MAF. In fact, California would use more than 5 MAF annually in the years before the US Supreme Court ruling in 1963. Temporarily, however, the agreement minimized a developing struggle between agriculture and growing metropolitan areas in Southern California that would revisit the region in the future.[76]

While Colorado River water finally flowed to agricultural interests in California, Los Angeles knew it would soon need a much larger share. In 1931 the Metropolitan Water District, in charge of providing water for Los Angeles, began plans to construct a canal that would bring Colorado River water to the city. The Bureau of Reclamation began construction of Parker Dam in 1934 to provide a reservoir for the aqueduct to Los Angeles.[77] While these plans were of utmost importance to Southern California, they seemed a threat to Arizona, which still refused to sign any apportionment agreements or to recognize California's claim to Colorado River water. In fact, Arizona had filed lawsuits against California three times during the 1930s to establish its own water rights and minimize California's, all to no avail.

As construction on Parker Dam began in 1934, Arizona governor Benjamin Baker Moeur decided to do more than protest to the media. As had his predecessor, George P. Hunt, he adamantly opposed California's share of the Colorado River and any attempts to construct an aqueduct. To show his determination, he mobilized Arizona's National Guard and sent it to the dam site. Unfortunately, this action ended up as a joke to many observers because the troops became stranded on an old steamboat and were rescued by the Los Angeles Department of Water and Power. The *Los Angeles Times* sent a reporter it jokingly called a "war correspondent" to cover the dispute.[78] Arizona was determined to prevent the Bureau of Reclamation from anchoring the dam to the Arizona side of the river, and at this point the secretary of the interior decided to halt the project. He referred the issue back to the US Supreme Court, which promptly ruled that indeed the Boulder Canyon Project Act had not authorized a dam at Parker. Congress quickly drafted the necessary legislation to make the dam legal, and in 1935, Arizona had to stand back and watch as construction resumed.[79] In 1941 the California Aqueduct carried its first water delivery from Parker Dam into Los Angeles. Suddenly, Los Angeles had plenty of water to allow it to grow, and between 1940 and 1970 the city's population nearly doubled.[80]

During most of the 1930s, President Franklin D. Roosevelt fully supported federal reclamation as part of his New Deal. At both ends of the Colorado River, the Bureau of Reclamation developed irrigation projects—one to address the needs of farmers in the Gila River Valley region near Yuma, and the other in Colorado to transport water from the western slope over the Continental Divide to Denver. The Gila Project, started in 1936, was not completed until after the interruption caused by World War II.[81] The project consisted of two sets of irrigation canals, one called the Wellton-Mohawk Division running east-west along both sides of the river, and the second a series of north-south canals called the Yuma Mesa Division. As farmers irrigated their fields along the Gila, over time the water and soil became more and more salty, making it impossible to grow crops. The canals carried Colorado River water to the area, and a series of drainage wells leached off excess groundwater from farms. Some of the canals brought in Colorado River water as early as 1943, but much of the Wellton-Mohawk Division was not ready until 1952 and was not fully completed until 1957.[82]

In 1937, work began on a large project at the north end of the river—the Colorado–Big Thompson Project, one of the largest ever constructed by the Bureau of Reclamation. It covers approximately 250 miles and includes dams, reservoirs, pipelines, power stations, pumping stations, and canals—all meant to transport the abundant water from the Colorado River on the western side of the Rockies to the arid eastern side. In spite of ongoing mistrust between the two parts of the state, western slope residents reluctantly agreed to pump their water to the eastern cities of Denver and Boulder.[83]

As with the Gila Project, construction on Colorado–Big Thompson began in 1937, slowed during World War II, and resumed until the project was completed in 1959. Both of these long-term projects diverted significant amounts of water from the Colorado River and used allotments in both the upper and lower river basins. The agreement for the Colorado–Big Thompson Project was made possible by the fear that if the upper basin did not begin using and storing its share of water, the lower basin would over-allocate and take more than its fair share. During the 1940s and 1950s, Colorado politicians such as Representative Wayne Aspinall would feel an urgency to develop more projects for the upper basin and Colorado in particular.[84]

While those in the western basin divided up the Colorado River and competed with each other for allocations, most users ignored any claims by Mexico (see chapter 7). For many years, Mexican officials and farmers became increasingly concerned and frustrated as Hoover Dam and the All-American Canal reduced the flow of water across the border. Both sides also disagreed on sharing the Rio Grande between Texas and Mexico. Over several decades the two nations talked, made offers, and finally reached an agreement in 1944. In the resulting treaty, Mexico allowed the United States to draw 350,000 acre-feet of water from the Rio Grande, and the United States agreed to deliver 1.5 MAF to Mexico from the Colorado River.[85] The document made no mention of water quality, creating much debate later, in the 1960s. The treaty also did not stipulate who would give up water in times of serious drought. These issues reappeared in the 1960s when diversions of agricultural runoff near the border made Mexico's allocation too salty to use.

By the end of World War II, the Colorado River had been dammed, divided, litigated, and transferred inside and outside the river basin. While more dams and canals would follow in the 1950s and 1960s, much of the work of turning the river into a complex plumbing system was complete. Most of the Colorado River's water was put to the "beneficial use" of agriculture, but although the Bureau of Reclamation continued to follow a policy of limiting irrigated agriculture, the ideal of individual farms gave way to a new reality of large-scale agribusiness. The story of the river in later decades became even more complicated, affected by a growing environmental movement and intensified disagreements between upper and lower basin states. Important issues such as Native American water rights, quality control, and urban demands versus agriculture characterized the story of the Colorado River in the second half of the twentieth century.

NOTES

1. An acre-foot is the amount of water it takes to cover one acre of land one foot deep, approximately 325,851 gallons. This is the primary unit of measure used for water flow in rivers in the United States. Eighteen acre-feet is approximately 1 miner's inch. Although miner's inches are still used in some parts of the American West, the acre-foot has been a much more common measure since the early twentieth century. The figure of 15 million acre-feet as the average water flow is somewhat limited in value. While current estimates over the past 100 years do place the average near 15 MAF (considerably less than the 1922 estimate of 18 MAF), recent tree-ring analysis indicates the existence of long drought periods over the past 500 years. This much longer-term analysis places an "average" flow well below 15 MAF and perhaps as low as 13 MAF. The evidence suggests the occurrence of a severe drought lasting multiple decades during the 1100s. See David M. Meko et al., "Medieval Drought in the Upper Colorado River Basin," *Geophysical Research Letters* 34, no. 10 (May 24, 2007): L10705.

2. Robert W. Adler, *Restoring Colorado River Ecosystems: A Troubled Sense of Immensity* (Washington, DC: Island, 2007), 27.

3. Western Water Assessment, Colorado River Streamflow, 2009, http://treeflow.info/lees /index.html (accessed December 30, 2012).

4. US Geological Survey, Climatic Fluctuations, Drought, and Flow in the Colorado River Basin, August 2004, http://pubs.usgs.gov/fs/2004/3062/ (accessed March 17, 2011).

5. Ibid. A more recent assessment using tree-ring data is Jeff Lukas (Western Water Assessment), Connie Woodhouse (University of Arizona), and Henry Adams (University of Arizona), Colorado River Streamflow, a Paleo Perspective, 2009, http://treeflow.info/lees /gage.html (accessed January 24, 2012). See also National Research Council of the National Academies, *Colorado River Basin Water Management: Evaluating and Adjusting to Hydroclimate Variability* (Washington, DC: National Academies Press, 2007).

6. NASA press release, Study Shows Desert Cuts Colorado River Flow, September 10, 2010, http://www.nasa.gov/topics/earth/features/colorado20100920.html (accessed March 17, 2011).

7. For information about the region's indigenous history, see Stephen Plog, *Ancient Peoples of the American Southwest* (New York: Thames and Hudson, 1997); Emil W. Haury, *The Hohokam, Desert Farmers and Craftsmen: Excavations at Snaketown, 1964–1965* (Tucson: University of Arizona Press, 1976). The best recent work is Craig Childs, *House of Rain: Tracking a Vanished Civilization across the American Southwest* (New York: Little, Brown, 2007).

8. For translated versions of Ulloa's journey, see Henry Raup Wagner, *California Voyages, 1539–1541: Translation of Original Documents* (San Francisco: J. Howell, 1925). See also Joyce Moss, *Spanish and Portuguese Literatures and Their Times* (Detroit: Gale Group, 2002).

9. Jack D. Forbes, "Melchior Díaz and the Discovery of Alta California," *Pacific Historical Review* 27, no. 4 (November 1958): 352.

10. An excellent study of the Spanish period is Michael C. Meyer, *Water in the Hispanic Southwest: A Social and Legal History, 1550–1850* (Tucson: University of Arizona Press, 1984).

11. See Richard W. Etulain, *Western Lives: A Biographical History of the American West* (Albuquerque: University of New Mexico Press, 2004); Robert Utley, *A Life Wild and Perilous: Mountain Men and the Paths to the Pacific* (New York: Henry Holt, 1997).

12. Richard Allen Chase, *The Pioneers of '47: Migration along the Mormon Trail* (Logan, UT: Watkins, 1997); Elliot West, *The Contested Plains: Indians, Goldseekers, and the Rush to Colorado* (Lawrence: University Press of Kansas, 1998).

13. Richard Griswold del Castillo, *The Treaty of Guadalupe-Hidalgo: A Legacy of Conflict* (Norman: University of Oklahoma Press, 1990). See also Treaty with Mexico, 1848, http://www.mexica.net/guadhida.php (accessed August 6, 2010).

14. For an excellent source on riparian water rights and the transition in the West to prior appropriation, see chapter 3 in Norris Hundley Jr., *The Great Thirst: Californians and Water—a History* (Berkeley: University of California Press, 2001 [1992]). See also Donald J. Pisani, *Water, Land, and Law in the West: The Limits of Public Policy, 1850–1920* (Lawrence: University Press of Kansas, 1996).

15. See Arthur Jay Sementelli, "Naming Water: Understanding How Nomenclature Influences Rights and Policy Choices," *Public Works Management and Policy* 13, no. 1 (July 2008): 4–11.

16. John C. Dernbach, *Stumbling toward Sustainability* (Washington, DC: Environmental Law Institute, 2002), 201.

17. John Wesley Powell, Eliot Porter, and Don D. Fowler, *Down the Colorado: Diary of the First Trip through the Grand Canyon, 1869* (New York: E. P. Dutton, 1969). See also Wallace Stegner, *Beyond the Hundredth Meridian: John Wesley Powell and the Second Opening of the*

West (Boston: Houghton Mifflin, 1954); Edward Dolnick, *Down the Great Unknown: John Wesley Powell's 1869 Journey of Discovery and Tragedy through the Grand Canyon* (New York: HarperCollins, 2001).

18. John Wesley Powell, *Report on the Lands of the Arid Region of the United States* (Boston: Harvard Common Press, 1983 [1878]), 6.

19. Ibid., 3. See also Donald Worster, *A River Running West: The Life of John Wesley Powell* (New York: Oxford University Press, 2000).

20. Powell, *Report on the Lands of the Arid Region*, 13.

21. Lisi Krall, "US Land Policy and the Commodification of Arid Land (1862–1920)," *Association for Evolutionary Economics* 35, no. 3 (September 2001): 662.

22. Ibid.

23. For an excellent recent work on the California Gold Rush, see Leonard L. Richards, *The California Gold Rush and the Coming of the Civil War* (New York: Alfred A. Knopf, 2007). For studies of the transcontinental railroad, see Stephen Ambrose, *Nothing Like It in the World: The Men Who Built the Transcontinental Railroad, 1863–1869* (New York: Simon and Schuster, 2000); David Haward Bain, *The Old Iron Road: An Epic of Rails, Roads, and the Urge to Go West* (New York: Viking, 2004).

24. Coy F. Cross, *Go West, Young Man! Horace Greeley's Vision for America* (Albuquerque: University of New Mexico Press, 1995); Dennis W. Johnson, *The Laws That Shaped America: Fifteen Acts of Congress and Their Lasting Impact* (New York: Routledge, 2009).

25. Frederick Jackson Turner, "The Significance of the Frontier in American History," essay presented to the American Historical Association, 1893 (Ann Arbor: University Microfilms, 1966 [1893]).

26. Frederick Jackson Turner, *The Frontier in American History* (New York: Holt, Rinehart, and Winston, 1962 [1893]). See also Rebecca Edwards, *New Spirits: Americans in the "Gilded Age," 1865–1905* (New York: Oxford University Press, 2011); and an older but excellent classic, Robert Wiebe, *The Search for Order, 1877–1920* (New York: Hill and Wang, 1967).

27. United States Congress, "An Act to Provide for the Sale of Desert Lands in Certain States and Territories," *United States Statutes at Large* 19, ch. 107 (Washington, DC: Government Printing Office, 1877), 377.

28. Donald J. Pisani, *Water and American Government: The Reclamation Bureau, National Water Policy, and the West, 1902–1935* (Berkeley: University of California Press, 2002), xiv.

29. James R. Kluger, *Turning on Water with a Shovel: The Career of Elwood Mead* (Albuquerque: University of New Mexico Press, 1992).

30. Donald J. Pisani, *To Reclaim a Divided West: Water, Law, and Public Policy, 1848–1902* (Albuquerque: University of New Mexico Press, 1992), 273–98.

31. Barbara Ann Metcalf, "Oliver M. Wozencraft in California, 1849–1887," MA thesis, University of Southern California, Los Angeles, 1963. See also the reproduction of Blake's

report in William Phipps Blake and Harry Thomas Cory, *The Imperial Valley and the Salton Sink* (Charleston, SC: Nabu, 2010).

32. A miner's inch was the amount of water that could flow through a 1-inch-square hole in a miner's sluice at a set pressure level. In general, this amount is close to a flow of 1.5 cubic feet per minute, but the measurement varied by state. See "Convert Miner's Inches to Other Values," February 24, 2012, http://www.western-water.com/water-calculators /convert-miners-inches-to-other-values.

33. William Oral Hendricks, "Guillermo Andrade and Land Development of the Mexican Colorado River Delta, 1874–1905," PhD diss., University of Southern California, Los Angeles, 1967. Other sources on Andrade and the development of Baja California are David Piñera Ramírez, *Los Orígenes de las Poblaciones de Baja California: Factores Externos, Nacionales y Locales* (Mexicali, Baja California México: Universidad Autónoma de Baja California, 2006); Marco Antonio Samaniego López, coordinador, *Breve Historia de Baja California* (Mexicali, Baja California México: Universidad Autónoma de Baja California, 2006).

34. George Kennan, *The Salton Sea: An Account of Harriman's Fight with the Colorado River* (Ithaca, NY: Cornell University Press, 2009 [1917]); Pat Laflin, *The Salton Sea: California's Overlooked Treasure* (Indio, CA: Coachella Valley Historical Society, 1999), 17–30.

35. Worster, *Rivers of Empire*, 196.

36. Laflin, *Salton Sea*, 21. See also Reisner, *Cadillac Desert*, 122–23.

37. Joseph E. Stevens, *Hoover Dam: An American Adventure* (Norman: University of Oklahoma Press, 1988), 13–15.

38. Kennan, *Salton Sea*, 32–63.

39. Worster, *Rivers of Empire*, 197.

40. For a recent study on eutrophication and its impact on water resources, see Abid A. Ansari et al., eds., *Eutrophication: Causes, Consequences and Control* (New York: Springer, 2011).

41. Imperial Irrigation District, 2009 Annual Water Report, June 2012, http://www.usbr .gov/lc/hooverdam/faqs/lakefaqs.html (accessed December 30, 2012).

42. See Donald Worster, *Under Western Skies: Nature and History in the American West* (New York: Oxford University Press, 1992).

43. William E. Smythe, *The Conquest of Arid America* (Charleston, SC: Nabu, 2010 [1905]); Peter Wild, *The Opal Desert: Explorations of Fantasy and Reality in the American Southwest* (Austin: University of Texas Press, 1999); Worster, *Rivers of Empire*, 118–25.

44. Bureau of Reclamation, Hoover Dam and Lake Mead, June 2012, http://www.usbr .gov/lc/hooverdam/faqs/lakefaqs.html (accessed December 30, 2012). See also US Geological Survey, Evaporation from Lake Mead, Arizona and Nevada, 1997–1999, Scientific Investigations Report, 2006-5252, http://pubs.usgs.gov/sir/2006/5252/pdf/sir20065252 .pdf (accessed January 26, 2012).

45. Many in California celebrated what the Newlands Act promised for Western farmers. See Hamilton Wright, "Millions of New Acres for American Farmers," *National Magazine* 23, no. 1 (October 1905): 197–203, http://www.archive.org/stream/nationalmagazine23brayrich #page/199/mode/1up (accessed August 7, 2010).

46. C. J. Blanchard, "Home-making by the Government: An Account of the Eleven Immense Irrigating Projects to Be Opened in 1908," *National Geographic Magazine* 19 (April 1908): 250–87.

47. Frederick H. Newell, "What May Be Accomplished by Reclamation," *Annals of the American Academy of Political and Social Science* 33, no. 3 (May 1909): 174–79.

48. A recent work on Theodore Roosevelt and his conservation and social planning ideas is Douglas Brinkley, *The Wilderness Warrior: Theodore Roosevelt and the Crusade for America* (New York: HarperCollins, 2009).

49. William D. Rowley, *The Bureau of Reclamation: Origins and Growth to 1945,* vol. 1 (Denver: Bureau of Reclamation, 2006), 47–60.

50. Pisani, *Water and American Government*, 56.

51. Ibid., 29.

52. Pisani, *To Reclaim a Divided West*, 312.

53. David Clark, *Uncompahgre Project* (Denver: Bureau of Reclamation History Program, 1994), http://www.usbr.gov/projects/Project.jsp?proj_Name=Uncompahgre%20Project &pageType=ProjectHistoryPage (accessed August 8, 2010).

54. Robert A. Sauder, *The Yuma Reclamation Project: Irrigation, Indian Allotment, and Settlement along the Lower Colorado River* (Reno: University of Nevada Press, 2009); Eric A. Stene, *Yuma Project and Yuma Auxiliary Project* (Denver: Historic Reclamation Projects, Bureau of Reclamation, 1996), http://www.usbr.gov/projects//ImageServer?imgName =Doc_1271086556202.pdf (accessed August 8, 2010).

55. Pisani, *Water and the American Government*, 56–62.

56. For more on the revolution, see Michael J. Gonzales, *The Mexican Revolution, 1910– 1940* (Albuquerque: University of New Mexico Press, 2002); Héctor Aguilar Camín and Lorenzo Meyer, *In the Shadow of the Mexican Revolution: Contemporary Mexican History, 1910–1989,* trans. Luis Alberto Fierro (Austin: University of Texas Press, 1993).

57. See Hundley, *Great Thirst*, 208–11.

58. David P. Billington, Donald C. Jackson, and Martin V. Melosi, *The History of Large Federal Dams: Planning, Design, and Construction* (Denver: US Department of the Interior, Bureau of Reclamation, 2005), 142–44.

59. Stevens, *Hoover Dam,* 26–27.

60. For more on Herbert Hoover's role, see George H. Nash and Kendrick A. Clements, *The Life of Herbert Hoover* (New York: W. W. Norton, 1983); Herbert Hoover, *The Memoirs of Herbert Hoover* (New York: Macmillan, 1965).

61. Hundley, *Great Thirst,* 211–15. See also Colorado River Compact Commission, Minutes of Colorado River Compact Commission, 1922, Arizona Historical Foundation Collection, Hayden Library, Arizona State University, Tempe.

62. Hundley, *Water and the West,* 204–5.

63. For a complete discussion of the compact, see ibid.; Pisani, *Water and Government.* For the full text of the compact, see http://www.usbr.gov/lc/region/pao/pdfiles/crcompct.pdf (accessed August 8, 2010).

64. Roderick Frazier Nash, *Wilderness and the American Mind* (New Haven: Yale University Press, 2001 [1967]), 27.

65. Hundley, *Great Thirst,* 214.

66. For more information on the Colorado River Compact and other court cases relating to prior appropriation rights, see Daniel Tyler, *Silver Fox of the Rockies: Delphus E. Carpenter and Western Water Compacts* (Norman: University of Oklahoma Press, 2003); Peter E. Black, *Conservation of Water and Related Land Resources* (Boca Raton: Lewis, 2001); David Berman, *Interstate Compact Water Commissions: Selected Case Studies: A Report* (Washington, DC: Washington Center for Metropolitan Studies, 1962).

67. Hundley, *Great Thirst,* 220–21. See also Hundley, *Water and the West,* 276–77.

68. For an excellent study on Yuma water politics, see Evan R. Ward, "Crossroads on the Periphery: Yuma County Water Relations, 1922–1928," MA thesis, University of Georgia, Athens, 1997. Ward argues convincingly that Yuma County shared more in common with the Imperial Irrigation District in California and supported the Swing-Johnson Bill.

69. For a recent work on Hoover Dam, see Michael A. Hiltzik, *Colossus: Hoover Dam and the Making of the American Century* (New York: Free Press, 2010). For information on Harry Chandler and William Randolph Hearst, see Perry J. Ashley, *American Newspaper Journalists, 1926–1950* (Detroit: Gale Research, 1984); David Nasaw, *The Chief: The Life of William Randolph Hearst* (Boston: Houghton Mifflin, 2000).

70. House Resolution 5773, "Boulder Canyon Project Act," Public-no. 642, 70th Congress, December 21, 1928, http://www.usbr.gov/lc/region/pao/pdfiles/bcpact.pdf (accessed August 8, 2010).

71. Bureau of Reclamation, Herbert Hoover and the Colorado River, September 10, 2004, http://www.usbr.gov/lc/hooverdam/History/articles/hhoover.html (accessed August 8, 2010). See also Sarah T. Phillips, *This Land, This Nation: Conservation, Rural America, and the New Deal* (New York: Cambridge University Press, 2007).

72. Bureau of Reclamation, Hoover Dam, January 20, 2005, http://www.usbr.gov/lc/hooverdam/History/essays/concrete.html (accessed September 1, 2011).

73. See Bureau of Reclamation, Colorado River and the Hoover Dam, Facts and Figures, January 2012, http://www.usbr.gov/lc/region/pao/brochures/faq.html (accessed December 30, 2012). For more information, see Donald C. Jackson, *Building the Ultimate Dam: John S.*

Eastwood and the Control of Water in the West (Norman: University of Oklahoma Press, 1995).

74. Hoover Dam is featured in more than a dozen movies, including *The Amazing Colossal Man* (1957), *Viva Las Vegas* (1964), *James Bond—Diamonds Are Forever* (1971), *Superman* (1978), *Lost in America* (1985), *Universal Soldier* (1992), *Fools Rush In* (1996), *Vegas Vacation* (1996), and *Transformers* (2007).

75. Evan Ward, *Border Oasis: Water and the Political Ecology of the Colorado River Delta, 1940–1975* (Tucson: University of Arizona Press, 2003), 32–37.

76. Full text of the Seven Party Agreement, 1931, US Bureau of Reclamation, http://www .usbr.gov/lc/region/g1000/pdfiles/ca7pty.pdf (accessed August 8, 2010).

77. Billington, Jackson, and Melosi, *History of Large Federal Dams*, 180–81. See also Bureau of Reclamation, Parker-Davis Project, May 11, 2011, http://www.usbr.gov/projects /Project.jsp?proj_Name=Parker-Davis%20Project (accessed December 30, 2012).

78. Other newspapers also covered the event. See "Arizona Troops Ready for Orders," *New York Times,* November 11, 1934; "Arizona Troops Get Order to Halt Dam Work," *Lima* [OH] *News*, November 12, 1934; "Arizona Expeditionary Forces Ordered to Return Home, Bloodless Victory Won, Work on Dam Project Is Halted," *Nevada State Journal* [Reno], November 15, 1934.

79. Reisner, *Cadillac Desert,* 257–59.

80. Hundley, *Great Thirst,* 231.

81. US Bureau of Reclamation, "Gila Project," January 30, 2012, http://www.usbr.gov /projects/Project.jsp?proj_Name=Gila+Project (accessed December 30, 2012).

82. Hundley, *Dividing the Waters*, 172–74. See also http://www.wmidd.org/history .html (accessed July 13, 2010).

83. Robert Autobe, Bureau of Reclamation, Colorado–Big Thompson Project, http:// www.usbr.gov/projects//ImageServer?imgName=Doc_1303159857902.pdf (accessed August 8, 2010). See also Daniel Tyler, *The Last Water Hole in the West: The Colorado–Big Thompson Project and the Northern Colorado Water Conservancy District* (Niwot: University Press of Colorado, 1992).

84. For an excellent work on Aspinall, see Steven C. Schulte, *Wayne Aspinall and the Shaping of the American West* (Boulder: University Press of Colorado, 2002).

85. Treaty between the United States of America and Mexico, signed in Washington, DC, February 3, 1944, http://www.usbr.gov/lc/region/pao/pdfiles/mextrety.pdf (accessed August 8, 2010). See also Charles T. Meyers and Richard L. Noble. "The Colorado River: The Treaty with Mexico," *Stanford Law Review* 19, no. 2 (January 1967): 367–419; United States Congress, *Water Treaty with Mexico: Hearings before the Committee on Foreign Relations, United States Senate, Seventy-Ninth Congress, First Session, on Treaty with Mexico Relating to the Utilization of the Waters of Certain Rivers, January 22–26, 1945* (Washington, DC: US Government Printing Office, 1945); Helen M. Ingram et al., *Divided Waters: Bridging the U.S.-Mexico Border* (Tucson: University of Arizona Press, 1995).

2

Farming the Desert

Agricultural Water Demands

THE PRIMARY USE of the Colorado River is and has always been for agriculture. In recent times, metropolitan applications of its water have made users forget where most of the water is actually consumed. Yet in spite of rapid urban growth of the major cities, most of the river still goes to farming. Today, an estimated 78 percent of the annual flow is used on fields to grow crops, watering approximately 1.8 million acres with almost 4 trillion gallons. The Colorado River basin produces approximately 15 percent of the crops and 13 percent of the livestock for the entire nation.[1] This agriculture generates approximately $1.5 billion in revenue each year.[2] While water marketing and farm-to-urban transfers will eventually lower this percentage, agriculture is still the primary purpose for which the river is used.

Agriculture and Its Impact on Water

Although only 16 percent of US farmland today is irrigated, that 16 percent generates almost half of all farm

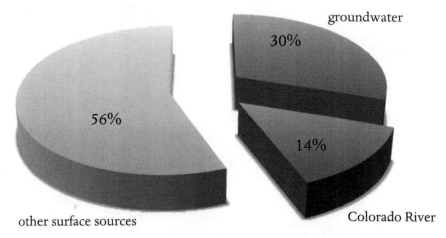

groundwater

30%

56%

14%

other surface sources

Colorado River

FIGURE 2.1 *California water sources. Created by author with data from the Colorado River Water Users Association, http://www.crwua.org/ColoradoRiver/MemberStates/California .aspx, and UCLA Institute of the Environment and Sustainability, http://www.environment .ucla.edu/reportcard/article.asp?parentid=4870.*

crops nationwide. Because of this fact, more than 80 percent of water from any source in the United States is used by agriculture. In most of the Southwest, between 60 percent and 96 percent of farm sales result from large-scale irrigated agriculture. In California and Arizona, these large-scale farms account for almost 66 percent of irrigation water used. Smaller farms, which make up about 81 percent of irrigated farms in the West, use the remaining 34 percent.[3]

While much of the West still relies on groundwater (see figures 2.1, 2.3, 2.5), surface water amounts to an average of 2.6 acre-feet per acre throughout the region. Pumping costs for irrigation water vary from state to state, with areas in Montana costing only around $11 per acre while California and Arizona see costs of more than $62 per acre.[4] According to slightly outdated numbers compiled in 2000 by the US Geological Survey, irrigated land in Arizona equaled around 976,000 acres. Of this acreage, 779,000 was irrigated by flood irrigation, 18,000 by sprinkler, and only 14,000 by micro-irrigation that includes systems either close to the surface or under the soil. According to these same statistics, California was the highest water user in the Colorado River basin, withdrawing a range of 15 billion to 31 billion gallons per day. Next in water use were Arizona and Colorado, both in the 5 billion to 15 billion gallons per day range. The lowest water users for irrigation in the basin were Nevada, Utah, and Wyoming in the 1 billion to 5 billion gallon range.[5]

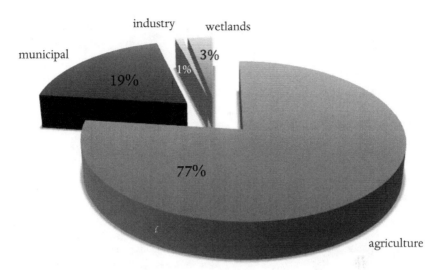

municipal
19%

industry
1%

wetlands
3%

77%

agriculture

FIGURE 2.2 *California water use. Created by author with data from Colorado River Water Users Association, http://www.crwua.org/ColoradoRiver/MemberStates/California.aspx, and UCLA Institute of the Environment and Sustainability, http://www.environment.ucla .edu/reportcard/article.asp?parentid=4870.*

In spite of recent agriculture-to-urban transfers, the most recent numbers show that agriculture is still the largest user of Colorado River water. In 2007 the Colorado River Water Users Association (CRWUA) provided updated agriculture statistics. California receives the largest share of Colorado River water in the basin and irrigates approximately 10 million acres throughout the state (see figure 2.2). The largest user of this water is the Imperial Irrigation District, which waters more than 1 million acres of land. The district's farmers produce many products, including alfalfa, carrots, lettuce, sugar beets, cantaloupes, onions, asparagus, and livestock. Overall, irrigation water districts have rights to 3.85 million acre-feet (MAF) per year of California's 4.4 MAF allotment of Colorado River water.[6]

Colorado irrigates around 3.4 million acres of land, producing corn, hay, and wheat, as well as fruits, vegetables, and livestock (see figure 2.4). Although its share of Colorado River water is the second largest of any user state, at 3.9 MAF, agriculture consumes approximately 12.8 MAF, or 90 percent, of all the state's surface and groundwater. Colorado uses an average of 3.76 acre-feet of water per acre and is the third-highest user behind Wyoming (the highest) and Arizona (the second highest). In terms of amounts of water per acre, California comes in a distant fourth.[7]

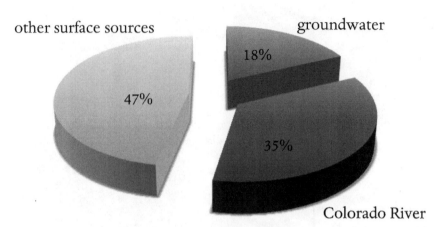

FIGURE 2.3 *Colorado water sources. Created by author with data from Colorado River Water Users Association, http://www.crwua.org/ColoradoRiver/MemberStates/Colorado.aspx, and Colorado Water Conservation Board, http://cwcb.state.co.us/water-management /water-supply-planning/Pages/main.aspx.*

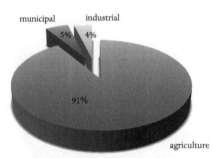

FIGURE 2.4 *Colorado water use. Created by author with data from Colorado River Water Users Association, http://www.crwua.org /ColoradoRiver/MemberStates/Colorado.aspx, and Colorado Water Conservation Board, http://cwcb.state.co.us/water-management /water-supply-planning/Pages/main.aspx.*

Arizona's share of the Colorado River is 2.8 MAF. Before the turn of the twentieth century, agriculture in the state started with water from the Salt River, a tributary of the Gila River. Today, with the Salt River Project and the Central Arizona Project (CAP) canal, Arizona waters almost 900,000 acres of land, with an aggregate income of nearly $1.8 billion per year from produce and livestock. The major crops include cotton, hay, wheat, barley, corn, sorghum, potatoes, lettuce, onions, cauliflower, broccoli, carrots, various melons and citrus, and grapes. Of total water used in Arizona, 25 percent comes from the Colorado River, 80 percent of which goes to agricultural uses (see figure 2.6).[8]

Utah receives 1.7 MAF yearly from the Colorado River and irrigates approximately 4.3 million acres of farmland throughout the state. About 400,000 acres receive

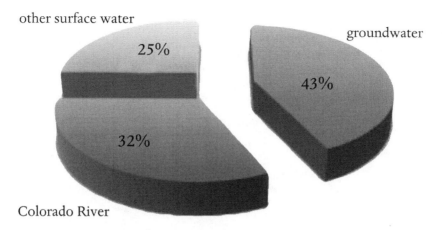

other surface water

25%

groundwater

43%

32%

Colorado River

FIGURE 2.5 *Arizona water sources. Created by author with data from Colorado River Water Users Association, http://www.crwua.org/AboutUs/Welcome.aspx, and Arizona Department of Water Resources, http://www.azwater.gov/AzDWR/PublicInformationOfficer/documents /supplydemand.pdf.*

Colorado River water. Although cattle, sheep, and poultry make up a large portion of the state's agriculture, Utah also raises alfalfa, barley, oats, corn, potatoes, onions, tomatoes, and fruits such as cherries, apricots, apples, and peaches. Wyoming receives one MAF per year, irrigating approximately 5 million acres. In spite of its smaller share of the river, Wyoming irrigates more acres than any of the other basin states and ranks ninth in the entire country in irrigated acres. Farming and grazing occupy almost 35 million acres, more than half of the entire area of the state. Livestock, alfalfa, and other small grains make up the bulk of the products produced.[9]

Both of the last two states that share Colorado River water—New Mexico and Nevada—irrigate a significant amount of farmland. New Mexico receives only 850,000 acre-feet of river water from the main stream of the Colorado but uses the nearly 2.5 MAF that come from the San Juan River, the Colorado River's largest tributary. Nevada, the driest state in the nation in terms of rainfall, receives only 300,000 acre-feet of Colorado River water and uses all of it for urban and industrial purposes. Agriculture receives its water from other sources, which irrigate approximately 2.3 million acres in the state. Nevada grows potatoes, onions, and garlic, among other vegetables, but mostly produces livestock and alfalfa.[10]

Although the Colorado River's water provides the necessary irrigation for all this farming, there are significant and long-term impacts on the water quality. One of the most significant problems is salinity. There are natural sources for this salt; according

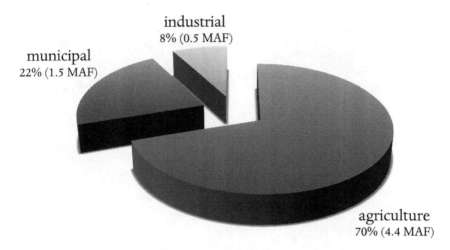

industrial
8% (0.5 MAF)

municipal
22% (1.5 MAF)

agriculture
70% (4.4 MAF)

FIGURE 2.6 *Arizona water use. Created by author with data from Arizona Department of Water Resources,"Securing Arizona's Water Future," http://www.azwater.gov/AzDWR /PublicInformationOfficer/documents/supplydemand.pdf.*

to geologists, the region that comprises the Colorado River watershed was once sub-merged below saltwater inland seas. As a result, the soil in the basin contains sodium chloride and calcium sulfate that leach into the water as it flows down through the watershed. These naturally occurring salts make up approximately 50 percent of the river's salinity. As this salty water is used for irrigation, the salt levels increase sig-nificantly. Around 37 percent of salt sources for the river is a result of irrigation, 12 percent results from concentration caused by evaporation at reservoirs, and another 4 percent comes from mining and industrial sources.[11]

Both using the water for irrigation and depleting the reservoirs increase salt loads in the river. According to recent reports from the Department of the Interior, irriga-tion runoff creates 3.4 million tons of salt every year, making up the 37 percent of the river's overall saline content cited in the previous paragraph. As irrigation water runs over the basin's already saline soils, it leaches salts into rivers and streams. Water in the upper basin region has lower salinity concentrations; however, as this water is diverted, the remaining water that runs into the lower basin is much saltier.[12] During the process of irrigation, excess water that is not consumed by plants or evaporated then seeps down through soils that are saline, draining higher salt levels into the river.[13] These high levels of salt make irrigated farmland more and more unsuitable for most crops, requiring farmers to flush their fields free of excess salts with large amounts of water. It becomes a self-perpetuating cycle. Added to the high saline levels are other chemical contaminants from pesticide use, including Dacthal (DCPA), Dieldrin, Toxaphene, Chlordane, and Dchloro diphenyl dichloroethylene (DDT). While all but one of these pesticides have been banned for use in the United States, high levels are still found in fish samples in the Colorado River basin because of heavy past usage.[14]

Agriculture has thus been both the greatest beneficiary of the river's water and its largest pollutant. In the first two decades after World War II, issues directly relating to agriculture dominated negotiations for Colorado River water. Although the 1922 Colorado River Compact and the 1944 treaty with Mexico seemed to permanently divide the water, conflict arose whenever individual states tried to use river allocations. California reaped the benefits from Hoover Dam's electricity and diverted water through aqueducts to the Imperial Valley and Los Angeles for agriculture and urban usage. Since few other states were using Colorado River water, California had all it could take from the main stream. The upper basin states felt left behind. They worried that unless they built dams and reservoirs soon, California would claim prior appropriation rights to much more than its share. For its part, California saw no reason to limit itself to the allocations if no one else was putting the water to beneficial use.

Eventually, Arizona and Nevada joined the upper basin in its concern about California's water use. Arizona, which hoped to develop more agriculture and reduce over-pumping of groundwater, planned for a pipeline to bring the river to its central desert. California saw CAP as a threat to its supply and argued that Arizona's Gila River water should count as part of the state's allocation. The issue of prior appropriation water law was wrapped around all of the negotiations and conflicts in these years, revealing the policy's fatal flaw. Instead of a fair distribution of water, it encouraged the basin states to grab as much water as possible because of a "use it or lose it" mentality. Taken to its most logical conclusion, this interpretation of the law resulted in over-allocation and wasteful uses of the river.

The 160-Acre Limit

At the end of the nineteenth century, explorer and geologist John Wesley Powell argued that only small farming was sensible in the arid West with its scarce water. Bureau of Reclamation officials did not follow all of his suggestions, but they did grasp Powell's argument and turned it into a vision: the arid West would be made suitable for small farmers through federal reclamation. The entire focus was to make the vast, unused West conform to Turnerian views of the region: a safety valve for overcrowding in the East and a haven for the small farmer. Coupled with this idealized agrarian dream was the realization of what subsidized water might do. Speculators could develop a large monopoly of land that had suddenly increased in value—land the small farmer would never be able to afford. Inexpensive water could lead to wasteful irrigation. Limiting federal water to farms of 160 acres became a major reclamation policy for the West.[15]

As discussions over water rights began in earnest between Arizona and California in the years after World War II, the 160-acre limit became a major tool in the fight. Arizona found the limit and its many violations a perfect weapon to shake at California, but for much of Southern California the discussion was moot. The Imperial Valley was already watering 120,000 acres by some accounts when Rockwood's canal gates broke in 1905.[16] Since that water was privately developed, federal limits did not apply. In the 1940s, however, farming had expanded north into the Coachella Valley, which was supporting more than 12,000 acres of farmland as early as 1946.[17] Predictions were that proposed extensions of the All-American Canal into the valley would irrigate an additional 75,000 acres.[18] However, the Imperial Irrigation District insisted on federal exemptions from the 160-acre limit and was reassured that it would keep them. For everyone else, however, the limit applied.

When construction on the All-American Canal began in the early 1930s, Arizona and other Colorado River users recognized how much water could potentially be drawn from the annual flow. With an almost 300-day annual growing season, farmers could grow two crops per year and bring onions, peppers, tomatoes, corn, spinach, carrots, eggplant, beans, peas, melons, grapes, and many other fruits and vegetables to American tables year-round.[19] Other states such as Utah also looked to increase agricultural production from irrigated farming. Utah developed extensive plans for a Central Utah Project. This project would bring Colorado River water to irrigate enough land to raise the state's annual agricultural income from $150 million to $240 million by 1949.[20] Arizona also sought to develop its share as work began on the Wellton-Mohawk Project in Yuma, which started in the 1930s to relieve farmers plagued by depleted groundwater sources and high levels of salt. The Bureau of Reclamation planned to send enough water for the more than 11,000 acres of farmland.[21]

To further its support of farming in the West, the Department of the Interior announced in 1948 that it would work throughout the West to more than double the number of irrigated acres by 1954.[22] The aim was to create more than 24,000 individual plots to encourage World War II veterans and others to establish small farms. Planners hoped that more than 100,000 people would thus be encouraged to move west.[23] The question of how to manage some of these plans led to an interesting discussion about regional planning. Some western water developers saw Franklin Roosevelt's Tennessee Valley Authority (TVA) as an excellent model for managing western rivers. In the TVA, water diversion, storage, irrigation, electricity generation, and flood control were all the responsibility of one government body that sought to allocate resources fairly. The developers argued that such a plan in the West might help eliminate some of the long struggles over western water about to erupt into yet another court case between California and Arizona.

In his State of the Union Address in early 1949, President Harry Truman held up the TVA as an example for western water development.[24] As newspaper editorials began to sound off on the subject, many argued vehemently against the idea of "valley authorities" in the West. Some journalists asserted that such valley authorities were created and paid for completely by the government, which then also ran and controlled them—something they viewed as both unconstitutional and un-American.[25] The National Reclamation Association (formed in 1932 and later renamed the National Water Resources Association) was a collection of western politicians and water users who regularly lobbied Congress for federal assistance. In 1949 the organization campaigned against the idea of using the "valley authority" model in the West and met with the president to protest the idea.[26] While Truman argued in favor of the model, it very quickly faded in importance compared with other water issues.

Arguments over the 160-acre limit appeared again in the press in 1949. During election campaigns in California, for example, sitting senator Sheridan Downey opposed candidate Helen Gahagan Douglas, who supported the limit. Douglas argued that violating the 160-acre limit deprived small farmers of their livelihoods. Downey insisted that while her argument made sense, it was not enforceable and it was a waste of time and money to try to enforce it. The reality was big agriculture for California, and the state government should accept that reality and work with it.[27]

Amid all the debate, changes in restrictions for growing cotton posed another problem for the Southwest. Long a staple, cotton had become king in the Southwest during the first half of the twentieth century because of too much moisture and infestation of the boll weevil elsewhere in the country. The Southwest was free of both of these problems, and its aridity provided an additional advantage. In the Southeast, cotton cultivation required a lot of labor in the spring to rid fields of weeds and grass; the Southwest, however, could easily handle weeds and grass with cultivators.[28] The development of mechanized cotton harvesters also enhanced cotton growing in the Southwest, speeding up production and saving labor costs. Because overproduction during the postwar years led to dropping prices and a market glut, government regulations had limited cotton production nationwide to 10 million bales per year but in

1951 raised the limit back to 16 million. This change in regulation spawned speculation in the newspapers over what impacts increased cotton production would have on Arizona water. Some analysts predicted (correctly) that Arizonans would dig more wells and seriously deplete the groundwater in times of drought. Some justified substantial water use because high production was good for showing strength in times of war (the Cold War and the hot war in Korea).[29]

Other discussions about cotton agriculture began to focus on new ways to irrigate. Up to the early 1950s, most cotton farmers watered their crops using flood irrigation, but some farmers began experimenting with sprinkler irrigation. According to one farmer's experience, sprinklers grew cotton just as well and used only a little more than two acre-feet of water compared with flooding, which used four to five acre-feet to grow a full crop.[30] Talk of water conservation was important, although not consistent enough to make a huge difference. Growing cotton in the Southwest yields approximately sixty pounds of lint for each inch of water used.[31] Economist Robert Adler has argued, "If water from Bureau of Reclamation projects was not so heavily subsidized, this economically inefficient use of water would not be possible."[32] This unsustainable relationship between agriculture and the Colorado River began a long time ago, but no one in the 1940s and 1950s was paying attention to water efficiency, only to state allocations and political struggles among themselves. Although not the thirstiest of the southwestern crops, cotton is still the largest water user in Arizona in the twenty-first century.

In 1952 the Colorado Cattleman's Association held its annual convention and publicly blasted price supports and the 160-acre limit as communism. One attendee, a former district attorney from New York who had moved to Colorado, argued that the acre limit rule "spells out a picture of the socialistic tendency to break down and divide those institutions that have made this country great."[33] Arizonans argued that a large cotton crop meant good income for the state, but, as one journalist stated, it was also important for "the clothing industry, defense plants which require the fiber, and . . . the economy of the nation generally."[34] Another journalist editorialized even more adamantly:

> Colorado River water is needed by Central Arizona not merely to keep this fertile region from reverting to the desert but because the United States before long will be in need of all the food it can produce. The world has seen how wars can bring a nation to the verge of starvation . . . We are helping to feed the world now but scientists point out that if the population of America continues to grow, there must come a time when the supply of food will be inadequate . . . Selfish interest cannot much longer be allowed to stand in the way of such a self-liquidating development as the Central Arizona Project.[35]

Similar arguments filled the editorial pages of Arizona newspapers throughout the months leading up to the filing of the *Arizona v. California* US Supreme Court case.

Meanwhile, California not only protested proposals for CAP but also complained that new efforts to enforce the 160-acre limits in California's Central Valley were unfair. No one, Californians argued, was worried about the size of farms to be watered by the Central Arizona Project. Supporters of CAP countered that its water would in fact preserve small, 160-acre farms and protect other small landholders by preserving tenuous groundwater sources.[36]

US secretary of agriculture Charles Brannan had written a report in 1949 about the small family farm and how the 160-acre limit protected this icon of American life. Brannan sought to put teeth into the Bureau of Reclamation's acreage limit by proposing that the US Congress officially ban any federal dollars, credits, and insurance for farms that violated the limit.[37] In spite of Brannan's ongoing efforts in 1952 to bring the law up for debate, Congress did not want to get involved in such legislation. Meanwhile, California upheld the irrigation rights of a farmer in the San Joaquin Valley who sought water for 190 acres. California was not willing to enforce Bureau of Reclamation laws, especially for what it saw as a minor violation.[38]

By mid-1953 western states were clamoring for a repeal of the 160-acre limit. William H. Harrison, a great-great-grandson of the ninth president of the United States, served as a Republican representative to Congress in the early 1950s. In 1953 he argued to the House Reclamation Subcommittee for repeal of the 160-acre limit. He and others asserted that such a limit was not enough land to allow farmers and ranchers to actually make a living and that enforcing the rule would "upset the existing, established farm economy of the area."[39] Meanwhile, the Bureau of Reclamation continued to try to enforce the rule and limit the acreage that could be planted in cotton, a policy that particularly affected Arizona. In 1953 the amount of land in Arizona dedicated to cotton totaled around 645,000 acres, but federal reduction requirements reduced that amount to slightly less than 288,000 in 1954.[40] California argued all the more strongly that Arizona did not need water for agriculture, especially for unsupportable crops such as cotton. For its part, Arizona did begin to diversify its crops but firmly maintained agriculture (and cotton) as the primary source of the state's income.[41]

Arizona v. California and CAP

While the ongoing discussion about acre limits continued, Arizona authorities became convinced that they would have to sue California over rights to water from the Colorado River to fill the proposed CAP canal. California was seeking more and more water to serve a growing urban population, but most of its Colorado River water share went to lucrative agricultural enterprises. Arizona believed the only route to increasing the state's economic status was to follow California's example by increasing irrigated agriculture. The only way to increase this acreage was to obtain more water, and the only source for more water was the Colorado River. Arizona needed to have a court settle the argument over how much water, and then federal monies, would be needed to help build the canal. These issues eventually led to the longest Supreme Court case in US history.

To understand the origin of *Arizona v. California*, one has to reach back to the early twentieth century.[42] The struggle between California and Arizona began a few years before the 1922 Colorado River Compact was put in place, when California requested assistance from the federal government for construction of the All-American Canal. Arizona worried that if the government stepped in to help California divert water from the Colorado River before any kind of apportionment was agreed to, California stood to gain much of its water through prior appropriation. At another fundamental level, Arizona protested the diversion from a states-rights perspective. Clinging to the attitude that a state should be in complete control of all of its natural resources, many Arizona politicians disagreed with any federal intervention and especially with any law that gave control of Arizona water to a body of states. Arizona did not support regional planning, while California was generally a willing partner in such agreements.

During the 1920s, Arizona lacked any real unity on water issues other than the assertion that California was grabbing too much water. While California eagerly participated in the Colorado River Compact and negotiated the 1928 Boulder Canyon Project Act, Arizona stood to the side, refusing to sign and ratify either agreement.[43] When California eventually agreed to limit itself to 4.4 MAF of Colorado River water annually, it eased some of the concerns of states such as Colorado and Nevada.[44] However, Arizona contended that neither compact did anything to protect water for Arizona. The 1922 compact did not guarantee Arizona a specific amount of water and placed water control outside the state. In addition, the 1928 act gave power to Los Angeles and the All-American Canal to the Imperial Valley at the expense of other states.

Finally, in 1930, Arizona filed suit in the US Supreme Court, protesting that both the Colorado River Compact and the Boulder Canyon Project Act were unconstitutional.[45] The suit was swiftly dismissed, and construction began on Boulder Dam and the All-American Canal. As the rest of the decade passed, there was still no official division of water among the lower basin states because of this ongoing disagreement. Arizona's governor Benjamin Baker (B. B.) Moeur refused to attend the dedication of Hoover Dam in 1935 because of the lack of a clear division. Hoover Dam would help California take more than its rightful share of water, in Arizona's estimation. Time had passed, Hoover Dam had been built without Arizona's signature, and an agreement with Mexico was the only progress being made in lower basin allocation agreements.

The fears voiced by Arizonans as early as the 1930s indicated the struggles to come as the state's politicians began to argue for federal help so Arizona could take its share of Colorado River water. In a December 15, 1938, interview with the newspaper *Californian,* Arizona governor Rawghlie Clement Stanford argued that "development of the Colorado river was a possible step in 'returning the nation to economic normalcy.'" In his view, irrigating Arizona's deserts could be one of the solutions to the Great Depression by opening up more land for unemployed farmers. "Arizona beckons to the world," he stated. "We have millions and millions of acres, but we need water, and that water should come from the Colorado River."[46]

The next day, an article in the *Albuquerque Journal* made a similar argument, citing Governor Stanford's statement about Arizona. The reporter agreed with Stanford and added that New Mexico also had a lot of land and water allocations that had never been put to use.[47] Soon, the issue of bringing Colorado River water to both states became the clarion call of the new decade. One editorial in the *Chandler Arizonan* in late 1940 illustrates this growing agenda among many state politicians and other boosters:

> Over on the coast, when their underground water began to run short, they went up and got the Owens River water; that wasn't enough, so they came to Arizona and got the Colorado River. Smart boys, those Los Angeles business men. But why doesn't Arizona do likewise? Sure, it will cost money ... Why doesn't this state get busy now, before the underground water supply is exhausted? There isn't any immediate danger in Arizona ... but ... the future of Arizona's vast agricultural industry depends largely on getting water from the Colorado River—and don't let anyone persuade you differently.[48]

Others continued the same argument as the United States began extending lend-lease aid to Great Britain. Even before the attack at Pearl Harbor, New Mexico and Arizona were arguing that developing Colorado River resources would help the war effort. In reality, they were very worried about California's growing aircraft industry and the power demands it required. If California needed more electricity, the state would soon demand Arizona's share, not just its own share of water.[49] If Arizona didn't hurry, nothing would be left. New Mexico heard what Arizona was saying and began its own call for action in its newspapers, pleading that the state must act quickly lest Arizona be the only one to benefit.[50]

Finally, by 1940 Arizona began to develop what one could call a unified stance on, or approach to, water issues. The states-rights argument was set aside as it became very clear that the region needed federal help to develop water resources from the Colorado River and that California had made enormous strides ahead of everyone else since it had already recognized this fact. Along with the developing war and California's expansion of Colorado River water usage, Arizona experienced a severe drought that would extend through most of the decade. Its population was expanding rapidly, with an increase of 67 percent between 1920 and 1940, followed by a veritable explosion of another 200,000 residents between 1940 and 1945.[51] This expanding population needed water and more electricity. It was time to cooperate with regional compacts and the Bureau of Reclamation.

As the war years unfolded, others argued that Arizona needed water to provide reclaimed land for returning veterans once the war was over. The prediction was that Maricopa County, then supporting a population of approximately 200,000, would easily grow and support 500,000 people if Colorado River water were obtained.[52] By the time the war ended, it seemed everyone was getting water except Arizona. By 1946 the Colorado River boasted four major dams (Hoover, Parker, Imperial, and Laguna), with Davis Dam under construction and serious plans being laid for Glen Canyon Dam and a dam in the Grand Canyon. The dam in Glen Canyon would eventually be the cornerstone of the 1956 Upper Colorado River Basin Storage Act. The proposed dam in the Grand Canyon was supposed to provide Arizona with water storage and electricity.[53]

Not everyone agreed, however, that Arizona or anyone else needed to develop more dams and diversion projects on the Colorado River. On February 22, 1947, a long article on this subject appeared in the *Saturday Evening Post*, replete with Bureau of Reclamation photos of Hoover Dam and a map of the river. Journalist Edward Churchill voiced the opinions of some observers regarding federal spending, the desert Southwest, and the potential end of its resources. While he was not completely critical, the title "Shall We Spend $2,000,000,000 More on the Colorado?" and his opening paragraph certainly implied that further spending was fruitless:

> One of these days, probably toward the end of the second generation hence, somebody is going to cross the Rockies heading in a southwesterly direction and make history . . . If he comes by train no band will meet him at the station; if by car no parade will gather at the city limits; if by plane no dignitaries will greet him as he alights. No one will recognize him for what he is—the last person the resources of the Southwest can sustain.[54]

Sounding like John Wesley Powell, Churchill went on to describe the costs of Hoover and the other three dams already in existence on the Colorado River. Listing the estimated costs of planned projects, he wondered whether all the money was worth it. Although he talked about expense versus return and silt problems, the most poignant point he made was regarding flow. Even though the Bureau of Reclamation would not allow itself to admit to over-allocations of the Colorado River for almost another decade, Churchill clearly argued that allocations divided up a mythical 10.9 million acre-feet per year while actual flows in 1947 were running at around 8.5 MAF.[55]

People as far away as Denver began to worry about California and Arizona conflicts over water, fearing that all resources would soon be dried up, with none left for the upper basin. A provocative cartoon that originally appeared in the *Arizona Republic* (Phoenix) was reprinted in the *Denver Post* in September 1947. In it, a large, menacing sailor, representing California, controls access to a barrel of drinking water six other small sailors desperately need as they ride on a tipsy raft on a choppy sea. California uses his strong muscles to keep the others away and rations the water, saying, "What are you guys worried about? I ain't started drinkin' YOUR SHARE—yet."[56]

Meanwhile, California and Arizona engaged in an editorial shouting match regarding who deserved Colorado River water and how much they should get. One editorial in the *Arizona Times* argued that the newspapers were fanning the flames and creating a false sense of urgency. Apparently, journalists had been referencing an ongoing drought, using it as the reason Arizona needed the proposed Central Arizona Project pipeline. In contrast to the story in the *Saturday Evening Post*, this writer argued that Arizona would be fine, with or without Colorado River water:

> There is absolutely no danger that this valley wonderland of ours, wrung from the desert, will dry up and revert to the desert, the current hysteria to the contrary notwithstanding. If worst comes to worst, we will have to taper off our agriculture a little. But we still have our incomparable climate. We still have our constantly growing tourist trade. We still have cattle. We still have great untapped potentialities for light industry. Arizona, and all the wondrous things which the name connotates, are here to stay.[57]

While this editorial sounded a positive note for Arizona, the emotional exchanges increased as 1948 brought yet another year of lingering drought.

Another important worry that gained voice in the region was the need for more power generation. Nevada needed more power and hoped to obtain some from Davis Dam, still under construction.[58] Soon, Arizona was also complaining about power shortages and predicting brownouts. More power generation was needed from the Colorado River, and Arizona meant to have its share.[59] Finally, in late 1948, Arizona and the Bureau of Reclamation finished plans to bring power to Arizona from Hoover and Davis Dams at a cost of almost $55 million. Maps showing the power lines stretching out from the Colorado River followed the general pathway the Central Arizona Project canal would eventually take.[60]

Still, power generation took a backseat to simple access to water for municipal and agricultural purposes. Arizona had reason to worry, as did the other Colorado River states, because California was indeed looking eagerly at the river's water. In the years just before the last of three Supreme Court cases Arizona filed against California, the latter's population had exploded in the southern part of the state. According to newspaper stories, the population of Los Angeles had increased by more than 1 million people between 1941 and 1948. The immediate future for it and other cities in Southern California was dependent upon the Colorado River.[61]

In a speech to the US Congress on April 6, 1948, Congressman Norris Poulson stated California's case poignantly:

> Without it [the Colorado River] vast areas of Southern California would be barren desert, burning and unproductive as they were through the thousand years before man came to them. Without this dependable water, there could be no great cities, no immense groves and vineyards, no fine pastures, no industry. And without it there could be no assured future for millions yet unborn. Progress and development would soon reach a saturation point, and California would of necessity say to those who wished to live there and build: You cannot come here, for we cannot provide for you . . . That is why Southern California must be assured that in its aqueducts and canals, water must always flow from the Colorado River. The water from the Colorado River is Southern California's lifeblood. We must never relax our vigilance in safeguarding it. We must never fail to preserve it. We must always be on guard against those who would take some of it from us. For without it we cannot exist.[62]

These strong words made it clear to everyone that California was likely to grab all of the water in the river unless it was stopped. Colorado and the other upper basin states paid attention and hastily drew up their own apportionment plans. California looked to the Columbia River as a possible supplement, while Arizona continued its battle of words.[63]

Central to all these conflicts was the issue of the CAP, designed to draw off Arizona's share of the Colorado River if the actual amount was ever agreed upon. As early as the mid-1880s, territorial governors of what would eventually be the state of Arizona had talked about the need for canal projects to bring Colorado River water into the desert. Nathan Oakes Murphy, territorial governor from 1892 to 1893 and again from 1898 to 1902, first proposed a canal that would run from a dam at the end of the Grand Canyon down to Phoenix.[64]

In 1943 Senator Carl Hayden quietly put forward a proposal to Congress for the Bureau of Reclamation to conduct a survey of the lower river basin. He argued that it would be good to identify beneficial projects to employ veterans when they began returning from the war. Congress approved the survey, and the bureau subsequently identified several such projects; one was for a dam at Bridge Canyon to provide hydroelectricity and a possible canal to central Arizona. Several possible locations for such a project were investigated, and a plan to start a canal at Parker Dam proved the most affordable. Arizona formed the Arizona Power Authority and the Arizona Interstate Stream Commission to support developing plans for a canal.

In 1946, boosters in Arizona formed the Central Arizona Project Association to lobby Congress and convince the public that such a canal was needed. In a 1947 feasibility report, Secretary of the Interior Julius Albert Krug argued that Arizona desperately needed the water—and soon. Phoenix was rapidly expanding, and Tucson was suffering water shortages as groundwater levels dropped severely.[65] By 1949 the association's efforts were creating deep concern in California. Los Angeles newspapers began calling Arizona's plans a "water grab" and started a publicity campaign against the proposed legislation. The real fear was that there was not enough water in the river for the CAP, and its construction might require California to give up some of its own water. The Colorado River Association of California began its own campaigns through newspaper ads that highlighted Arizona and claimed that one state was demanding federal resources that rightfully belonged to all western states equally. California claimed in full-page advertisements that Arizona was asking for more than 700 million federal dollars—almost as much as eleven western states combined had received in the past ten years.[66]

Another series of advertisements asked Californians to fill out form letters to their senators and members of Congress protesting Arizona's plans for CAP. "Why should your Senators and your Congressman vote for something that would cost you money and give you no benefits?" one advertisement asked.[67] Other headlines encouraging Californians to oppose CAP asserted: "Cotton and Alfalfa Will Thrive in Arizona While Electric Bills Go Up in California."[68]

A pair of interesting editorials gives an idea about the depth of the suspicion between California and Arizona. In one editorial from Palm Springs, California, in 1949, the author wrote:

> Our water life-line from the Colorado River is being threatened by the attempts of Arizona promoters to grab more and more of the precious stuff . . . If the Arizona "grabbers" succeed with their scheme, the result will be a drastic cut in the amount of Colorado River water and power vitally needed to serve the southland . . . Property values will be affected. Employment opportunities will be restricted. Industry and business

will stop growing. Agricultural expansion will be finished. It is an economic life and death struggle for terrific stakes and we Californians can't afford to lose.[69]

The language in Arizona was equally dramatic: "California will spend any amount of money and stoop to any trick to cheat Arizona of Arizona's own share of the Colorado River . . . California twists truths and half-truths, juggles figures, makes comparisons that are not comparisons at all, tries to frighten everyone, tries to bull-doze and fabricate a way to get the whole of the Colorado river."[70]

Other states soon joined California in protesting the Central Arizona Project. Nevada claimed it would not receive water it needed for defense industries and important air force bases.[71] Arizona countered that important defense needs would not be met if drought-plagued Arizona could not grow its crops. In a letter to the editor of the *Arizona Republic* in 1951, one reader argued that California should want Arizona to feed California as it concentrated on growing its defense industry sector.[72] Farmers in Colorado argued that irrigated agriculture was as necessary as bombs to defend freedom in the midst of the Cold War.[73]

As the feud in the lower basin became more vehement, the upper basin states also began to voice concern. Editorials in Utah and Colorado began to urge upper basin planning and water diversion before no water was left. One Utah article argued that the upper basin needed a plan to put before Congress that would utilize all of its water within a decade.[74] Words were much stronger in Colorado:

> Before many years, we in Colorado are going to be fighting California and Arizona for our water. We should be putting this water to a beneficial use as they are doing . . . If this water is not used by us, someone else will use it and we will lose our rights . . . There are too many dry spots in this state. Why? We're just waiting for it to rain—and four rivers full of water [are] going to the Hoover Dam.[75]

This kind of rhetoric did not abate over the next fifteen years.

While the debate continued in the press, Arizona quietly put forward legislation for CAP, first in 1946 and again in 1947. Neither bill got very far, but support was building. In 1950 the US Senate passed a version of a bill for CAP, much to Arizona's delight. The bill stalled in the US House of Representatives, however, as California flooded the committee with arguments as delaying tactics. The bill was again tabled. When it was reintroduced in both houses the following year, the bill died quickly as it became clear that California and Arizona would not agree on water alloca-tions. Congress declared that a settlement had to be reached before CAP could be approved. Arizona again filed suit in the Supreme Court in early 1952.[76]

Through the suit's fifty-four pages of motions and complaints, Arizona lawyers argued that the state's primary source of income—agriculture—desperately needed

Colorado River water to survive. The population was growing, and its urban centers would soon need water almost as badly as agriculture did. Because the lawyers did not want to recognize California's prior appropriations of Colorado River water, Arizona's legal team based its arguments on demands for equity. All the states in the river basin should share its waters fairly, regardless of prior appropriations or legal compacts.[77]

During the first few years of the court case, attorney John P. Frank served as Arizona's chief litigator. His opponent was the experienced California water lawyer Northcutt Ely. At first, these two men and their respective teams made arguments in front of a special master appointed by the Supreme Court to hear testimony and form an opinion on the case before it was heard by the court. The first special master, George I. Haight from Chicago, began hearing preliminary motions in 1954, but later that year he died suddenly. His replacement, Simon Rifkind from New York, began hearing testimony in 1955. The following year formal hearings began, but they were delayed when Rifkind suffered a heart attack in early 1957.[78]

During the early hearings and testimony, Arizona lawyers argued that the prior appropriation policy should be set aside for the good of all states in the basin. California, on the other hand, kept carefully to its interpretation of the law, bringing forward witness after witness to establish water rights through prior appropriation.[79] It seemed that Arizona did not have a legal basis to stand on. Many in the state began complaining that attorney Frank had no coherent strategy and that the case was going badly. California argued its case slowly and methodically, asserting prior appropriation rights to more than 5.3 MAF of Colorado River water annually. Any assumptions that California should only have 4.4 MAF came from negotiations over the Boulder Canyon Project Act in 1928, which Arizona had long refused to sign and whose legality the state continued to debate. California intended to appropriate as much water as possible while the court case dragged on, sticking firmly to already established water rights law. Arizona seemed to be pleading that the law did not matter or should not apply to the Colorado River. Arizona papers published stories about how the case was being lost.[80]

By the end of 1957, it seemed that California would eventually win the court case, limiting Arizona to perhaps as little as 1 MAF per year of mainstream Colorado River water—not enough to justify building the CAP canal. Wisely, Arizona decided to try a new approach by shaking up its legal team. Mark Wilmer, originally from Wisconsin, had been practicing law in Phoenix since 1931. This local man proved to be Arizona's salvation as he took over the reins of an almost lost court battle. Wilmer saw immediately that Arizona had been taking the wrong approach to the case from the beginning and took a huge risk by changing the entire game plan. While Frank had argued that prior appropriation law and other compacts should be set aside for the sake of equality and fairness, Wilmer used the law as the basis of the case.[81]

On August 5, 1957, Wilmer filed the "Amended and Supplemental Statement of Position by Complainant, State of Arizona." In this document, Wilmer embraced prior appropriation law and argued that Arizona had appropriated and proven its claim to all of the Gila's waters prior to the 1922 compact. The compact had referred to "perfected rights" that should not in any way be affected by the new compact document. In other words, all states could keep the tributary waters they were already using prior to 1922. The compact meant only to divide the waters not already so appropriated—the main stream of the Colorado River. Wilmer further argued that according to precedent, the federal government had the right to further apportion the river water. Previous cases had affirmed that whoever built a reservoir had the right to distribute that water. In this case, the entity that held the right was the federal government.[82]

Thus instead of trying to argue against the 1922 compact or the 1928 act or even against prior appropriation, Wilmer illustrated over the next several years that Arizona had a legal right to its water and to at least 2.8 MAF of mainstream water per year. California, in spite of its demands for 5.3 MAF through prior appropriation, was violating legal divisions of water that limited California to 4.4 MAF. Part of the problem was that California had reluctantly agreed to that limit in 1929 to obtain ratification of the 1928 Boulder Canyon Project Act. Arizona intended to hold California to this limit and assert its rights to 2.8 MAF of mainstream water, as outlined in the 1928 act. It was indeed a "shift from a case based on equity to a case based on statutory apportionment."[83]

By 1960, California had made all the rebuttal arguments it could but recognized that Arizona had completely turned the case around. Special Master Rifkind wrote his opinion siding with Arizona and sent it on to the Supreme Court. In 1962 the Court heard arguments and finally ruled in Arizona's favor on June 3, 1963.[84] The implications of this Court case for river users were many. First, the road was now open for Arizona to again seek legislation for the Central Arizona Project. Second, California would now be forced to seek additional water from other sources and find ways to reduce its Colorado River water usage to its legal limits. Third, federal recognition of American Indian water rights was reaffirmed, with the burden resting on states in which an Indian reservation resided. This aspect of the ruling placed a very heavy burden on Arizona, which held the largest proportion of reservation land entitled to Colorado River water (see chapter 6). Finally, the Court ruling firmly placed federal control over interstate waters, putting states-rights issues to rest at long last. Of course, none of this impact meant the Colorado River would now be apportioned peacefully. While most of the legal battles had finally ended in the courtrooms, they simply moved to the legislative branch of the government.

Upper Basin Projects

While the lower basin argued over allocation, the 160-acre limit, and the Central Arizona Project, the upper basin watched with concern. California was using its full 4.4 MAF of water per year, benefiting from Hoover Dam electricity, Parker Dam, the All-American Canal to the Imperial Valley, and the Colorado River Aqueduct to Los Angeles. It was also planning other water diversions, such as the Central Valley Project, to bring water from the northern part of the state.[85] The upper basin feared the lower basin would soon claim the entire river by prior appropriation.

If John Wesley Powell had been around, he would have been surprised that the upper basin eventually got *any* irrigation projects. The contrast in climate between the upper and lower basins is fairly extreme. There is nearly a mile difference in elevation for much of the upper basin compared with the lower basin, as well as a much shorter and colder growing season. To lower basin farmers, irrigation agriculture made sense in California and Arizona, where one could grow many vegetables and fruits in high demand, but the only crops and livestock that could do well in the upper basin were hay, alfalfa, corn, and cattle. Why should the upper basin be allocated as much water as the lower basin? This is yet another example of how prior appropriation law and state politics encouraged people in the Southwest to ignore the long-term consequences of their use of natural resources. If states had managed some kind of basin-wide approach, as Powell tried to do, numerous water distribution issues would look different. Both the upper and lower basins were too busy grabbing what water they could instead of seeking a sustainable relationship with that water.

Seeking to utilize their half of the Colorado River as soon as possible, representatives from the upper basin states met in 1948 and hammered out an agreement, as required before any water could be diverted.[86] The 1948 Upper Colorado River Basin Compact gave percentages of water to the four states instead of set amounts in acre-feet, recognizing that the amount of water that might actually be available would likely be much different from estimated flows used in the original 1922 Colorado River Compact. In the 1948 agreement, Colorado was allotted 51.75 percent of the water, Utah 23.00 percent, Wyoming 14.00 percent, and New Mexico 11.25 percent. Northern Arizona was given 500,000 acre-feet of the water annually, and the act also established a permanent commission to settle any disagreements that might occur. This important law of the river made it possible for the upper basin states to appeal to Congress and the Bureau of Reclamation for help in developing their water shares.[87]

At about this same time, the bureau developed an important concept that would make irrigation projects in the upper basin possible: the cash register dam. In fact, the bureau had already built its first such dam. Although significant irrigation came from the water storage in Lake Mead, Hoover Dam's electricity generation was its primary value, especially for the growing metropolis of Los Angeles.[88] Once the Great Depression was over, the Bureau of Reclamation could find the financial resources to build dam projects that would generate large amounts of electricity. These dams would pay not only for themselves but also, through creative accounting, for expensive and unprofitable irrigation projects.

In the late 1940s, Bureau of Reclamation commissioner Michael W. Straus loved building huge dams. The upper basin worked quickly to cultivate his support for irrigation diversion projects, and Straus responded with a unique plan he called River Basin Planning, or River Basin Accounting.[89] Informed by the Tennessee Valley Authority example, Straus believed building large dams to generate and sell electricity could fund irrigation projects for farmers who would never make enough profit to pay for them. He felt such large projects were justified by the accounting of the entire basin, not the individual pieces of the plan. This idea was what it would take to have any irrigation projects built in the upper plains where, if measured individually, they would never be cost-effective.

What the upper basin lacked in good, irrigable farmland, it had in abundance in good, deep-river canyon locations for large dams. Since this kind of project most appealed to Straus anyway, he soon became a friend of the upper basin. In 1953, just before he was removed from office by the incoming Eisenhower administration, Straus put before Congress his large plan: the Upper Colorado River Basin Storage Project.[90] Some politicians understood that expenditures for such projects were foolhardy. Their farmers were being paid not to grow the kinds of crops the upper basin would grow poorly with bureau projects. Western politicians, however, argued that they deserved the projects. There were also powerful members of Congress from the West, such as Wayne Aspinall of Colorado, who eventually controlled the House Interior and Insular Affairs Committee with an iron fist.[91]

Aspinall's rise to power in the US House of Representatives is a fascinating story of a man who believed first and foremost in the benefits of growth in the West. Having grown up in a family that supported itself with a peach orchard, Aspinall learned very early the importance of water to Colorado. After studying law and serving for a decade in the state legislature, Aspinall won election to the US Congress in 1948, serving until 1973. He spent most of those years chairing the powerful House Interior and Insular Affairs Committee. Aspinall remembered the impact water resources had had on his family orchard and had witnessed the remarkable rise in the family's fortune as a result of irrigation water.[92]

Several important issues and events converged to convince Aspinall that Colorado needed federal reclamation projects. Experiences drawn from the Great Depression and New Deal policies demonstrated the benefits of federal projects for agriculture that Aspinall believed could and should be applied to the West. He, like many fellow western politicians, believed the rest of the country had benefited from federal projects and it was now the West's turn. Specifically, Aspinall believed California had already received more than its share of reclamation projects: the Hoover Dam, the All-American Canal to the Imperial Valley, and the Colorado River Aqueduct to Los Angeles.

During the war years, Colorado had seen dramatic growth around Denver and the eastern slope of the Rocky Mountains. Aspinall and others used Cold War rhetoric to argue the need for more federal support for Colorado industry, and that meant water. He also shared with many others the belief that science and technology could solve all problems. All society had to do was apply new research and solutions to create long-term growth.[93] For Aspinall, the environment and water were resources to be used. The West's natural resources could and should push western states into a place of importance at long last. It was the West's turn to lead, and Colorado needed to participate.

Western politicians from the lower basin, such as Carl Hayden of Arizona, also supported the upper basin projects, hoping those states would be allies in Arizona's development plans once it solved its dispute with California. In many ways, it was all the western states versus California; they all stood to lose their water to California if it was not used as soon as possible. Even though California had been granted a specific amount of Colorado River water, it was already using more than that share, and the other basin states feared laws of prior appropriation might be used to give that water to California permanently.

The 1956 project included plans for six major dams: the Navajo Dam on the San Juan River in New Mexico, Cross Mountain Dam on the Yampa River in Colorado, Flaming Gorge Dam on the Green River in Utah, Echo Park Dam at the confluence of the Green and Yampa Rivers in Colorado, Curecanti Dam on the Gunnison River in Colorado, and Glen Canyon Dam in northern Arizona.[94] Congressman Aspinall made an impassioned speech in support of the project in early 1956, arguing that although some conservationists were upset about dams that might flood national park areas, the project was necessary. "Put yourself in that farmer's shoes," he urged. "Can you say to that man, 'sorry, but I refuse to do anything for you?' All this man asks is the right to work hard, to put to use a God-given resource . . . can you deny him that opportunity?"[95] As environmentalist voices grew louder in protest over dam proposals, Aspinall convinced Congress that without the act, western farmers in the upper basin would not survive.

In many ways, the years leading up to the passage of the project bill indicated a shifting of attitudes toward both natural resources and human interaction with the environment. No one talked about sustainability yet. Very few admitted that some of the so-called triumphs of technology and human ingenuity had serious drawbacks or unanticipated consequences. Yet times were changing, and soon a new kind of power player emerged in the political struggles over the Colorado River. Although faintly at first, people began to hear the river's voice.

NOTES

1. Colorado River Water Users Association, Agriculture, 2007, http://www.crwua.org /ColoradoRiver/RiverUses/agriculture.aspx (accessed August 12, 2010).

2. Save the Colorado Association, Agriculture, 2010, http://www.savethecolorado.org /river.php (accessed March 22, 2011).

3. US Department of Agriculture, Economic Research Service, The Economics of Food, Farming, Natural Resources, and Rural America, July 20, 2004, http://www.ers.usda.gov /data/westernirrigation/ (accessed July 21, 2010).

4. Ibid.

5. US Geological Survey (USGS) Irrigation Withdrawals by Source and State, 2000, http://pubs.usgs.gov/circ/2004/circ1268/htdocs/figure07.html (accessed July 21, 2010).

6. For graphs illustrating the impact of irrigated agriculture in the West as of 1998, see Western Water Policy Review Advisory Commission, *Water in the West: The Challenge for the Next Century* (Boulder: Western Water Assessment, June 1998), appendixes A1–A6.

7. Estimated Water Use in the United States in 2000, http://pubs.usgs.gov/circ/2004 /circ1268/htdocs/table07.html (accessed July 21, 2010).

8. Agriculture, 2007, http://www.crwua.org/ColoradoRiver/RiverUses/agriculture.aspx (accessed July 21, 2010).

9. Estimated Water Use in the United States in 2000, http://pubs.usgs.gov/circ/2004 /circ1268/htdocs/table07.html (accessed July 21, 2010).

10. Nevada Department of Agriculture, Agriculture in Nevada, http://agri.nv.gov /AgInNevada.htm (accessed September 15, 2012).

11. US Bureau of Reclamation, 2003 Quality of Water Colorado River Basin, http://www .usbr.gov/uc/progact/salinity/pdfs/PR21.pdf (accessed March 24, 2011).

12. Ibid., 15–16.

13. Robert W. Adler, *Restoring Colorado River Ecosystems: A Troubled Sense of Immensity* (Washington, DC: Island, 2007), 52.

14. USGS, Water Resources of Colorado, Fact Sheet, Distribution and Concentrations of Selected Organochlorine Pesticides and PCBs in Streambed Sediment and Whole-Body Fish in the Upper Colorado River Basin, 1995–1996, http://pubs.usgs.gov/fs/fs167-97/ (accessed March 24, 2011).

15. See Charles V. Moore, "Impact of Federal Acreage Limitation Policy on Western Irrigated Agriculture," *Western Journal of Agricultural Economics* 7 (December 1982): 301–16; Milton R. Copulos, "Enforcement of an Anachronism: The 160 Acre Limitation," December 15, 1977, Policy Archive for Non-Partisan Public Policy Research, http://www.policyarchive .org/handle/10207/bitstreams/8434.pdf (accessed August 12, 2010).

16. Pat Laflin, "The Salton Sea: California's Overlooked Treasure," *Periscope* (Indio, CA: Coachella Valley Historical Society, 1999), http://www.sci.sdsu.edu/salton /PeriscopeSaltonSeaCh5-6.html (accessed September 10, 2010).

17. *Imperial Valley Press* [El Centro, CA], June 19, 1946, Box 3, Folder 1947, Colorado River Board Collection (hereafter CRBC), Water Resource Center Archives (hereafter WRCA), University of California, Berkeley.

18. "Coachella Valley Vote Insures Water Supply Bureau of Reclamation Contract for Colorado River Distribution Laterals Approved," *Los Angeles Times*, December 18, 1947, Box 5, Folder Coachella, CRBC, WRCA.

19. *Riverside Enterprise*, August 9, 1947, Box 3, Folder 1947, CRBC, WRCA.

20. *Las Vegas Review*, July 6, 1948, Box 17, Folder 2, CRBC, WRCA.

21. *Yuma* [AZ] *Sun and Sentinel*, November 30, 1949, Box 3, Folder 2, CRBC, WRCA. The bureau anticipated growth and designed the Wellton-Mohawk Project to serve 75,000 acres. According to 2004 numbers, the project provided irrigation for just under 63,000 acres.

22. "Dreams May Come True," *Arizona Republic* [Phoenix], November 18, 1948, Box 5, Folder Colorado River 1948, CRBC, WRCA.

23. See Brian Q. Cannon, *Reopening the Frontier: Homesteading in the Modern West* (Lawrence: University of Kansas Press, 2009).

24. *Inglewood* [CA] *News*, January 6, 1949, Box 5, Folder Colorado River, CRBC, WRCA.

25. "Siding with Reclamation," *Deseret News* [Salt Lake City], October 30, 1949, Box 13, Folder 1, CRBC, WRCA.

26. Collection of newspaper articles, Box 13, Folder 1—National Reclamation Association, 1947–1950, CRBC, WRCA. See also Clayton R. Koppes, "Public Water, Private Land: Origins of the Acreage Limitation Controversy, 1933–1953," *Pacific Historical Review* 47, no. 4 (November 1978): 607–36.

27. *Los Angeles Times*, December 7, 1949, Box 1, Folder 1, CRBC, WRCA.

28. Moses S. Musoke and Alan L. Olmstead, "The Rise of the Cotton Industry in California: A Comparative Perspective," *Journal of Economic History* 42, no. 2 (June 1982): 389. See also Cameron Lee Saffell, "Common Roots of a New Industry: The Introduction and Expansion of Cotton Farming in the American West," PhD diss., Iowa State University, Ames, 2007.

29. Westbrook Pegler, "Too Much Cotton Endangers Arizona," *Arizona Republic*, May 22, 1951, Box 7, Folder Crops, CRBC, WRCA.

30. "Cotton under Sprinklers," *Arizona Farmer* [Phoenix], September 1, 1951, Box 7, Folder Crops, CRBC, WRCA.

31. "Cotton and Natural Resources," Cotton Today, 2011, http://cottontoday.cottoninc .com/natural-resources/water/ (accessed March 25, 2011).

32. Adler, *Restoring Colorado River Ecosystems*, 253.

33. Ralph Partridge, "Cattlemen Rap 'Socialist Trend' in Government," *Denver Post,* May 24, 1952, Box 5, Folder Colorado, CRBC, WRCA.

34. "Proving Value of Water," *Arizona Republic*, August 10, 1952, Box 7, Folder Crops, CRBC, WRCA.

35. "Food for the Future," *Arizona Republic*, March 6, 1952, Box 2, Folder Arizona Propaganda, CRBC, WRCA.

36. Editorial, *The Mirror* [Los Angeles], August 25, 1952, Box 1, Folder 1, CRBC, WRCA.

37. *Denver Post*, September 19, 1952, Box 1, Folder 1, CRBC, WRCA.

38. *Los Angeles Daily News*, October 3, 1952, Box 1, Folder 1, CRBC, WRCA.

39. *Arizona Republic*, April 21, 1953; *Los Angeles Times*, April 21, 1953, both in Box 1, Folder 1, CRBC, WRCA.

40. *Arizona Farmer-Ranchman* [Phoenix], December 5, 1953, Box 1, Folder 2, CRBC, WRCA. In 1954, Arizona irrigated approximately 1.17 million acres of farmland. California irrigated a little more than 7 million acres. See Henry Dworshak, "Remarks on Agriculture and Irrigation," US Senate Testimony, 1959, Box 1, Folder 8, Elmer K. Nelson Papers, WRCA.

41. For a discussion of federally subsidized water for western agriculture, see Terry L. Anderson and Donald R. Leal, "Priming the Invisible Pump," in Peter G. Brown and Jeremy J. Schmidt, eds., *Water Ethics: Foundational Readings for Students and Professionals* (Washington, DC: Island, 2010), 91–104.

42. *Arizona v. California,* 373 U.S. 546 (1963). The best work on this court case is Jack L. August Jr., *Dividing Western Waters: Mark Wilmer and Arizona v. California* (Fort Worth: Texas Christian University Press, 2007).

43. Residents of Graham County, Arizona, Oppose the "Swing-Johnson Boulder Canyon Dam Bill," petition, January 14, 1927, Colorado River Central Arizona Project Collection (hereafter CRCAPC), Arizona State University, Tempe. Digitized for Western Waters Digital Library, http://repository.asu.edu/items/12285 (accessed August 13, 2010).

44. See page 5 of Transcript, Boulder Canyon Debate, December 11, 1928, CRCAPC, http://repository.asu.edu/items/12277 (accessed August 13, 2010).

45. *Arizona v. California et al.*, Papers of Delph E. Carpenter and Family, Water Resources Archive, Colorado State University, Fort Collins. Digitized copy at http://digitool.library .colostate.edu:80/R/?func=dbin-jump-full&object_id=80767 (accessed August 13, 2010).

46. Article, author unknown, *Californian* [Bakersfield], December 15, 1938, Box 1, Folder 7, CRBC, WRCA.

47. Article, author unknown, *Albuquerque Journal*, December 16, 1938, Box 1, Folder 13, CRBC, WRCA.

48. Author unknown, "Arizona Must Have Water from the Colorado," *Chandler Arizonan*, December 13, 1940, Box 1, Folder 7, CRBC, WRCA.

49. Miscellaneous articles, esp. in *Chandler Arizonan*, January 10, 1941, Box 1, Folder 1941, CRBC, WRCA.

50. Author unknown, *Las Vegas, NM, Optic*, April 6, 1942, Box 1, Folder New Mexico: 1939–1945, CRBC, WRCA.

51. August, *Dividing Western Waters*, 48.

52. Franklin D. Lane, Editorial, *Phoenix Gazette*, October 6, 1943, Box 1, Folder Arizona, CRBC, WRCA.

53. Author unknown, "Celebration Today Marks 10th Anniversary of Giant Project," *Los Angeles Daily News*, October 23, 1946, Box 12, Folder Misc. 1946, CRBC, WRCA.

54. Edward Churchill, "Shall We Spend $2,000,000,000 More on the Colorado?" *Saturday Evening Post,* February 22, 1947, 28–29, Box 12, Folder Misc. 1946, CRBC, WRCA.

55. Ibid.

56. Reg Manning, artist, "Over-Weaning" (cartoon), *Denver Post*, September 22, 1947, Box 5, Folder Colorado/Big Thompson Project, CRBC, WRCA.

57. Editorial, author unknown, *Arizona Times*, May 29, 1947, Box 1, Folder 1947, CRBC, WRCA.

58. *Reno Gazette*, April 6, 1948, Box 1, Folder 3, CRBC, WRCA.

59. *Arizona Republic*, May 4, 1948, Box 2, Folder Arizona Power, CRBC, WRCA. See also *Arizona Republic*, August 5, 1948, Box 12, Folder 2, CRBC, WRCA.

60. *Arizona Star* [Tucson], October 30, 1948, Box 2, Folder Arizona Power, CRBC, WRCA.

61. "Heavy Drain on Water from Colorado Cited," *Los Angeles Times*, July 15, 1948, Box 5, Folder Colorado River, CRBC, WRCA.

62. Norris Poulson to Congress, April 6, 1948; reprinted in *Los Feliz Hills News* [Los Angeles], July 22, 1948, Box 5, Folder Colorado River 1948, CRBC, WRCA.

63. *Los Angeles Times*, July 15, 1948, Box 5, Folder Colorado River, CRBC, WRCA; *Pasadena Star News,* December 28, 1948, Box 1, Folder 1, CRBC, WRCA; *Phoenix Times*, December 6, 1948, Box 1, Folder 1, CRBC, WRCA.

64. August, *Dividing Western Waters,* 31.

65. Julius Albert Krug, "Report on Feasibility Bridge Canyon Route Central Arizona Project," Project Planning Report no. 3-8b.4-1, February 1947, CRCAPC, http://repository .asu.edu/items/12263 (accessed September 18, 2012).

66. *Arizona Republic,* August 14, 1949, Box 6, Folder 1, CRBC, WRCA.

67. *Phoenix Times,* August 19, 1949, Box 6, Folder CRA, CRBC, WRCA.

68. *Los Angeles Times,* June 2, 1949, Box 6, Folder 5, CRBC, WRCA.

69. "Not a Drop to Drink," *Limelight News* [Palm Springs, CA], March 15, 1949, Box 3, Folder 1949, CRBC, WRCA.

70. "Official Reveals Fortune Is Spent on Propaganda," *Arizona Republic,* August 14, 1949, Box 6, Folder 1, CRBC, WRCA.

71. *Los Angeles Examiner,* March 18, 1951, Box 13, Folder NE, CRBC, WRCA.

72. Letter to the Editor, *Arizona Republic,* March 29, 1951, Box 7, Folder Crops, CRBC, WRCA.

73. *Denver Post,* February 21, 1951, Box 7, Folder Conservation, CRBC, WRCA.

74. "Ten Years and the Colorado River," *Deseret News,* September 17, 1949, Box 17, Folder Utah, CRBC, WRCA.

75. "Cash down the River," *Denver Post,* July 15, 1949, Box 5, Folder 49–50, CRBC, WRCA.

76. The full records of *Arizona v. California* are available in Record Group 95, subgroups 4–11, Archives and Public Records, Arizona State Library, Phoenix. For a good narrative on CAP proposals and events leading to the 1952 Supreme Court case, see August, *Dividing Western Waters,* 83.

77. August, *Dividing Western Waters,* 83.

78. Ibid.

79. *Arizona v. California,* 26 California Exhibits, Witnesses List, 1958, J. Willard Marriott Library, University of Utah, Salt Lake City, http://content.lib.utah.edu/cdm4/document .php?CISOROOT=/wwdl-azvca&CISOPTR=7133 (accessed August 13, 2010). The entire collection of court documents for *Arizona v. California* can be found in Record Group 95, *Arizona v. California,* SG 8 US Case Material (1890–1967), Archives and Public Records, Arizona State Library, Phoenix.

80. August, *Dividing Western Waters,* 74; "Water Case Shift," *Arizona Republic,* April 3, 1957.

81. Norris Hundley Jr., "Clio Nods: *Arizona v. California* and the Boulder Canyon Act—A Reassessment," *Western Historical Quarterly* 3, no. 1 (January 1972): 23–24.

82. "Arizona Changes Claim in Colorado Water Suit: Gila Rights Held Separate," *Arizona Republic,* August 6, 1957.

83. August, *Dividing Western Waters,* 79–84.

84. Supreme Court of the United States, no. 8, Original, *State of Arizona, Plaintiff, v. State of California et al.*, Defendant's Decree, March 9, 1964, http://www.usbr.gov/lc/region /pao/pdfiles/supctdec.pdf (accessed August 15, 2010).

85. Although a great deal of planning for this project dragged on through the 1950s, it was not officially approved and begun until the early 1960s.

86. Text of Upper Colorado River Basin Compact, 1948, http://www.usbr.gov/lc/region /pao/pdfiles/ucbsnact.pdf (accessed August 12, 2010).

87. "Colorado Basin Board Pushes Work for Pact," *Los Angeles Times,* October 5, 1948, 15; "The Southwest," *New York Times,* January 23, 1949, E6.

88. Hydropower at Hoover Dam, February 2009, http://www.usbr.gov/lc/hooverdam /faqs/powerfaq.html (accessed August 12, 2010).

89. Reisner, *Cadillac Desert*, 135–36.

90. Text of Upper Colorado River Basin Storage Project, Public Law 485, http://www .usbr.gov/lc/region/g1000/pdfiles/crspuc.pdf (accessed August 12, 2010).

91. See Stephen C. Sturgeon, *The Politics of Western Water: The Congressional Career of Wayne Aspinall* (Tucson: University of Arizona Press, 2002).

92. Steven C. Schulte, *Wayne Aspinall and the Shaping of the American West* (Boulder: University Press of Colorado, 2002), 13.

93. Ibid., 53.

94. Upper Colorado River Grass Roots, Inc., The Colorado River Storage Project, Wayne Aspinall Papers, Box 40, Folder 9, Penrose Library, University of Denver, Denver, CO.

95. Wayne Aspinall, Speech, House of Representatives, February 1956, Wayne Aspinall Papers, Box 40, Folder 6, Penrose Library, University of Denver, Denver, CO.

3

Saving the River

The Environmental Movement

IN 1963, THE ELEVEN-YEAR Supreme Court case *Arizona
v. California* finally ended, seemingly resolving the last
major conflict in a forty-year struggle over sharing the
Colorado River. Yet before Arizona could finally tap the
main stream, a young and growing environmental movement challenged the entire policy of western water development. The river itself was becoming a power player in
the struggle over water as its deteriorating riparian ecosystem began demanding attention. Western water users
reluctantly moved away from big dams and project building toward preservation and restoration policies. One of
the sharpest realities water users gradually faced was the
fact of scarcity. For all the divisions and negotiated allocations, there was simply not enough water.

Altered River Environment

Over many millennia, the powerful flow of the Colorado
River carved out many canyons along its path on its nearly

1,500-mile journey to the sea. At first, the only canyon anyone knew much about was the Grand Canyon, first sighted by Europeans in the sixteenth century. Later, in the 1920s, Bureau of Reclamation engineers paid considerable attention to Black Canyon and Boulder Canyon along the present-day California-Arizona border as potential sites for the first large dam on the river. There are many more canyons, however, all of them beautiful in their own ways.

The first canyon beneath the river's headwaters in the Rocky Mountains of Colorado is Little Yellowstone Canyon. Named after a similar canyon in Wyoming's Yellowstone National Park, the river carved this canyon out of volcanic ash and lava layers. Then the river passes through a string of other canyons—Byers, Gore, Glenwood, De Beque, Ruby, and Westwater—before reaching Canyonlands National Park and then Lake Powell, the now flooded Glen Canyon. Below Glen Canyon the river enters Marble Canyon just before entering the Grand Canyon and then exiting through Boulder Canyon, now part of Lake Mead. Hoover Dam stands astride Black Canyon, and below it the river naturally ran through a lower course and historically wandered over a larger floodplain.

Although today the river is tightly channeled and reduced to a small stream by the time it reaches the US-Mexico border, the Colorado once flowed into a large delta area full of rich marshes that provided habitat for many species of fish and birds. The part of the delta that remains must be artificially enhanced with water, usually salty agricultural runoff. In earlier times, the delta regularly experienced a tidal bore or wave when tides in the Gulf of California sent seawater partway up the mouth of the Colorado River around Montague Island at the mouth of the delta. This island is an active breeding ground for at least seven species of waterbirds. The delta region as a whole was home to a large variety of wildlife species, but today some of them are gone, and at least five are listed as endangered: the desert pupfish, Yuma clapper rail, delta clam, vaquita porpoise, and bobcat.

Along its length, humans have harnessed the Colorado with seven major dams on the main stem and dozens more on its tributaries. From north to south, these dams are the Glen Canyon, Hoover, Davis, Parker, Imperial, Laguna, and Morelos. Above Glen Canyon in the upper basin, the Aspinall Unit on the Gunnison River has three dams: Blue Mesa, Morrow Point, and Crystal. On the Green River in northeastern Utah is Flaming Gorge Dam and its large reservoir. Navajo Dam bridges the San Juan River in New Mexico. While many other tributaries and their dams all contribute to the storage and diversion of Colorado River water, these major dams are the ones that have contributed the most to the changing river ecosystem.

Probably the most significant impact of damming the Colorado River is on its fish species. Ecologically isolated from other watersheds, the Colorado was once home to perhaps sixteen species found nowhere else.[1] These fish are usually not considered valuable or desirable for human consumption, as are trout and salmon, but some were heavily fished and appreciated as recently as the 1930s. The Colorado pikeminnow (previously named squawfish) is the largest fish in the river and at one time served as a major food supply for native communities and later European arrivals.[2] The pikeminnow is at the top of the food chain, and its decline indicates two primary problems. First, if a top predator declines, it indicates that the overall river environment's health is in serious trouble. Second, the decline of an important predator means uncontrolled populations of species further down the food chain. It also encourages an increase in non-indigenous species introduced into the river, such as carp.[3]

The other major fish species in the river and on the endangered list are the humpback chub, bonytail, and razorback sucker. Dams have endangered native fish populations by blocking their natural migration patterns. While scientists continue to debate whether fish ladders actually help mitigate this effect, the first fish ladder (at the Redlands Diversion Dam on the Gunnison River in Colorado) was not constructed until 1996.[4] In 2008 another fish ladder was installed at the Grand Valley Diversion Dam, about forty-four miles north of Grand Junction, Colorado.[5] Fish species in the river need a variety of depths and gravel beds for different times in their life cycles. With the presence of dams, the fish cannot migrate to these places.

One of the major changes to the river that affects fish species is the alteration of water flow. As humans have dammed, diverted, and stored Colorado River water, the amount of average flow downstream has been significantly reduced. Diversions out of the river take out nearly one-third of the flow, and another 1 million acre-feet (MAF) evaporate every year from the reservoirs.[6] Even more significant is the elimination of annual flooding, which signaled the time for certain fish behaviors, such as spawning. With a much more even flow in the stream, important sweeping of gravel beds by spring floods and accumulation of sandbar habitats are prevented. Since the early 1990s, scientists and dam operators at Glen Canyon have attempted mitigation by creating artificial floods (see chapter 4), but it is still unclear whether these efforts will have favorable results in the long term. Daily fluctuations in water levels because of hydroelectric dam operations have also damaged fish habitat.

One of the most difficult results to mitigate is the change in water temperature. Before the river was dammed, water temperatures fluctuated, sending important signals that it was time to spawn. Colder river water is now thought to inhibit the development of juvenile fish species.[7] In addition, dams trap important nutrients that no longer flow downstream with the silt. Before the dams were constructed, scientists estimate that an average of 374 metric tons of sediment washed downstream each day.[8] Studies have determined that suspended sediment loads in the river were reduced by

70 percent to 80 percent after the construction of Glen Canyon Dam. Such barriers also obstruct leaves, twigs, and other plant materials—all sources of important nutrients—from flowing downstream. Lower levels of nitrogen and increased levels of phosphorus directly affect phytoplankton productivity, an important producer of oxygen necessary for maintaining freshwater food systems.[9] Sediment also provided important nutrients for riparian habitat such as cottonwood, mesquite, and willow trees. Damming the Colorado River has also trapped various pollutants that used to be washed downstream. Mining pollution in the river basin has leached into groundwater and flowed in runoff into the main stream. Some of these pollutants are uranium, radium, lead, pesticide chemicals, vanadium, and high levels of the natural chemicals salt and selenium—causing severe damage to the river environment.[10]

Most of the individuals involved in the early environmental movement of the 1960s and 1970s had little understanding of some of the effects dams were having on the river environment. Instead, they were most concerned about the changing aesthetics of the country's natural landscape. The roots of this modern environmental movement that would impact the Colorado River can be traced as far back as World War II. According to historian Samuel P. Hays, major social changes that took place after the war, such as an increased interest in outdoor recreation during the 1950s and awareness of pollution problems in later decades, encouraged a larger concern about the environment. Instead of merely concentrating on productivity and wealth accumulation in the modern age, society became more interested in quality of life. The idea that nature's aesthetics were important to life's quality began to inform and encourage those who argued for wilderness preservation.[11]

As living standards and education levels increased in the second half of the twentieth century, more and more people came to value the natural environment as an added aspect of quality of life. The bulk of a person's income was spent on basic survival prior to the twentieth century, and no real concern about natural environments existed except as resources to use for economic benefit. With higher living standards, however, came much more leisure time and an appreciation for the finer things in life. Specific features of the American landscape came to symbolize natural values that modernity had endangered and that must be preserved. Hays argued that these landscapes, whether of land or water, shaped a "sense of belonging, a mechanism for focusing a feeling of shared responsibility for one's surroundings."[12]

The development of color photography after World War II helped encourage attention to, and action on behalf of, nature. Photographers like Ansel Adams spread awareness through what we now call "coffee table books," which provided dramatic visual images of nature across the country. The impact of the growing environmental movement in the Colorado River basin is directly attributable to such photographic books, which raised awareness of canyons threatened by dams and the reservoirs destined to flood them.[13] Beginning in the 1950s, the history of the river became intertwined with that of the American environmental movement through the work of David Brower and the Sierra Club. Founded by John Muir in 1892, the Sierra Club worked to increase public awareness of the natural environment and to preserve it from negative human interaction. A major force behind the creation of Yosemite, Glacier, and Yellowstone National Parks, the Sierra Club also fought a hard battle to prevent the flooding of the Hetch Hetchy Valley in California in the early 1900s.[14] In many ways, the specter of Hetch Hetchy served as a constant reminder to preservationists across the country of the importance of their cause.

Other environmental organizations, such as the Audubon Society, had been around almost as long as the Sierra Club. Others like the National Wildlife Federation and the Wilderness Society were formed during the New Deal era in the 1930s. Although most of the Sierra Club's membership was on the West Coast in its early years, it began to expand nationally in the 1950s, mostly because of David Brower—the club's executive director from 1952 to 1969 and founder of Friends of the Earth in 1970 and of the Earth Island Institute in 1982. Through his experiences as a mountain climber and editor for the University of California Press, Brower became an expert publicist for environmental issues and a passionate activist.[15]

Saving the Grand Canyon

The issue that pushed both Brower and the Sierra Club into the national spotlight was the 1950 battle over Echo Park Dam. In 1909 paleontologist Earl Douglass discovered an enormous dinosaur fossil bed in a canyon area along the Colorado River, a little bit west of present-day Grand Junction, Colorado. The fossils became famous, and in 1915 President Woodrow Wilson declared the area a national monument to protect them. Other than some scientists interested in dinosaurs, few people knew of or visited the area. As the Bureau of Reclamation made plans to build dams for the upper Colorado River basin in the late 1940s, a proposal called the Colorado River Storage Project (CRSP) went before Congress that called for building a series of dams above Hoover, including two in the Grand Canyon. The Sierra Club and other conservationists did not immediately oppose any of the dams as first proposed. One of these dams was planned for Echo Park Canyon, which was inside Dinosaur National Monument.[16]

Although initially the proposal for Grand Canyon dams failed in Congress, plans for the other dams began to draw attention, particularly the one at Echo Park. The National Park Service protested first, upset that a dam would flood a major portion of Dinosaur National Monument. Interior Department secretary Oscar Chapman listened to what the Park Service had to say but decided that the dam was important enough to flood one national monument area and promised that he would not make the practice a precedent.[17] Soon, other voices joined in the protest over Echo Park. Bernard DeVoto, a western historian and novelist, wrote an article for the *Saturday Evening Post* blasting the Department of the Interior for planning to flood a national park area.[18] He asserted strongly that a dam and reservoir in the park would "degrade the great vistas." He added, "No one has asked the American people whether they want their sovereign rights, and those of their descendants, in their own publicly reserved beauty spots wiped out."[18] Others quickly joined the crusade to save the park, including the Sierra Club's new president, David Brower. As an editor, he believed the power of books, especially those with beautiful photographs, would help convince the general public to join the fight to save Echo Park.[19]

In 1953, just before Dwight Eisenhower was to take office, Secretary Chapman and commissioner of the Bureau of Reclamation Mike Straus struggled hard to pass one last large reclamation bill that would leave something of a legacy. The CRSP, with its many parts and smaller dams, also called for two very large dams and reservoirs: one on the Green River at Echo Park and another on the main stem of the Colorado at Glen Canyon. It seemed an appropriate answer for the upper basin states, which felt they had no way to divert their share of Colorado River water. While the plan sounded good to western governors and members of Congress, Brower and other conservationists were horrified. Brower ran three boat trips down the Green River and through Dinosaur National Monument in 1953, hoping to draw attention to the beauty of the canyon that would be buried under water if the Echo Park Dam were built.

Even though Chapman and Straus were not able to shepherd the CRSP bill through Congress before their tenure was over, the Republican appointees who followed them (Douglas McKay and Wilber Dexheimer) continued to move the plan forward. Brower's boat trips and DeVoto's article, reprinted in *Reader's Digest,* had stirred up public interest in the park. Once hearings on the CRSP began in early 1954, Congress received thousands of letters defending the beauty of Echo Park and protesting the dam.[20] Many conservationists testified against the dam at the congressional hearings, including David Brower. He suggested that raising the height of the proposed Glen Canyon Dam would enable the project to store all the water needed without a dam and reservoir at Echo Park. Bureau of Reclamation engineers had argued that raising Glen Canyon would not be enough, but Brower successfully demonstrated in front of everyone present that the engineers were wrong. Using the loss of Hetch Hetchy as his example, Brower argued against the loss of yet another valley so many people had recently discovered.[21]

After another round of hearings in the US Senate, both houses of Congress seemed ready to debate and approve the CRSP with the Echo Park Dam, but letters kept pouring in and Brower kept working. He persuaded historian Wallace Stegner, also a staunch environmentalist, to edit a photo-filled book about the canyon that would be flooded if the project moved forward. Titled *This Is Dinosaur*, the book appeared in early 1955. Brower sent copies to every senator and representative in Washington, as well as to the editors of all major western newspapers.[22] As debate moved first through the Senate and then the House, major newspapers published story after story about how a beautiful canyon would soon be lost. Those who believed in the new emphasis on the aesthetics of nature used the appeal of beauty to raise public support for campaigns against dams, and the general public responded. Many people were beginning to see nature and its beauty as a public good, one that was necessary for a high quality of life.[23]

Wayne Aspinall and other supporters of the CRSP bill realized that they could probably afford to delete Echo Park from their plans, and finally, in 1956, Congress passed the CRSP without Echo Park. The importance of the victory for Brower and other environmental activists was that land set aside as a national monument was protected from development projects. Although Brower was also worried about Glen Canyon Dam and the flooding of the surrounding canyons, he and many others had never traveled there or seen what would be flooded. It was not that Brower did not care about Glen Canyon; he simply realized that he could probably win only one battle at this point, and that battle should be to protect federally preserved lands.[24] Later, when Brower did visit Glen Canyon, he mourned its loss, as his predecessors had mourned the loss of Hetch Hetchy.

Activist groups such as the Sierra Club were not the only opponents of the CRSP and its dams. California was very much opposed to the entire project, mostly because it felt the dams would cost far too much and withhold water the lower basin deserved. California's lawyer, Northcutt Ely, was busy arguing his state's case in its lawsuit with Arizona, begun in 1952. However, he was not too busy to ignore plans for Glen Canyon Dam and reservoir. He stated California's opposition clearly in a speech to the press on July 22, 1955:

> We now know that the Upper Basin cannot use half of the water apportioned to it with-out billions of dollars in Federal subsidies, and that with like subsidies the Lower Basin could use several times its apportionment . . . The pending Colorado River Storage Project as passed by the Senate is a scheme . . . The project turns the Colorado compact upside down. Instead of using 7,500,000 acre feet and releasing the remainder to the Lower Basin, some Upper Basin states would give us 7,500,000 and keep the balance

. . . Water projects depend upon permanent water rights. Your title is threatened when works adverse to yours are authorized, and that is the time to resist the encroachment. It is suicide to wait until your adversary's gigantic works are built at the cost of hundreds of millions and have impounded your water behind them.[25]

From a twenty-first-century perspective, it seems surprising that Ely either could not or would not think in terms of sustainable development. In Ely's words, there appeared to be no real concern that both basins could deplete the resource. In fact, this attitude among politicians from all the basin states illustrates the drive to use as much water as possible before the other side could grab it. Such thinking created the inevitability of shortage in later years.

In spite of opposition from California and belated opposition from David Brower of the Sierra Club, construction on Glen Canyon Dam began in 1956. By the time the gates closed to fill the dam in 1963, Brower realized that a canyon much more beautiful than Echo Park had been lost forever. Very few people had ever visited the canyon, including Brower, who finally did so on three separate trips before the gates closed. In 1966 the Sierra Club published another picture book with photos taken before Glen Canyon was filled with water.[26] Ironically, while it took seventeen years for Lake Powell to fill, river users in the early twenty-first century would see that over-allocation and drought could empty it in much less time.

Echo Park was not the only national monument area threatened by the CRSP. There was a big danger that once the gates of Glen Canyon Dam had closed, the water might threaten Rainbow Bridge National Monument in Utah. This natural arch carved out of sandstone is 275 feet across, 33 feet wide, and 290 feet tall. First seen by Europeans in the 1800s, in 1910 the place came to the attention of President William Howard Taft, who declared it a national monument. After Lake Powell began to fill in 1963, concerned activists and the National Park Service worked hard, in cooperation with the Navajo Nation, to keep the waters from flooding the national monument. Ironically, the lake made visitor access to the park much easier, leading to later conflicts over park management and respect for native sacred sites.[27]

In spite of conservationist protests and laments over the flooding of Glen Canyon, most members of Congress, advocating for their respective states, sided with development and not preservation. During the 1940s and 1950s, the Bureau of Reclamation and its supporters worked to further develop western water resources, especially those along the Colorado River. States in both basins of the river worked hard to divert and use their allocations lest another state try to take their share. Glen Canyon was the last big dam to bridge the main stream of the river, but one last diversion project remained: the Central Arizona Project canal.

CAP: The Last Big Diversion Project

When the long battle between Arizona and California finally ended, the struggle for the Central Arizona Project (CAP) resumed. Stewart Udall, secretary of the interior in 1961, strongly supported a regional water plan for the lower basin to mirror the one developed by the upper basin.[28] He believed such regional planning would be the only way to ensure water for the region in the future and to finally obtain CAP approval. To fund CAP and other regional projects, Udall recommended building two dams in the Grand Canyon: a high dam above the national park at Marble Canyon and a low dam at Bridge Canyon below the park. He, along with other water engineers, believed these two cash register dams would do no damage to the park itself, would only flood parts of canyons the public never saw, and would provide huge amounts of funding from hydroelectric power generation.

Reflecting the changing times, Udall was torn by contradicting interests. On one hand, he had been impacted by the growing attention to the aesthetics of nature and fully supported the new consideration paid to environmental issues such as clean air and water and preserving wild places.[29] On the other hand, he was a westerner who supported his home state and its needs for water. In his book *The Quiet Crisis*, published in 1963 and reprinted in 1964, Udall argued for a new, modern approach to the environment to mitigate the environmental consequences of industrialization.[30] In this frame of mind, Udall proposed one of the first regional plans he hoped would coordinate water usage and preservation throughout the arid West. The plan was comprehensive and ambitious, and it met with much opposition. By attempting to straddle the fence between state water needs and conservation, Udall ended up pleasing no one.

Within the Southwest region, states' attitudes were also at odds. California's political liberalism supported Udall's ideas for a regional water plan, but Arizona's much more conservative politics meant strict adherence to the concept of state authority over water. No one in Arizona was willing to give up any control of internal resources to a regional board. Although they differed, the attitudes of both states ignored the idea of sustainable resources. When Aspinall and other upper basin politicians also protested against Udall's plans or later variations, they did so because they feared the lower basin would take their share of water.[31]

Finally ready to unveil his regional plan, Stewart Udall put forward a large $8 billion proposal in January 1963. Called the Pacific Southwest Water Plan, Udall's vision included not only the Grand Canyon dams and CAP but also plans for California water projects and a regional fund for pooling resources. Stewart's brother, Arizona representative Morris Udall, thought this plan might help CAP pass through the US House, but public opinion in Arizona was less positive. Stories in the *Arizona*

Republic criticized the plan, arguing that it would put control of CAP into the hands of others outside the state.[32] Many Arizonans saw the Udall plan as something that would destroy their plans for CAP.

Senator Carl Hayden was one of those who thought Udall's plan was bad for Arizona. The day after the US Supreme Court gave its judgment, the senator submitted his plan for CAP. Called the Lower Colorado River Basin Project (LCRBP), Hayden's plan included 340 miles of pipeline, with pumps and impoundments to carry approximately 1.2 MAF of water from the Colorado River down to central and southern Arizona.[33] To pay for the project, Hayden included plans for the Bridge Canyon Dam. Most Arizona lawmakers understood the importance of at least one Grand Canyon dam to make CAP funding possible. Although many Arizonans were strongly opposed to the dams, the public was largely unaware of the details of the plans when first proposed. However, pro-conservation members of Congress were already opposing the Grand Canyon dams and CAP approval.[34] While economic interests still trumped most decisions about the environment, members of Congress began to respond to a growing public appreciation for preserving nature for its own sake.

Although many criticized plans for a dam at Marble Canyon, people were less concerned about the strip of canyon north of Grand Canyon National Park. Bridge Canyon Dam would back up water seventeen miles into the national park from the south all the way through Grand Canyon National Monument (connected to the lower end of Grand Canyon National Park). Morris Udall and other supporters of the CAP bill argued that the water would never be visible to most park visitors unless they were hikers who went off the main rim roads and trails. Without the dams, Arizona politicians and Stewart Udall believed CAP would never be constructed. After fighting so long against California, no Arizonan would tolerate that outcome.[35]

Twice before, in the early 1950s, Arizona senator Carl Hayden had put forward bills for CAP and lost. After Arizona won the Supreme Court case against California, though, Hayden once again created a bill he thought would finally receive the support needed for passage. However, such a bill would have to pass through the House Interior and Insular Affairs Committee, chaired by Representative Wayne Aspinall of Colorado. Although Aspinall had been favorable to Arizona's plans for CAP, he had developed concerns about how much water was actually available in the river for implementation of the plan.[36] There were good reasons for this concern, and many researchers had long suggested that the river was already over-allocated. To address these questions, Aspinall asked Secretary Udall for an assessment of Colorado River water flows and the feasibility of the plan for CAP. Udall argued that there was plenty of water in the river to implement CAP.

Even though Udall tried to assure Arizonans that he supported CAP—it was, after all, a part of his proposal—Hayden saw the bill as a betrayal. While in principle he was not opposed to a broader regional plan, he believed CAP needed prompt and separate approval to protect it from opponents. While he vigorously argued for his bill from 1963 to 1968, other plans were proposed and debated. To gain California's approval for CAP, some legislators formulated a plan to augment the Colorado River with water from the Columbia River. In spite of Stewart Udall's and Bureau of Reclamation commissioner Floyd Dominy's assurances that there was plenty of water, California argued that its fears were legitimate. Any attempts to legislate an inter-basin water transfer, however, were thwarted by the chair of the Senate Committee on Interior and Insular Affairs, Henry Jackson of Washington.

During the five years of debate, counterproposals, and amendments, the two most difficult issues to conquer turned out to be proposals for a dam in the Grand Canyon and California's demand for priority water rights. Regarding California, Hayden at first refused to negotiate. In 1965, however, he reluctantly agreed to provisions that would guarantee California priority rights first for twenty-five years, then for twenty-seven. In the final bill, California was granted priority water rights in perpetuity. This aspect of the CAP bill essentially overturned Arizona's gains from the Supreme Court case. Hayden was ultimately more interested in obtaining funding and approval for CAP than in anything else.[37] Water from CAP would be used to supplement shortfalls other states might experience, especially California. Although some thought he was selling out Arizona, Hayden reluctantly came to believe it was a price he had to pay.[38]

While CAP supporters were trying to gain California's approval, environmentalists stirred up other trouble over the Grand Canyon dams. Both Udall and Hayden believed cash register dams were necessary to fund the proposals. As early as 1964, the Sierra Club and David Brower began lobbying Congress against Udall's bill and later Hayden's, starting a media campaign to change public opinion. First, the club published a coffee table book titled *Time and the River Flowing: Grand Canyon,* which featured beautiful photographs of the threatened sections of Colorado River canyons.[39] Ironically, Brower argued that funding for CAP could be found by building coal-fueled power plants instead of the dams. Although pollution from coal eventually became a hot topic within environmental groups, Brower was obviously more concerned with water and canyons.

While the Sierra Club and other groups protested the proposed dams, one surprising group actively supported Bridge Canyon Dam: the Hualapai Indians, on whose reservation the dam would be built. No one asked for their permission or approval, but the Hualapai saw the dam as a possible solution to poverty, providing jobs and, they hoped, drawing large numbers of tourists. They carefully crafted their support for Hayden's bill with provisions that would give the tribe the proceeds from hydro-electric generation and actively lobbied Congress to approve their amendments to the bill. This unexpected support created a stumbling block for environmentalists who argued against the dams.[40] They could no longer use the inundation of native lands as a reason against building them.

Still, the voices of the preservationists were much louder than those of the Hualapai people, and the latter's needs and arguments were soon drowned out by a loud public outcry. By the time serious debate on the LCRBP bill began in the US Congress, Morris Udall and other supporters were being barraged by a letter-writing campaign. Although Morris challenged Brower rather eloquently, stating that coal-powered electric plants might pollute the Grand Canyon much more than damming the Colorado River would, his arguments fell on deaf ears. The public believed the Grand Canyon was a national treasure that must be protected from money-grabbing politicians who cared little for nature. Many people took part in the "Save the Canyon" campaign David Brower created.[41]

The fight against Grand Canyon dams became an epic battle between western government water interests and the environmental movement. In March 1966 the Sierra Club and *Reader's Digest* teamed up to hold a workshop at El Tovar Lodge on the south rim of the Grand Canyon. At this meeting, anti-dam invitees made speeches and held discussions about how to prevent the terrible tragedy of building the dams. Pro-dam interest groups were generally kept out of the meetings, leading to loud protests from Morris Udall and others. One surprising and uninvited voice at the workshop was that of former Arizona senator Barry Goldwater. In an impromptu speech to the group, he declared his love for the Grand Canyon but argued that a dam at Bridge Canyon would not harm the canyon's beauty in any way. Many in attendance found his speech moving and convincing.[42]

Yet the Sierra Club continued its campaign, and in June 1966 it took out full-page ads in the *New York Times*, the *Washington Post,* and other major newspapers calling for the general public to help "Save the Grand Canyon." The *Times* ad asked, "Should we also flood the Sistine Chapel so that tourists can get nearer the ceiling?"[43] The ads encouraged readers to mail coupons provided with the advertisements to various members of Congress in protest.[44] The media blitz was successful, and the Sierra Club gained large numbers of members immediately after the campaign. Morris Udall was angry, but there seemed little he could do to silence the public outcry. The day after the ads ran, the Internal Revenue Service (IRS) visited the Sierra Club, telling the group it might lose its nonprofit status because it was directly lobbying Congress. Brower accused Morris Udall of tipping off the IRS; he vehemently denied having done so.[45] In the end, the Grand Canyon dams were removed from the Senate and House versions of the bill by early 1968.

While opposition from California and environmental groups threatened to derail Hayden's hopes for CAP approval, Wayne Aspinall, who tightly controlled the House Interior and Insular Affairs Committee, was another formidable roadblock. The Colorado representative distrusted California, but he also feared CAP would take water away from upper basin projects. To protect his interests and gain Arizona's support for his own plans, Aspinall drafted a revision of the bill that dropped the word "Lower" and added five Colorado water projects. He argued that these projects would give something to everyone and increase support in Congress. Arizona representative Morris Udall and others argued that Aspinall was "blackmailing" Arizona with his Colorado projects. They were all in his district, and none had the approval or support of the Bureau of Reclamation. The bureau did eventually create new financial studies to help Aspinall's projects pass muster in Congress because it favored a bill authorizing CAP.[46]

Although they were asking for more water projects, western lawmakers were beginning to understand that the Colorado River was over-allocated. Studies in the years following the 1922 compact clearly demonstrated that this bill and subsequent ones divided up more water than was actually available.[47] Most analysts believed that for everyone to receive the water shares promised them, the Colorado River would have to be augmented. It would not simply be Arizona's CAP that would be tapped to meet needs, but major augmentation of the river might be needed in the foreseeable future just to meet treaty obligations with Mexico. However, Senator Henry Jackson from Washington state chaired the Senate Committee on Interior and Insular Affairs, which would need to approve the 1968 bill, and he adamantly refused to support any augmentation plans from the Columbia River. Therefore, the revised bill specifically placed a ten-year moratorium on river augmentation plans that would come from the Columbia River or any other natural river.[48] The bill did authorize federal funding and oversight for future augmentation projects, if needed.

Another interesting provision of the bill that appeared in the final version was a restatement of the treaty obligation to Mexico, signed in 1944, that promised to deliver 1.5 MAF of Colorado River water to Mexico each year. Not only does the section reaffirm this treaty commitment, but it also states that meeting this obligation would be a "national responsibility," not just a regional one.[49] Arizona wanted to make sure that in times of drought, the burden of meeting the treaty obligations would not fall completely on its or the lower basin's shoulders. By making it a national obligation, the Bureau of Reclamation could demand water from the upper basin and spread the responsibility around. While some saw this provision in the bill as an unfair shifting of responsibility from the states to the nation, the lower basin wanted this stipulation, and it slipped through without a major challenge.

On September 30, 1968, the bill finally passed both houses, and at long last CAP was approved. The Colorado River Basin Project Act became the last major piece of legislation dividing up the waters of the Colorado River. The central provision of the bill—approval of the Central Arizona Project aqueduct—would be the last major diversion of the river and would ensure that Arizona would receive its designated 2.8 MAF of water each year, provided there was enough water in the river to meet everyone's needs.[50] To gain California's support, the bill dictated that in times of drought, Arizona's CAP held the most junior water rights to the river and would bear the brunt of any cutbacks.[51] Other basin states received project approval and funding in the bill: Colorado for five small projects, Utah for two, and New Mexico for one. Nevada was addressing its water needs in a separate bill, and Wyoming was not asking for anything. In the end, Arizona's project was the costly and controversial piece, but Aspinall's other small projects helped create at least the illusion of a larger, regional plan to garner the necessary approval in Congress.

David Brower, members of the Sierra Club, and other concerned citizens breathed a sigh of relief that no dams would be built in the Grand Canyon. The preservationists had won a battle against unrestrained western reclamation. In many ways, this victory was less about saving the Grand Canyon from any real damage than it was a cooling salve following the failure to preserve Glen Canyon years before. In 1985, seventeen years after passage of the Colorado River Basin Project Act, Arizona opened a major section of its CAP aqueduct and began taking its share of Colorado River water for the first time. Coal was strip-mined from the Hopi Reservation and used to fuel a coal plant on the Navajo Reservation.

Because of two conflicting sections of the bill—one making national funding for Colorado River augmentation possible and one banning such augmentation from other rivers—an unexpected outcome was experiments with "weather modifications."[52] These experiments primarily involved cloud seeding with silver iodide to encourage rainfall. From 1970 to 1975, the federal government subsidized these projects along the San Juan mountain range in southwestern Colorado. The environmental outcomes of these experiments are unknown, but most scientists eventually agreed that there was little or no evidence that cloud seeding actually worked.[53] These efforts to modify the weather demonstrated the limits of science to expand the water supply. Once this fact was fully recognized, gradual support for sustainable use could develop.

Environment versus Development

Even though there was a growing awareness of water shortages by the 1960s, the Bureau of Reclamation and Colorado River basin states continued project building. One of the most important forces behind this effort was Floyd Dominy, bureau commissioner from 1959 to 1969. Dominy was one of the last big dam builders and oversaw much of the completion of the Colorado River Storage Project for the upper basin while the lower basin fought its battles over allocations and CAP. He was a force within the Bureau of Reclamation with whom few wished to tangle. Dominy believed strongly in water management and control; as the last great project builder, he fought relentlessly against environmentalists who tried to halt construction on various parts of the CRSP throughout the 1960s.[54] In many ways he was a man whose time had passed, as the public began to view big dam projects as unnecessary tax burdens and terrible blights on the beauty of nature. As the West urbanized in the 1960s and beyond, it became harder and harder to argue for expensive water projects meant mostly for the irrigation of farmland.

Dominy, however, held firm to his belief that the bureau and its projects were vital to the West's prosperity and success. He saw no value in leaving rivers in their natural state. He had grown up on a homestead in Nebraska and, as a boy, watched his grandfather and then his father farm the land through great struggle and backbreaking work. Dominy believed a river undammed was a river wasted. He maintained this attitude throughout his entire career, once stating in a public speech that the Colorado River without dams was "useless to anyone" and "I've seen all the wild rivers I ever want to see."[55] On a regular basis, he argued for funding and support for reclamation projects as necessary for national growth, particularly that of the western states.[56]

Through careful politicking, Dominy managed to obtain congressional support for his project and worked actively to develop relationships with two powerful western politicians: Aspinall of Colorado, who ran the House Interior and Insular Affairs Committee, and Hayden, who chaired the Senate Appropriation Committee. One of the major water projects connected to the CRSP finally began near the end of Dominy's tenure. The Central Utah Project (CUP) was a large diversion project that would do for Utah what CAP was intended to do for Arizona. Designed to water a little over 400,000 acres of farmland, its seven units would also serve industry and municipal uses. Construction began on the first unit in 1959, and the last major piece of the system was finished in 1987.[57] As with other parts of CRSP, CUP has never been fully completed and was challenged by environmental groups during the late 1960s and the 1970s. Although it functions as a vital part of the water sources for central Utah and Salt Lake City, it remains controversial, mostly because of the high costs.[58]

Although Utah and other western states begged regularly for funding to complete CUP, Congress continued cutting funding for western water projects in general.[59] The Grand Canyon controversy exemplified an ongoing debate about the aesthetics of nature and its importance to the quality of life. Preservationists celebrated the saving of the Grand Canyon while lamenting the lost Glen Canyon. Yet even though the Colorado River continued to flow freely through the Grand Canyon, the battle to save this section from damming was not completely over. Proposals to dam the canyon continued to appear from time to time, although they did not pass.[60] Finally, in 1975, Congress significantly enlarged Grand Canyon National Park to end the controversy of dam building there once and for all.[61]

As the environmental movement began to build up steam in the mid-1960s, Secretary Udall supported what became known as the National Wild and Scenic Rivers Bill. For over four years, he and others pushed for the federal legislation but were met with strong opposition from western politicians such as Aspinall.[62] Finally signed by President Lyndon Johnson in 1968, the bill created a system to protect undeveloped rivers in the United States. Motivated by the Grand Canyon dam controversy and patterned after the 1964 Wilderness Act and its National Wilderness Preservation System, the new act eventually protected 252 free-flowing rivers or sections of rivers. Eighty-six of these rivers were added in 2009 and signed into law by President Barack Obama.[63]

Another piece of federal legislation that would impact the Colorado River was the National Environmental Policy Act (NEPA), signed in 1969.[64] The most important part of this legislation was the process for conducting environmental impact assessments on all federal projects. Once proposed, any project such as a dam or an aqueduct had to pass extensive studies to make sure the environment was minimally impacted. In 1970 President Richard Nixon signed a bill creating the Environmental Protection Agency (EPA). This federal agency consolidated all oversight of environmental issues, including water and air quality. From this point on, Bureau of Reclamation operations became far different from those in previous decades. Instead of project building, the bureau slowly turned toward concentrating on improving operations and alleviating adverse environmental impacts of existing water and power projects.[65]

Although it initially focused on air-quality issues, the EPA began to tackle water-quality issues in 1972. Eleven years earlier, the Bureau of Reclamation had built a long drainage canal to help local farmers in the Wellton-Mohawk Project east of Yuma, Arizona. Long-term over-pumping of groundwater for irrigation had created such high levels of salt that farms began to suffer greatly, and they asked the bureau for help. The drain was a way to channel salty irrigation runoff away from the fields and down to the Colorado River. Of course, this action meant that water flowing across the border to Mexico became too salty for its use, and Mexico accused the United States of violating the 1944 treaty that guaranteed delivery of 1.5 MAF each year (see chapter 7).[66] Stewart Udall came under serious pressure from the president and the US State Department to come up with a solution quickly. He received no cooperation from western politicians such as Hayden, however, who thought the treaty contained no obligation regarding water quality. Arizona politicians in particular had disliked the Mexican treaty when it was first signed and argued later that any efforts to manage salinity issues for Mexico's sake would damage the local economy.[67]

Finally, in 1973 the United States and Mexico reached an agreement to reduce salinity in the Colorado River, and Congress passed the Colorado River Basin Salinity Control Act in 1974.[68] Just two years earlier, in 1972, several amendments to previous legislation had become known as the Clean Water Act, regulating the flow of wastewater into water sources. In 1968 Congress had appointed a National Water Commission to study US water resources; the group reported the results to Congress in 1973.[69] In its report, the commission recommended that better attention be paid to the environment and that some kind of federal coordination be implemented for all water policy nationwide, which encouraged the EPA to turn its attention to water quality. The commission report also recommended that water be connected to pricing to both pay for infrastructure and encourage conservation. However, while many of the recommendations are still discussed today, very little in the form of federally coordinated water policy has ever been implemented.

Other pressures, not immediately obvious to people in the Southwest, continued to strain the Colorado River. The nation faced a growing energy crisis in the 1970s following the Yom Kippur War, a twenty-day conflict that began when Egypt and Syria led a collection of Arab armies in a surprise attack against the state of Israel. The United States quickly came to Israel's aid, and, after the Arab states were defeated, many Middle East leaders blamed their failure on the United States. To retaliate for US involvement in the war, members of OPEC, the Organization of Petroleum Exporting Countries, declared an oil embargo. Lasting from October 1973 to March 1974, when Israel agreed to withdraw from the Sinai Peninsula, the embargo created widespread fear and economic instability. In response to what became known as the "energy crisis," Americans began searching for untapped US energy resources in an effort to reduce dependency on Middle East imports.[70]

Some analysts believed one of the possible alternatives to dependency on Middle East oil was to extract more natural gas, coal, and oil shale—much of which could be found in Colorado, Wyoming, Utah, and New Mexico. Colorado governor John D. Vanderhoof spoke out in protest when he heard about plans to mine oil shale in his state. The problem, he argued, was that any of the extraction processes used to mine the shale required very large amounts of water that would have to come from the Colorado River. "We're not going to cut off water to our cities and farms," he announced to the Colorado River Water Users Association in November 1973. "The potential impact of the demand for energy production on available water supplies of the Colorado River can become enormous," he continued. If the upper basin became the new source for energy, the Colorado River Compact would have to be renegotiated. To support his argument, he quoted statistics that estimated a need for 27,000 acre-feet of water to produce 1 million kilowatts of electricity.[71]

To help cope with the energy crisis, the Bureau of Reclamation pushed for increased funding for the Central Utah Project to allow the allocation of more water for energy resources. While Vanderhoof was leery of such developments, Utah and Wyoming welcomed them. Wyoming governor Stanley Hathaway proclaimed in 1974:

> About 40 percent of the world's coal is in the United States, and the United States is blessed with oil shale, which represents 17 times the known crude oil reserves in the world. The greatest danger facing the United States is that of failing to make any decision at all, after lengthy studies, to avoid making mistakes in use of our resources. Somebody has to start making decisions.[72]

Some western states saw the oil embargo as an opportunity to push for extraction industries in their home states.[73] Mining oil shale, however, requires a lot of water. In a commonly used process called aboveground retorting, from one to five barrels of water are required to extract one barrel of shale oil.[74] In this process, the shale is heated until oil vapors and gas condense for collection. Most of the oil shale in the United States is found in the Green River deposit in Utah and Colorado. The industry never really became successful, however. Although interest in shale had faded by the early 1980s, recent prices and worries about shortages have restimulated interest. The impact of mining and processing the shale, however, will continue to cause serious concerns about local groundwater and the Colorado River.[75]

While mining oil shale had supporters and detractors from the beginning, one mining industry already posed serious threats to the water quality of the Colorado River. In 1952 geologist Charlie Steen discovered what would prove to be the largest uranium deposit ever found in the United States. Used in nuclear-generating plants as well as weapons production, this uranium was in very high demand during the Cold War years. Soon, Steen's Uranium Reduction Company was processing an average of 1,400 tons of uranium per day. Radioactive waste from the mine, which was positioned on a bank of the Colorado River, was stored in a bulldozed basin right beside the mine.[76]

Operating until 1984, the mine produced an estimated 16 million tons of tailings and contaminated soil. While no one knows for certain what exactly leached into the river, some environmental activists assert that around 110,000 gallons of radioactive groundwater have seeped into the Colorado River every day since the mine was closed. After the mine closed and its former owners failed to clean up the tailings, the US Department of Energy took over the project and decided to move the contaminated material away from the river to a location near Crescent Junction, Utah, thirty miles north of the current site. In 2000 the government made plans to move the pilings that covered around 130 acres, but funding for this difficult cleanup project is expected to exceed $1 billion. Finally, in 2009 transportation of the pilings began, and by June 2011, 4 million tons had been moved, about 25 percent of the total.[77] As of February 2012, workers were transporting tailings four days a week, with the remaining days devoted to security and maintenance tasks. One train of thirty-six railcars is shipped per day, Monday through Thursday; each trainload contains approximately 5,000 tons of tailings. With reductions in funding, the final completion time is unknown, but estimates say that if no cuts are made to the annual budget of $90 million, the entire pile could be moved by the end of 2019.[78]

By the end of the 1970s, attitudes toward the environment had made a dramatic shift, from Teddy Roosevelt's conservation approach to a modern environmental movement. Based on a new appreciation for aesthetics and the benefits to human life of preserving nature's beauty, this movement evolved into one that advocated preservation and mitigation of the ill effects of human use. In many ways, the movement of the 1960s and 1970s failed to go beyond attitudes of equal access to clean water and air, beyond preserving redwood forests and preventing catastrophic oil spills. While those were important starting points, the environmental activists during those decades advanced slowly beyond the interest in preservation and cleaning up nature toward the concept of sustainability that would appear near the turn of the twenty-first century.

NOTES

1. Robert W. Adler, *Restoring Colorado River Ecosystems: A Troubled Sense of Immensity* (Washington, DC: Island, 2007), 27.

2. The name was changed from squawfish to pikeminnow by the Names of Fishes Committee of the American Fisheries Society at the request of native communities. See *Denver Post*, March 28, 1999; *Colorado Springs Gazette*, March 29, 1999.

3. Adler, *Restoring Colorado River Ecosystems*, 29.

4. Ibid., 30.

5. US Fish and Wildlife Service, Grand Junction Colorado River Fishery Project, March 15, 2011, http://www.fws.gov/grandjunctionfishandwildlife/management.html (accessed March 26, 2011).

6. Adler, *Restoring Colorado River Ecosystems,* 34.

7. Robert W. Clarkson and Michael R. Childs, "Temperature Effects of Hypolimnial-Release Dams on Early Life States of Colorado River Basin Big-River Fishes," *Copeia* 2000, no. 2 (May 8, 2000): 402–12.

8. Adler, *Restoring Colorado River Ecosystems,* 38. See also Harold M. Tyus and James F. Saunders III, "Nonnative Fish Control and Endangered Fish Recovery: Lessons from the Colorado River," *Fisheries* [Bethesda, MD] 25, no. 9 (2000): 17–24.

9. Larry J. Paulson and John R. Baker, "Nutrient Interactions among Reservoirs on the Colorado River," *Publications: Water Resources,* Paper 58, 1980, http://digitalscholarship.unlv.edu/water_pubs/58/ (accessed March 26, 2011).

10. Ibid.

11. Samuel P. Hays, *Beauty, Health, and Permanence: Environmental Politics in the United States, 1955–1985* (New York: Cambridge University Press, 1987), 3.

12. Ibid., 36.

13. Ibid., 37.

14. See Robert W. Righter, *The Battle over Hetch Hetchy: America's Most Controversial Dam and the Birth of Modern Environmentalism* (New York: Oxford University Press, 2005).

15. David Ross Brower, *For the Sake of the Earth: The Life and Times of David Brower* (New York: Peregrine Smith Books, 1990); Debby Anker, John De Graaf, and Antonio Castro, *David Brower: Friend of the Earth* (New York: Twenty-First Century Books, 1993).

16. Mark W.T. Harvey, *A Symbol of Wilderness: Echo Park and the American Conservation Movement* (Albuquerque: University of New Mexico Press, 1994).

17. Russell Martin, *A Story That Stands Like a Dam: Glen Canyon and the Struggle for the Soul of the West* (Salt Lake City: University of Utah Press, 1999 [1989]), 53.

18. Bernard DeVoto, "Shall We Let Them Ruin Our National Parks?" *Saturday Evening Post,* July 22, 1950, 17–19. See also James Lawrence Powell, *Dead Pool: Lake Powell, Global Warming, and the Future of Water in the West* (Berkeley: University of California Press, 2008).

19. For an interesting perspective on David Brower, see John McPhee, *Encounters with the Archdruid* (New York: Farrar, Straus and Giroux, 1971). See also Michael P. Cohen, *The History of the Sierra Club, 1892–1970* (San Francisco: Sierra Club Books, 1988).

20. Letters Protesting Echo Park Dam, Box 29 (M008), Folder 6, Wayne Aspinall Papers, Penrose Library, University of Denver, Denver, CO.

21. McGee Young, "From Conservation to Environment: The Sierra Club and the Organizational Politics of Change," *Studies in American Political Development* 22, no. 2 (2008): 183–203.

22. Wallace Stegner, ed., *This Is Dinosaur: Echo Park Country and Its Magic Rivers* (New York: Alfred A. Knopf, 1955).

23. For more on environmental aesthetics, see Allen Carlson, *Nature, Aesthetics, and Environmentalism: From Beauty to Duty* (New York: Columbia University Press, 2008).

24. Mark W.T. Harvey, "Echo Park, Glen Canyon, and the Postwar Wilderness Movement," *Pacific Historical Review* 60, no. 1 (February 1991): 43–67.

25. "Two Attacks on California's Colorado Water Rights . . . ," *The Commonwealth* 31, no. 31 (August 1955): 187, Hutchins Collection, Box 5, Folder 171, Water Resource Center Archives, University of California, Berkeley.

26. Eliot Porter, *The Place No One Knew: Glen Canyon on the Colorado,* David Brower, ed. (San Francisco: Sierra Club, 1966).

27. For more on the controversy over Rainbow Bridge, see Thomas Gary Smith, *John Saylor and the Preservation of America's Wilderness* (Pittsburgh: University of Pittsburgh Press, 2006), chapter 8, 142–57. See also Harvey, "Echo Park, Glen Canyon, and the Postwar Wilderness Movement."

28. Pacific Southwest Water Plan, AZ 372, Box 174, Folder 3, Morris K. Udall Papers, University of Arizona Library, Special Collections, University of Arizona, Tucson, digital format, http://content.library.arizona.edu/cdm/ref/collection/udallcoloradoAZU /id/3432= (accessed March 1, 2011). For a good look at Stewart and Morris Udall, see Henry Sirgo, *Establishment of Environmentalism on the U.S. Political Agenda in the Second Half of the Twentieth Century: The Brothers Udall* (Lewiston, NY: Edwin Mellen, 2004). For an overview of CAP, see Bureau of Reclamation, *Central Arizona Project* (Washington, DC: Department of the Interior, 1995).

29. Robert Dean, "Dam Building Still Had Some Magic Then: Stewart Udall, the Central Arizona Project, and the Evolution of the Pacific Southwest Water Plan, 1963–1968," *Pacific Historical Review* 66, no. 1 (February 1997): 86–87.

30. Stewart L. Udall, *The Quiet Crisis* (New York: Avon Books, 1964).

31. For a good discussion of water politics and motives in Arizona and California, see Wendy Nelson Espeland, *The Struggle for Water: Politics, Rationality, and Identity in the American Southwest* (Chicago: University of Chicago Press, 1998). See also Bonnie G. Colby and Katharine L. Jacobs, eds., *Arizona Water Policy: Management Innovations in an Urbanizing Arid Region* (Washington, DC: Resources for the Future, 2007).

32. For a discussion of varying newspaper opinions of Udall's plan, see James M. Bailey, "Reconsideration and Reconciliation: Arizona's 'Brothers Udall' and the Grand Canyon Dams Controversy, 1961–1968," *New Mexico Historical Review* 80, no. 2 (Spring 2005): 138.

33. Don Irwin, "Conservationists Oppose 2 Colorado River Dams: Projects Will Impair Grand Canyon Beauty, Add No Water to Supply, House Unit Told," *Los Angeles Times,* September 1, 1965.

34. Bailey, "Reconsideration and Reconciliation," 137.

35. Letter from Earnest W. McFarland to John Rhodes, March 8, 1956, Colorado River/Central Arizona Project Collection, Hayden Library, University of Arizona, Tempe, http://repository.asu.edu/search?q=March+8%2C+1956&col=Colorado+River+Central+Arizona+Project (accessed September 18, 2012).

36. See Stephen C. Sturgeon, *The Politics of Western Water: The Congressional Career of Wayne Aspinall* (Tucson: University of Arizona Press, 2002), 76; Steven C. Schulte, *Wayne Aspinall and the Shaping of the American West* (Boulder: University Press of Colorado, 2002), 188–89.

37. Carl Hayden and Barry Goldwater, S. 1658: A Bill to Authorize, Construct, Operate, and Maintain the Central Arizona Project, Arizona–New Mexico, and for Other Purposes, Box 22, Folder 1, Colorado River/Central Arizona Project Collection, Hayden Library, Arizona State University, Tempe. Also available in digital form through Western Waters Digital Library, http://repository.asu.edu/items/12307 (accessed September 18, 2012).

38. Sturgeon, *Politics of Western Water*, 114.

39. François Leydet and David Brower, *Time and the River Flowing: Grand Canyon* (San Francisco: Sierra Club, 1964).

40. Statement of Hualapai Indians of Arizona on Bridge Canyon Bill, Submitted to the Senate Committee on Interior and Insular Affairs, June 2, 1949, Colorado River/Central Arizona Project Collection, Hayden Library, University of Arizona, Tempe. Also available at ASU Digital Repository, http://repository.asu.edu/items/12276 (accessed September 7, 2012).

41. For a discussion of Brower's campaign against the dams and his use of photographs, see Finis Dunaway, *Natural Visions: The Power of Images in American Environmental Reform* (Chicago: University of Chicago Press, 2005).

42. Bailey, "Reconsideration and Reconciliation," 144–45.

43. "Should We Also Flood the Sistine Chapel So Tourists Can Get Nearer the Ceiling?" *New York Times*, June 9, 1966.

44. "Now Only You Can Save Grand Canyon from Being Flooded . . . for Profit," *Washington Post*, June 9, 1966.

45. Bailey, "Reconsideration and Reconciliation," 146–47.

46. Schulte, *Wayne Aspinall*, 192.

47. See Dustin Garrick, Katharine Jacobs, and Gregg Garfin, "Models, Assumptions, and Stakeholders: Planning for Water Supply Variability in the Colorado River Basin," *Journal of the American Water Resources Association* 44, no. 2 (April 2008): 384; H. G. Hidalgo, T. C. Piechota, and J. A. Dracup, 2000: Alternative Principal Components Regression Procedures for Dentrohydrological Reconstructions, Water Resources Research 36, 3241–49, http://www.u.arizona.edu/~gmgarfin/2008.garrick.jawra.pdf (accessed September 15, 2010).

48. Preference between Language of H.R. 3300 and S. 1004: Significant Differences, July 18, 1986, CM MSS-1, Box 18, Folder 6, Colorado River/Central Arizona Project Collection, Hayden Library, Arizona State University, Tempe.

49. Public Law 90-537, Colorado River Basin Project Act, September 30, 1968, Title II—Investigations and Planning, Section 202.

50. Public Law 90-537, Colorado River Project Basin Act, September 30, 1968, http://www.usbr.gov/lc/region/pao/pdfiles/crbproj.pdf (accessed August 16, 2010).

51. Preference between Language of H.R. 3300 and S. 1004: Significant Differences.

52. Letter from Ival V. Goslin, executive director, Upper Colorado River Commission, to Northcutt Ely, October 19, 1967; letter from Dallas E. Cole to Ival V. Goslin, November 28, 1967, both in Box M008.16.0288, Folder L-11b(1)I, Wayne Aspinall Papers, Penrose Library, University of Denver, Denver, CO.

53. Michael Garstang et al., "Weather Modification: Finding Common Ground," *American Meteorological Society* (May 2004): 647–55.

54. Floyd Dominy Oral History, by Brit Storey, April 6, 1994, in Boyce, VA, 75, Bureau of Reclamation Library and Archives, Denver, CO. See also Marc Reisner, *Cadillac Desert: The American West and Its Disappearing Water* (New York: Penguin Books, 1993 [1986]), 214–54.

55. Reisner, *Cadillac Desert*, 242.

56. Donald J. Pisani, Floyd E. Dominy, June 18, 2002, Waterhistory.org, http://www.waterhistory.org/histories/dominy/ (accessed March 26, 2011).

57. US Bureau of Reclamation, Central Utah Project, Bonneville Unit, May 27, 2009, http://www.usbr.gov/projects/Project.jsp?proj_Name=Central+Utah+Project+Bonneville+Unit (accessed August 16, 2010).

58. Adam R. Eastman, The Central Utah Project, *Historic Reclamation Projects Book* 26, May 27, 2009, http://www.usbr.gov/projects//ImageServer?imgName=Doc_1232656886173.pdf (accessed August 16, 2010). See also Jon R. Miller, "The Political Economy of Western Water Finance: Cost Allocation and the Bonneville Unit of the Central Utah Project," *American Journal of Agricultural Economics* 69, no. 2 (May 1987): 303–10.

59. Leo Perry, "Bonneville Project Is Pushed," *Deseret News* [Salt Lake City], March 11, 1965; "Utah's Appeal for Funds for Bonneville Unit Work," *Sunday Herald* [Provo, UT], May 16, 1965; Ann Shields, "Water Panel Votes to Ask $7.5 Million to Continue Central Utah Project," *Salt Lake Tribune*, April 5, 1966.

60. Steve Comus, "The Colorado River Could Provide L.A. with 5 Billion Watts of Power," *Herald-Examiner* [Los Angeles], December 4, 1973.

61. For an excellent discussion of the Grand Canyon controversy, see Roderick Frazier Nash, *Wilderness and the American Mind* (New Haven: Yale University Press, 2001 [1967]), 227–36.

62. Stewart Udall, Oral History Interview conducted by Joe B. Frantz, May 19, 1969, 27, Bureau of Reclamation Library and Archives, Denver, CO.

63. David Moryc, American Rivers Press Release, March 30, 2009, http://www
.americanrivers.org/newsroom/press-releases/2009/president-obama-signs.html (accessed
March 4, 2011).

64. Norman Miller, *Environmental Politics: Stakeholders, Interests, and Policymaking*,
2nd ed. (New York: Routledge, 2009). See also National Environmental Policy Act of 1969,
http://www.nps.gov/history/local-law/fhpl_ntlenvirnpolcy.pdf (accessed August 18, 2010).
For a discussion of the law and its applications, as well as guidelines for environmental
impact statements, see the NEPA webpage at http://www.epa.gov/compliance/nepa (accessed
August 18, 2010).

65. United States Environmental Protection Agency, Our Mission and What We Do,
http://www.epa.gov/aboutepa/whatwedo.html (accessed March 4, 2011). See also Hays,
Beauty, Health, and Permanence; Hays, *History of Environmental Politics since 1945*.

66. Treaty between the United States of America and Mexico, signed in Washington,
DC, February 3, 1944, http://www.usbr.gov/lc/region/pao/pdfiles/mextrety.pdf (accessed
August 8, 2010).

67. Udall Interview, 14.

68. International Boundary and Water Commission, United States and Mexico, Text
of Minute 242—Permanent and Definitive Solution to the International Problem of the
Salinity of the Colorado River, August 30, 1973, http://www.usbr.gov/lc/region/pao/pdfiles
/min242.pdf (accessed July 7, 2010). See also Lynne Lewis Bennett, "The Integration of
Water Quality into Transboundary Allocation Agreements: Lessons from the Southwestern
United States," *Agricultural Economics* 24, no. 1 (December 2000): 113–25, and "Colorado
River Basin Salinity Control Act," http://www.usbr.gov/lc/region/g1000/pdfiles/crbsalct
.pdf (accessed September 18, 2012).

69. Water Policies for the Future: Final Report to the President and to the Congress of the
United States, National Water Commission Records, 1969–1973, Box 10, Folder 1, Special
Collections, Stanford University Library, Palo Alto, CA.

70. See Kent Hughes Butts, "The Strategic Importance of Water," *Parameters* (Spring
1997): 65–83; James T. Bartis et al., *Oil Shale Development in the United States* (Santa
Monica: RAND Corporation, 2005).

71. Ray Hebert, "Peril to Colorado River Supply Seen in U.S. Quest for Energy," *Los
Angeles Times*, November 27, 1973.

72. Central Utah Water Conservancy District, "Water for Energy: District Meets Key
Area Water Needs," *News Report: Central Utah Project* (Winter 1974–75): 5.

73. For example, Aspinall became very involved in the oil shale business after he left office
in 1973. See stories in the *Denver Post,* July 19, 1979, and March 2, 1980; "Wayne N. Aspinall:
'I Fitted the Epoch,'" *Shale Country* (December 1982): 17.

74. Bartis et al., *Oil Shale Development in the United States.*

75. Department of Energy, Office of Petroleum Reserves, Oil Shale Water Resources, http://www.fossil.energy.gov/programs/reserves/npr/Oil_Shale_Water_Requirements .pdf (accessed March 4, 2011); Department of Energy, Office of Petroleum Reserves, Oil Shale and the Environment, http://fossil.energy.gov/programs/reserves/npr/Oil_Shale _Environmental_Fact_Sheet.pdf (accessed March 4, 2011).

76. Moab Uranium Mill Tailings Remedial Action (UMTRA) Project, History, 2011, http://www.moabtailings.org/history.htm (accessed September 11, 2010).

77. Minutes, Moab Tailings Project Stakeholders Committee Meeting, June 22, 2010, Grand County Council Chambers, Moab, UT, http://www.moabtailings.org/pdf/minutes _2010_06_22.pdf (accessed September 11, 2010). See also Moab, Utah, UMTRA Project website, http://www.gjem.energy.gov/moab/ (accessed January 27, 2012).

78. Moab, Utah, UMTRA Project, Site Operations and Maintenance, http://www.gjem .energy.gov/moab/general/site_ops.htm (accessed September 12, 2012).

4

Sharing the Shortage

A River in Control

By THE MID-1970S, the young environmental movement had significantly impacted American society and begun to shape federal policies. Yet western state governments and representatives to Congress still reflected a widespread reluctance to reduce spending for reclamation projects so popular with constituents. This period is marked by a strange disconnect between media portrayals of environmental awareness and actual government actions. President Jimmy Carter tried to rein in federal spending on water projects seen as boondoggles or pork barrel spending, but to little avail.[1] However, several drought cycles and the realities of water shortages eventually brought some change to the Colorado River basin. First, the Bureau of Reclamation began shifting its focus toward a more sustainable operation of its dams and reservoirs. Second, both urban centers and agriculture began making changes in the way they obtained and used water. Change was gradual, but by the end of the twentieth century, concerns about global warming had inspired a new era of cooperation and regional approaches to the environment.

Since the 1970s, two of the most important issues confronting Colorado River water users have been water quality and shortages. Those seeking possible solutions often use the term *sustainability* in place of older terms such as *conservation* and *preservation*. The origin of the term *sustainable development* is probably the 1983 United Nations (UN) resolution to study long-term environmental issues and to seek ways to obtain "sustainable development to the year 2000 and beyond."[2] Connected to this resolution, the UN formed the World Commission on Environmental Development (WCED) and appointed the dynamic Gro Harlem Brundtland its chair.

Brundtland was a physician who served as the Norwegian minister for environmental affairs from 1974 to 1979 and briefly as prime minister from February to October 1981.[3] She was largely responsible for making the term *sustainability* popular worldwide after 1987, when the WCED held hearings and published the report *Our Common Future*.[4] Usually referred to as the Brundtland Report, this document led directly to the 1992 United Nations Conference on Environment and Development held in Rio de Janeiro. At this important summit, the Commission on Sustainable Development was formed and the famous Agenda 21 was adopted, which helped focus world attention on problems of global warming and resource depletion in a comprehensive way.[5]

Although much of Brundtland's later work dealt more specifically with world health issues connected to poverty, she always believed that caring for the environment was an integral part of dealing with all kinds of global challenges, including poor health in developing countries. In a speech she gave while director of the World Health Organization, she made this statement:

> We cannot separate the people from their environment. Investing wisely in health means caring for our natural environment and ensuring that we endow future generations with that precious resource. If we manage, hundreds of millions of people now and in the future will be better able to fulfill their potential, enjoy their legitimate human rights and be driving forces in development. People would benefit. The economy would benefit. The environment would benefit. We have a long way to go to reach this goal.[6]

Brundtland thus challenged international organizations to address environmental issues as part of any solution to global issues such as health, population pressures, the treatment of indigenous peoples, and economic challenges in developing nations. Largely because of Brundtland's work, scholars and scientists now speak of sustainable use of resources such as the Colorado River.

Such policies, however, were slow to come to the American Southwest and in many ways are just beginning to take hold. As early as 1986, the publication of journalist Marc Reisner's *Cadillac Desert* alerted western states to the possible long-term implications of their unsustainable water use.[7] Later, in 1992, law professor Charles F. Wilkinson published *Crossing the Next Meridian: Land, Water, and the Future of the West*.[8] In this book, Wilkinson called for a new focus on sustainable use of the environment through community planning. He acknowledged that such planning was not something with which the western states had any real experience.[9] For many years, the concept of sustainability remained vague and in the background in the Colorado River basin. Policies of the Bureau of Reclamation did begin to turn toward a more balanced relationship with the environment, but it took time for such concepts to be incorporated into regional planning. During the late twentieth century, very few people spoke of sustainable resource use.

Defeat of Presidential Reform

Although the US environmental movement was well under way by the mid-1970s, most Colorado River users operated much as they had before. However, shortages caused serious concern when drought affected much of the nation and especially the western states in 1976–77. Although it was not as long as the drought that had plagued the region from the late 1950s well into the mid-1960s, many in the Southwest worried that their river lifeline would not meet demands, especially once the Central Arizona Project (CAP) canal finally began diverting water. For a short time, conditions in California were especially dire. Emergency regulations ordered restaurants to serve water only if customers specifically requested it. No one was allowed to hose down sidewalks or driveways, public fountains were turned off, and lawn watering was severely limited.[10] Although the restrictions were lifted when wetter conditions returned in 1978, California kept them on the books in case of a future water crisis.[11]

In the midst of these concerns, in 1977 the federal government passed amendments to the 1972 Clean Water Act.[12] This legislation was intended to reduce the discharge of pollutants into the nation's water from industrial and agricultural locations. In the Colorado River basin, pesticides and other pollutants were regularly found in the water, but the major problem confronting agriculture—the largest group of river users—was salt. Although the 1973 Salinity Control Act sought to address the quality of water sent across the border to Mexico, it was also meant to address salinity that affected users up and down the river.[13] The 1977 amendments to the national Clean Water Act addressed some of the issues the Colorado River basin experienced with enforcing the 1973 law. Measuring standards were simplified and made more flexible. However, these provisions still seemed too difficult to apply to the Colorado River and the problem of salinity.

Many studies of the Colorado River indicated that the high salt levels were at least 50 percent naturally occurring. The river ran through a landscape that naturally leached a high amount of salt into the water, and irrigation created even more salt. If the new 1977 standards were enforced on the Colorado, the zero-effluent or -return flows might result in increasing, rather than decreasing, salinity. Agricultural economics professor John Keith argued that none of the methods then used to reduce salinity, such as replacing flood irrigation with sprinklers or ponding the runoff for evaporation, would improve water quality for the Colorado. Instead, he asserted, people needed to realize that special conditions made it impossible to require zero-return flows in the arid West. A mere 10 percent decrease in salinity levels would actually require a "50 percent reduction in irrigated agriculture."[14] Nearly a decade later, scientists and scholars viewed salinity as one of the major problems still not being addressed in the Colorado River basin.[15]

One of the few discussions of comprehensive approaches to water shortages in the West came from the field of cultural anthropology. Using Clifford Geertz's theory of "involution," scholar Gary Palmer argued in 1978 that development in the basin should occur through careful planning and various conservation strategies rather than "competitive expansion."[16] Two of his most important recommendations were the concept of "urban agriculture" and an open water market.[17] Palmer asserted that producing food in and around urban areas could save approximately 3 million acre-feet (MAF) per year in the lower Colorado River basin alone.[18] While the idea of water marketing remains controversial, the concepts of growing food in urban areas and eating food produced locally have become more prevalent.[19] While the idea of growing food close to consumers surfaced in several places in the 1970s, it was largely ignored until recently.

In the midst of drought in the West, newly elected US president Jimmy Carter began to examine federal policy for the development of western water projects. Any president assuming office in the 1970s would have encountered the impact of the growing environmental movement. Although many planned projects in the Colorado River basin were either not yet started or unfinished, most people involved with western water understood that the great age of dam building and large diversion projects was almost over. Especially along the Colorado, everyone from politicians to engineers believed the river was almost completely allocated, and many suspected that the last great building project—the Central Arizona Project—would likely create shortages in times of drought.[20]

With an engineering degree from the US Naval Academy in Annapolis, Maryland, Carter was well equipped to analyze water issues. In 1973, as governor of Georgia, Carter carefully studied plans for a new dam project called Spewrell Bluffs, submitted by the Army Corps of Engineers. Because he was an engineer, he was able to understand the numbers and believed the proposal was an inaccurate assessment of costs and benefits for a $133 million project. Carter came to believe that dam projects themselves were destructive to the environment and damaging to the federal budget.[21]

After he became president, Carter took a similarly hard look at the Bureau of Reclamation and in 1977 announced plans to radically reduce its size and its programs. In the middle of the driest year ever recorded in California and in much of the West, Carter announced the cancellation of federal funding for nineteen water projects, including CAP. Since most water-building projects were in the western states, those governors and members of Congress began calling Carter's plans the "hit list" and the "war on the West."[22] Of the nineteen projects scheduled for complete cancellation, nine were west of the Mississippi, but only three were connected to the Colorado River basin (the Narrows Project, Fruitland Mesa Project, and Savory–Pot Hook Project).

Of five additional projects still to be funded but at severely reduced amounts, two were in the West, in both parts of the Colorado River basin (CAP and the Bonneville Unit for the Central Utah Project [CUP]). In fairness, nine projects were approved for full funding, three of which were in the West in parts of the Colorado River basin (the Dolores and Dallas Creek Projects in Colorado and the Lyman Project on the Green River in Wyoming).[23] As a firestorm erupted in Congress, where funding of water projects had long been an important political currency, Carter and his team identified even more projects for cuts or termination. Ending up with a list of almost eighty, Carter's advisers and Vice President Walter Mondale warned him that fighting Congress over so many projects would doom his presidency.[24]

Carter finally reduced the list to eighteen, but they were to be completely terminated. Among them were three in the upper basin of the Colorado River—Dolores, Fruitland Mesa, and Savory-Pot Hook—as well as the entire Central Utah Project and the Central Arizona Project. Congress responded by drafting its own version of the 1978 budget for Public Works, which restored funding for all but one of the projects on Carter's list and added even more. Faced with a likely veto override, Carter backed down and signed a compromise bill that removed nine projects from funding, but they were included in the budget the following year, and this time Carter did veto it. Congress drafted another version of an appropriations bill, but the president was in a political pinch for support over having returned the Panama Canal to Panama. Carter signed the bill, which not only continued heavy funding of water projects but also exempted Tellico Dam on the Tennessee River from the Endangered Species Act.[25] Although Carter had tried to change water reclamation policy, he had failed.[26] He later spoke about the decision to sign the compromise bill:

The compromise bill should have been vetoed because, despite some attractive features, it still included wasteful items which my congressional supporters and I had opposed. Signing this act was certainly not the worst mistake I ever made, but it was accurately interpreted as a sign of weakness on my part, and I regretted it as much as any budget decision I made as president.[27]

Although Carter had hoped to make a serious change in the way the Bureau of Reclamation operated and the projects that would be funded, he found himself bound by congressional politics. A radical change would not come from the executive office.[28]

Slow Change

In spite of the victory western politicians had won in restoring project funding, budgets were tight in all sectors of the economy. The era of big projects was indeed ending, and the Bureau of Reclamation was beginning to make a significant change in its overall focus, from project building to refined management planning. Bureau commissioner R. Keith Higginson, who served in that post during the Carter administration, admitted that although operations did not change much, the ideas were beginning to shift. Water shortages would be met with different conservation tools instead of dams and diversion canals. A little more than a decade after Carter had done so, Higginson agreed with most others involved in water policy in the West that new approaches were needed:

> So I think what's going to happen to take care of these future water shortages is conversion of uses, changes of uses from one to another. And I think you're going to see continual pressure to take water that is now being diverted for out-of-stream uses, like irrigation, and have these rights return to the stream, for reestablishing the stream flows for fish and wildlife.[29]

Higginson argued that the real problem was that western water law did not account for environmental uses. States would have to alter their laws and reassign water rights for true redistribution of water applications to occur.

In 1979 the Government Accounting Office (GAO) issued a report on the Colorado River basin and threats of water shortages. After declaring that "the Colorado River Basin is in trouble," the GAO outlined the basic problem of over-allocation and asserted that a lack of full-basin cooperation made the problem worse. Unless representatives from all seven basin states, the Bureau of Reclamation, and other private entities worked together, the region might face disaster.[30] While the Department of the Interior agreed with the GAO's basic recommendations, it argued that creating a large basin management group would never work because the states were not willing to relinquish any control of water in their jurisdictions.[31] In so many ways, this assessment illustrates one of the largest ongoing challenges in the Colorado River watershed area: the problems with regional planning and cooperation. Too much political intrigue and too many state-centered interests have kept the states from coordinating ever since the 1922 compact controversy. However, seemingly inevitable water shortages are gradually creating the possibility of cooperation to meet impending catastrophe.

One of the interesting details pointed out in the GAO report was an estimation of stream flow for the Colorado River. Bureau of Reclamation officials thought they were being conservative by using 14.8 MAF annually as the average flow, limiting projects to fit within that figure. The GAO, however, argued that 14.8 MAF was far too optimistic and that the number should realistically be closer to 13.7 MAF. If that estimate was correct, then each basin had much less than the 7.5 MAF allocated by the 1922 compact.[32] In response, the bureau prepared its own consumptive uses report for the river in 1981, examining trends from 1976 to 1980. In this parameter, the bureau estimated complete consumptive use of the river, including the allocation to Mexico, which amounted to 16.7 MAF in 1976 and as much as 22.0 MAF in 1980. Meeting these withdrawals meant drawing down the water levels in Lakes Mead and Powell.[33] In their next five-year report, bureau analysts estimated that total consumptive uses had risen steadily to 28.7 MAF in 1985, far exceeding the river's actual annual flow.[34]

In addition to measuring water consumption more carefully, the bureau made some changes through the 1982 Reclamation Reform Act. It raised the acreage limitation for reclamation water from 160 acres per person or 320 acres per family to 480 acres and 960 acres, respectively. While this increase was less than some farmers wanted, the bureau acknowledged that 160-acre farms were not financially viable in the arid West. Two other important changes included requiring each water district to develop a conservation plan and allowing the federal government to be party to lawsuits against states. This last feature enabled the federal government to assist Indian communities in their suits and settlements over water allocations (see chapter 6). Critics of the reform act argued that the acreage limitations were nothing more than smokescreens to hide the fact that huge farming enterprises in the West received very cheap, heavily subsidized water. Enforcing the acre limits was never very efficient, and so many exceptions were made that the limits did nothing to protect small farmers.[35] This unsustainable policy is undergoing considerable scrutiny in the twenty-first century as many analysts are concluding that the only way to reduce water consumption in the basin is to raise water prices for all users.

In the early years of the Reagan administration, most of the discussions in California revolved around impending shortages after CAP was completed. Southern

Californians and especially heads of the water districts of Los Angeles and San Diego feared shortages once Arizona began withdrawing its full 2.8 MAF annual allocation from the Colorado River.[36] Plans to build a "peripheral canal" that would bring Northern California water to Los Angeles failed because of environmental concerns over the Sacramento Delta. Once the plan was voted down, journalist Philip Fradkin warned that another "water war" would plague the West unless lawmakers created a solution.[37] Various conservation plans seemed the only real answer for urban areas such as San Diego, which was already using nearly 25 percent of the Colorado River water allotted to Los Angeles when it was entitled to only 15 percent.[38]

Arizona, on the other hand, was anticipating the opening of the CAP canal and trying to determine who would get the water. In the state's water plan for 1977, officials predicted that by 2010, CAP would be delivering approximately 400,000 acre-feet for urban uses, 100,000 acre-feet for Indian water settlements, and around 500,000 acre-feet for agriculture. Even with CAP water, Arizona would have to cut back on irrigation if there was to be enough water for the growing urban centers.[39] Anticipating future shortfalls, Arizona passed a groundwater act in 1980 to try to stem the overdraft of the aquifers.[40]

The year 1983 was unusually wet, with record snowpack and rainfall that led to widespread flooding up and down the Colorado River. When water began pouring over the floodgates of both Hoover and Glen Canyon Dams, operators opened the spillways to draw down the reservoirs to protect dam integrity. The resulting floods caused millions of dollars in damage in towns along the river, and many farmers as far south as Mexico had to evacuate their homes.[41] In late 1984, however, San Diego was negotiating with corporation owners in Colorado—the Galloway Group, Ltd.—to buy water. Although intrastate purchases were not totally unusual, this move was an interstate water sale, not yet legal in most states in the West. Los Angeles officials were also worried that San Diego's purchase might drive up the price of water for everyone.[42]

While California cities worried about the implications of Arizona's CAP, Arizonans defended the pipeline and emphasized its importance to the state. One of CAP's most vocal supporters was Arizona representative Burton W. Barr. The *Los Angeles Times* quoted him during one of his discussions about the importance of CAP to Arizona. He looked at the state's history and used the ancient Hohokam as an example of what would happen to Arizona without more water:

> You can go down there to the Hohokam . . . they had one of the most intellectual Indian groups, they were supposed to have the greatest water program in the world and all that's left there is a guy from the Department of [the] Interior with a brown hat on who says

"here's where they were." I tell the people that's what we'll have, the guy in the brown hat standing there.[43]

The specter of the Hohokam, whose culture had mysteriously disappeared, was an image all Arizonans understood. Unless they received CAP water, they were told, Phoenix and Tucson would end up like that ancient people—mysteriously gone. The water was also absolutely necessary for agriculture, one of the largest segments of the state's economy. People like Representative Barr breathed a sigh of relief when CAP pumped its first test waters in September 1985. The canal was dedicated in November, but construction would continue on CAP until its completion eight years later in 1993, at a total expense of $3.6 billion.[44]

A great irony missed by Arizonans at the time was the fact that irrigation had helped destroy the farmlands of the Hohokam by creating high salinity levels in the soil. Even if the water arrived and continued to flow, agriculture might disappear just as it had in ancient times because of the use of both ground and surface water. Another irony was that while CAP water was intended to keep farmers from over-drawing the groundwater, it instead went mostly to urban areas and to settle Indian water claims. The main reason for this change was cost. Only four years after water began flowing through the canal, CAP and its management organization, the Central Arizona Water Conservation District (CAWCD), found it hard to sell the water to farmers. To pay back the reimbursable part of the canal project, the CAWCD had to charge rates much higher than most farmers could afford. The rate was $38 to $48 per acre-foot in 1989. If farmers had more than the 960 acres allowed by the 1982 Reclamation Reform Act, then the cost of CAP water would be between $200 and $250 per acre-foot. Groundwater, on the other hand, cost only around $20 per acre-foot and up. California farmers were getting their federally subsidized reclamation water for as little as $3.50 per acre-foot, so Arizona still saw its expensive groundwater as the cheapest supply.[45]

During the late 1980s, Americans continued to pay attention to environmental hazards in both the air and the water. Although Ronald Reagan was less interested in the environment than Carter had been, activist groups and world congresses continued to draw US participation in sustainability efforts. In 1987 the US Congress passed amendments to the 1977 amendments to the Clean Water Act. Known collectively as the Water Quality Act, these amendments made further efforts to pinpoint sources of polluted effluents affecting both ground and surface waters. The interest in clean water impacted operations of the Colorado River system, mostly in salinity levels from agricultural runoff.[46]

As if worrying about water were not enough, scientists discovered first one and then two holes in the ozone layer, leading to the 1987 United Nations meeting in

Montreal and a document known as the Montreal Protocol. In that document, world nations vowed to phase out and eventually stop using harmful chemicals such as chlorofluorocarbons (CFCs) that depleted ozone. These harmful substances were used in everything from dry cleaning and hairsprays to insulating foams in refrigerators and water heaters. Encouraged by the successful negotiations in Montreal and other actions, the second Earth Day celebration took place in 1990, and the UN Conference on Environment and Development (Earth Summit) in Rio de Janeiro in 1992 pushed forward the notion of "sustainable development" through its Agenda 21. The Exxon *Valdez* oil spill in 1989 was another motivating event that encouraged worldwide attention to the environment.

In the American Southwest, however, the major concern continued to focus on water, as a drought began that would last through 1992. While this drought affected most states west of the Mississippi in some way, the lower Colorado River basin suffered the most. Already concerned about Arizona's CAP operations, California media treated the drought as a water crisis. At least one journalist, however, argued that the state should not worry, there was plenty of water. It was simply being managed poorly and allocated foolishly.[47] Other journalists and politicians asserted that responsibility for controlling water quality should be shared more equally among users, that the responsibility for addressing salinity issues had been placed unfairly on the upper basin. Both basins contributed to salinity and should share costs as well as water.[48]

As drought and water-quality issues continued to dominate all other issues in the West, the Bureau of Reclamation found itself working to sharpen its focus on operations management and environmental issues. The bureau had to find ways to continue meeting the 1944 US-Mexico Water Treaty obligations during times of drought, help manage salinity issues, settle Indian water claims, and continue upper basin development in spite of over-allocation.[49] Although priorities had shifted, bureau commissioners believed they played a very important role in western water, even after the great age of dam building had ended. C. Dale Duvall, bureau commissioner from 1986 to 1989, made this assertion in a 1993 interview: "The Bureau of Reclamation is the best neighbor that anybody has in the west . . . they're not the Corps of Engineers that comes in, builds a flood control project and then leaves town. When the Bureau come[s] in and builds a project, the Bureau becomes a member of that community . . . the Bureau is there."[50] In his view, the bureau was now a community partner in water conservation efforts.

Other bureau officials made similar assertions. Regional director for the upper Colorado basin Roland Robison (1989–1993) observed that the bureau was moving toward efficiency, safety, and protecting the environment. The bureau conducted environmental impact studies, and one for Glen Canyon Dam led directly to major operational changes, part of an agreement in the 1992 Grand Canyon Protection Act.[51] The act stipulated that the bureau would keep all state agencies and the US Congress apprised of any changes in dam operations and would conduct an environmental impact study to determine what those operations should entail.[52]

No matter how well the bureau managed its water projects, however, there was still a threat of shortages. When CAP was completed, the lower basin had completely developed its resources and was using close to its full allocation. California had once relied on "surplus" water from Lake Mead that Arizona was not yet using, but after 1985 that supply began to dwindle, and it had disappeared completely by the mid-1990s.[53] In 1992, Las Vegas projected that its needs would increase dramatically over the next decade. In 1989 the city's daily water usage equaled approximately 100 billion gallons per year. Most of that water came from the Colorado River, but Nevada's annual share of the river is only 97.75 billion gallons (300,000 acre-feet). Researchers predicted that Las Vegas would use the full allocation by 2006 and that the city would need at least 250 billion gallons of water per year by 2030.[54] The booming tourist industry makes Las Vegas successful, but there is a disconnect between tourists in Las Vegas and the desert environment. Tourists flock to the region to enjoy the weather, paying little attention to the desert environment that has such limited water resources. As historian Evan R. Ward noted, the reality of aridity is hard to remember when one is standing in front of the Bellagio's fountains or playing in the many hotel pools along the strip.[55] Not even Arizona worried about water shortages until a decade after the first water flowed from CAP. To hedge against future shortages, Arizona started a state water-banking plan by recharging groundwater with any unsold CAP water.[56]

Shortages in the region seemed inevitable based on increased population and water usage alone. Compounding these issues was the threat of climate change, just starting to enter the mainstream debate. In 1993, scientists predicted that a change in global temperature of just two degrees Celsius would reduce the flow of the Colorado River and its tributaries by up to 12 percent. If runoff from snowpack were reduced by only 5 percent, the salinity levels would be far above acceptable standards throughout the basin. They further warned that a reduction of only 10 percent of main-stream flow would reduce electricity generation by nearly 26 percent.[57] As these dire predictions began to surface, a Bureau of Reclamation study of consumptive uses of Colorado River water revealed troubling numbers that were steadily rising. In 1991 the total consumptive use amounted to more than 17.5 MAF, 3 to 4 MAF above actual stream flow. In 1993 consumptive use had increased to over 20.7 MAF. It was encouraging that the number went down to about 17.4 in 1995, but that amount was still far above the actual normal yearly flow.[58]

River Use at the Turn of the Twenty-First Century

Armed with these troubling statistics, scientists and decision-makers began to address future shortages. The topics of water banking and marketing entered the conversation with increasing frequency, along with recycling and reducing water use. Urban centers experimented with some of these methods, and various metropolitan areas educated the public on wise water use. Eventually, many municipal water districts saw the purchase of agricultural water as the answer to their shortages (see chapter 8). Although still the largest water user in the Colorado River basin, agriculture is slowly selling its water to urban areas. The growth of cities during the 1990s and the early twenty-first century placed heavy demands on water supplies, and the only place left to look for water was the farm. Small to medium-sized farmers are increasingly selling out to neighboring metropolitan water districts that buy them for their water rights and to accommodate ever expanding suburban sprawl.

Cities have continued to purchase water from farms, and the Salt River Project in Arizona provides an interesting example. Originally irrigating more than 250,000 acres of farmland in the state, the project was serving only 10 percent of that acreage for agriculture in 2007.[59] California converts approximately 40,000 acres of farmland to urban uses each year.[60] Throughout the entire nation, from 1992 to 1997, around 13.7 million acres were converted from farming to urban use—a 51 percent increase from the decade 1982–1992.[61]

While agriculture-to-urban transfers dominated western water activities in the 1990s, President Bill Clinton's administration worked to address environmental issues, including water. From 1993 to 2001, Bruce Babbitt oversaw Bill Clinton's environmental policy. As governor of Arizona from 1978 to 1987, Babbitt had worked to preserve Arizona's water, but as Clinton's secretary of the interior, he had to take a much larger view of water resources. While much of the environmental effort was focused on the Florida Everglades, Babbitt also oversaw the trimming and streamlining of the Department of the Interior and served as a broker in several water conflicts in the West. Through his efforts, California finally settled debates over the California Bay delta region, where urban and environmental uses conflicted.

Mostly, however, Babbitt saw it as his role to promote cooperation and new directions. He considered the 1973 Endangered Species Act the major turning point in sustainable water policies and mused about the future of water management:

> Our challenge is not to build more dams but to operate them then in a more river-friendly way. Our task is not to irrigate more lands but to promote more efficient use of water on lands now in production. Our task is not to develop new supplies but to make use of those that already exist. We do have allocation and distribution problems, but

they can be resolved through use of water markets, conservation, and other innovations. Our task in the coming century is to restore rivers, wetlands, and fisheries.[62]

To accomplish these goals, Babbitt argued frequently that bringing contending players' interests together was his most important task. "The only way you can fix a watershed is by creating partnerships between governments, between landowners large and small, among all the stakeholders," he asserted.[63] Using this philosophy, Babbitt tackled many sustainability issues, such as restoring habitat for the spotted owl in the Northwest, settling water claims for Indian reservations, and returning wolves to Yellowstone National Park. He also shepherded a regional partnership among the lower basin states, the Bureau of Reclamation, and the US Fish and Wildlife Service to develop a multi-species management and restoration plan known as the Lower Colorado River Multi-Species Conservation Program. Babbitt helped organize the operation of this basin-wide management group (see chapter 7).

One of the important projects for the Colorado River during Babbitt's administration was the effort to restore beach habitat in the Grand Canyon through simulated floods, which was partly an outgrowth of environmental impact studies conducted on Glen Canyon Dam operations. The Bureau of Reclamation conducted a simulated flood in March 1996 to try to restore the riparian ecosystem.[64] Two more high-flow experiments were conducted in 2004 and 2008.[65] Beaches and sandbars, created by naturally occurring sediment, provide vital habitat for young native fish, riparian vegetation, and camping sites for recreation. The natural flow of the river along with seasonal floods historically sustained these important conditions through the Grand Canyon, but Glen Canyon Dam trapped virtually all sediment in Lake Powell, leading to erosion and loss of habitat.

Many in the bureau believed these experiments were a significant move away from older patterns of development toward a new attention to environmental issues. In 1997, Pacific Northwest regional director John Keys stated, "Now it's up to all of us to ensure that the system provides water for all who need it, from fish to farmers, from Indians to Industry."[66] Although in theory the simulated floods were expected to restore sandbar habitats, scientists discovered that the sandbars that did form were very fragile and washed downstream within only a few days or, at best, a few months. Adjusting the timing of the floods helped conditions somewhat, but it remains unclear what the long-term results will be. During the temporary floods, native vegetation and wildlife suffer temporary damage, but river managers believe the damage will be temporary and offset by larger gains. Still, enhancing the river habitat for fish does help the native humpback chub downstream, but it also increases the number of nonnative rainbow trout upstream. These trout eventually crowd out the chub and other native species.[67]

For native communities, Babbitt and the State of Arizona negotiated a purchase of CAP water to augment the San Carlos Reservoir and maintain the lake fishery.[68] For Las Vegas, he opened the way for the lower basin states to share and transfer water rights. Arizona was allowed to bank water and then sell shares to Las Vegas from Lake Mead. Babbitt believed water banking and interstate transfers were good ways to meet shortfalls and ensure the best use of Colorado River water.[69] Another partial victory for Babbitt was his successful negotiation of a quantification agreement among Southern California users of the Colorado River. Since the 1930s, the Metropolitan Water District, Imperial Irrigation District, and Coachella Valley Water District had argued among themselves over the amount of river water each was entitled to. Although the Quantification Settlement Act was not fully completed until 2003 and was declared unconstitutional by the California Supreme Court in 2010, Babbitt's work in 1998 and 1999 enabled San Diego to begin to purchase water from Imperial Valley.[70] These agriculture-to-urban transfers continued as the ruling was appealed. In December 2011 the Third District Court of Appeals overturned the 2010 ruling. The controversy over who will pay for mitigation damage to the Salton Sea and local farmers continues.[71]

Other important projects in the 1990s included efforts to address the ongoing issue of salinity. Figures 4.1 and 4.2 illustrate the sources and recent estimates of the costs of salinity. By 2009, damage to land and infrastructure from salt was running upward of $350 million each year in the United States. In its most recent evaluation, the Bureau of Reclamation stated that salinity controls currently in place are reducing the amount of salt in the river by approximately 1.2 million tons per year. To keep water quality within current standards, the bureau estimates it will need to increase this control further, to 1.85 million tons by 2030.[72]

As the twenty-first century approached, Bureau of Reclamation policies began to reflect the values of sustainability. As law professor A. Dan Tarlock eloquently stated in 1997, "Sustainable use recognizes that artificial systems are permanent landscape features, but seeks to use science-based adaptive management."[73] Daniel F. Luecke, Rocky Mountain regional director for the Environmental Defense Fund, also articulated the issue with this perspective: "Describing and estimating the benefits of large dams require the treatment of rivers as commodities. Describing and assessing the environmental costs require viewing the rivers as integral systems."[74]

Some years earlier, in 1985, Governor Babbitt had formed the Grand Canyon Trust, an organization tasked with preserving and enhancing the Grand Canyon region.[75] The organization seeks a sustainable relationship between humans and the natural environment. In 1997 the trust conducted a Colorado River Basin Management Study and reported it to the bureau. In that report the authors stated: "The next 75

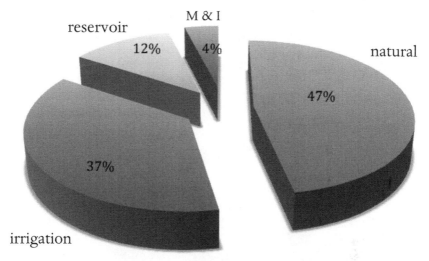

FIGURE 4.1 *Sources of salinity. Source: Bureau of Reclamation, Quality of Water Colorado River Basin Progress Report no. 23, US Department of the Interior, 2011.*

years of Colorado River Basin Management will be fundamentally different from the past 75 years."[76] Their recommendations included developing coordination among various management bodies throughout the basin and increasing the involvement of all stakeholders. While this goal has not yet been achieved, various groups and management teams are meeting, seeking this lofty goal of a basin-wide plan. One such symposium—held in Scottsdale, Arizona, in 2008—included scholars, lawyers, and representatives from the US Geological Survey, National Park Service, Bureau of Reclamation, and US Fish and Wildlife Service.[77]

One of the ongoing programs managed by the Fish and Wildlife Service is a study of the impact of high levels of selenium on endangered fish in the Colorado River basin. Selenium is a naturally occurring chemical element found in the various soils of the region. Although trace amounts are necessary for cellular function in animals and humans, too much selenium is toxic and kills fish and birds. Irrigation runoff can leach it (along with arsenic and other pesticide chemicals) into rivers and lakes. Evaporation, a constant ongoing condition in any water impoundment or canal, concentrates the levels of selenium, making plants and insects in those waters toxic to fish and waterfowl.[78]

In 1982, naturalists discovered significant numbers of deaths and deformities among waterfowl in the Kesterson National Wildlife Refuge in California. When tests showed the cause was high levels of selenium, the US Geological Survey, US

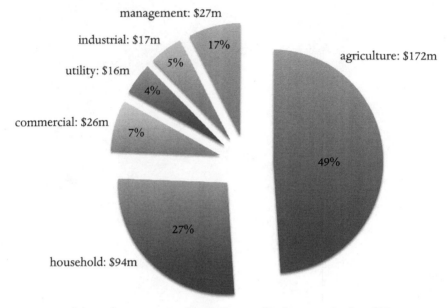

FIGURE 4.2 *Salinity damage per year. Source: Bureau of Reclamation, Quality of Water Colorado River Basin Progress Report no. 23, US Department of the Interior, 2011.*

Fish and Wildlife Service, Bureau of Reclamation, and Bureau of Indian Affairs teamed up to seek solutions. In 1985 the US Department of the Interior, under the auspices of the US Geological Survey, created a program to examine selenium contamination in western waters. Now called the National Irrigation Water Quality Program, the project has worked to study western regions and find ways to reduce harmful chemical levels.[79] Several key sites selected as part of the program include one on the Gunnison River in the Grand Valley in west-central Colorado at the spot where it joins the Colorado River. Another site is the Dolores–Ute Mountain Project area in Colorado along the Dolores River. One large study area includes most of the lower part of the main-stream Colorado River, from Davis Dam almost to Imperial Dam near Yuma, Arizona. Other major sites that affect tributaries of the Colorado include one on the Green River in Utah south of Flaming Gorge Dam and the San Juan River in New Mexico.[80]

Although most of these studies were suspended in 2005 because of federal budget cuts, ongoing efforts include partnerships between the Bureau of Reclamation and state groups, such as the Grand Valley Selenium Task Force in Colorado. In 2007 this group evaluated various ways to reduce selenium levels. One option includes diverting runoff water and increasing clean water to dilute levels of the chemical along important backwater habitats where endangered fish lay their eggs.[81] Other options include diluting the water in major drain areas, treating water with high levels of selenium, and creating replacement habitat at other locations. The fish species threatened in this portion of the upper Colorado River include four on the Endangered Species list: the Colorado pikeminnow, razorback sucker, bonytail, and humpback chub.[82]

Although many challenges to the Colorado River remain, recent events are encouraging. The most important element for any equitable and sustainable use of the river and restoration of its watershed ecosystem is the cooperation of multiple interests. The new high-flow release protocol plans for Glen Canyon Dam operations illustrate some of this new cooperation. The planning of the protocol included all major federal agencies (National Park Service, Bureau of Indian Affairs, US Fish and Wildlife Service, US Geological Survey, and Western Area Power Administration), the Arizona Game and Fish Department, and the upper Colorado River commissions, as well as active involvement of the Hopi, Hualapai, and Pueblo of Zuni communities.[83]

In 2004 the nonprofit environmental activist group American Rivers named the Colorado River "America's most endangered river."[84] In 2005 a large group calling itself the Clean Colorado River Alliance held meetings in Arizona to discuss water-quality issues. The major problems this large group of more than twenty organizations, municipal suppliers, and government offices identified included these pollutants: mercury, ammonia, boron, pesticides, uranium, perchlorate, and selenium.[85] These chemicals, along with high levels of salt, make water quality in the river an ongoing challenge. In 2010 the upper Colorado River remained on American Rivers' most endangered rivers list, at number six.[86] The organization declared 2011 the Year of the River, with plans to remove dams and restore riparian habitats along many rivers from Washington state to Maine. To better address Colorado River issues, American Rivers opened a new office in Denver.

In an ongoing effort to address the endangered condition of the humpback chub and other fish native to the Colorado River, the Bureau of Reclamation has experimented with nonnative fish removal. In 1988 the governors of Colorado, Utah, and Wyoming, along with federal organization leaders, formed the Upper Colorado River Endangered Fish Recovery Program.[87] Rainbow and brown trout prey on the humpback chub and other native species, so this group and others have conducted experiments with fish removal since 2003. In January 2011 the Bureau of Reclamation released a new assessment plan for nonnative fish control that again called for fish removal. The interesting factor in this latest report is a discussion of Native American interest in the project. When the removal plans were first publicized, the Hopi, Kaibab Band of Paiute, Hualapai, and Zuni Tribes protested the experiment unless some beneficial human uses were made of the euthanized fish. Part of the way the bureau addressed this concern and similar protests from the Pueblo of Zuni in 2009 was to move the operation away from a sacred area near the confluence of the Colorado and Little Colorado Rivers, near the top end of the Grand Canyon, farther upstream to the Lees Ferry area.[88]

Although there is little chance of restoring the Colorado River to its natural state, several groups have advocated radical mitigation efforts, such as the removal or circumvention of Glen Canyon Dam. Since 1996, the nonprofit organization Glen Canyon Institute has worked for restoration of Glen Canyon. In its Citizen's Environmental Assessment, completed in 2000, the institute argued that the costs of the dam far outweighed any benefits from storage of river water in Lake Powell.[89] Some of the recommendations include lowering the level of Lake Powell to 237 feet and drilling bypass tunnels to allow water to flow naturally around the dam. The group believes the dam would become a major tourist attraction as a relic of an earlier time and a representation of a new sustainable relationship with the river.[90] Yet it remains to be seen what kinds of future changes will be made to dam operations that might affect many groups of people both upstream and downstream. Any real success in creating a sustainable relationship with the Colorado River will require both widespread cooperation and an unflagging commitment to find solutions.

NOTES

1. Scott A. Frisch and Sean Q. Kelly, *Jimmy Carter and the Water Wars: Presidential Influence and the Politics of Pork* (Amherst, NY: Cambria, 2008), 40–43.

2. United Nations Resolution 38/161, Process of Preparation of the Environmental Perspective to the Year 2000 and Beyond, 1983, http://www.un-documents.net/a38r161 (accessed July 21, 2010).

3. She later served two more terms as prime minister, from 1986 to 1989 and again from 1990 to 1996, before serving as director-general of the World Health Organization from 1998 until 2003. For more, see her autobiography, Gro Harlem Brundtland, *Madam Prime Minister: A Life in Power and Politics* (New York: Farrar, Straus and Giroux, 2002).

4. Gro Harlem Brundtland, *Report of the World Commission on Environment and Development: Our Common Future* (New York: United Nations, 1987).

5. United Nations Environment Programme, Agenda 21, 1992, http://www.unep.org /Documents.Multilingual/Default.asp?documentid=52 (accessed July 21, 2010). For a recent assessment, see Joachim H. Spangenberg, Stefanie Pfahl, and Kerstin Deller, "Towards Indicators for Institutional Sustainability: Lessons from an Analysis of Agenda 21," *Ecological Indicators* 2, nos. 1–2 (November 2002): 61–77.

6. Gro Harlem Brundtland, speech to International Consultation on the Health of Indigenous Peoples, November 23, 1999, http://gos.sbc.edu/b/brundtland.html (accessed March 26, 2011).

7. Marc Reisner, *Cadillac Desert: The American West and Its Disappearing Water* (New York: Penguin Books, 1993 [1986]).

8. Charles F. Wilkinson, *Crossing the Next Meridian: Land, Water, and the Future of the West* (Washington, DC: Island, 1992).

9. Ibid., 299–302.

10. Benedykt Dziegielewski, Hari P. Garbharran, and John F. Langowski Jr., *Lessons Learned from the California Drought (1987–1992): National Study of Water Management during Drought* (Darby, PA: Diane, 1997), 50–56.

11. Richard Simon, "Water Conservation Pushed—Before Drought Returns," *Los Angeles Times*, February 5, 1981.

12. The law was amended again in 2002; http://epw.senate.gov/water.pdf (accessed July 28, 2010).

13. Salinity Control Act, 1973, http://www.usbr.gov/lc/region/pao/pdfiles/crbsalct.pdf (accessed July 28, 2010)

14. John Keith, "Water Quality Management and Irrigated Agriculture: Potential Conflicts in the Colorado River Basin," *American Journal of Agricultural Economics* 59, no. 5 (December 1977): 952.

15. James L. Wescoat Jr., "Impacts of Federal Salinity Control on Water Rights Allocation Patterns in the Colorado River Basin," *Annals of the Association of American Geographers* 76, no. 2 (June 1986): 157–74.

16. Gary B. Palmer, "Water Development Strategies in the Colorado River Basin: Expansion versus Involution," *Anthropological Quarterly* 51, no. 2 (April 1978): 99.

17. See also B. Delworth Gardner, "Water as an Equity Problem," *Center Magazine* (November-December 1980): 32–41.

18. Palmer, "Water Development Strategies," 112.

19. Ibid., 103–4. See also the works of German economist E. F. Schumacher, including *Small Is Beautiful: Economics as If People Mattered* (New York: Harper and Row, 1975).

20. "Is U.S. Running Out of Water?" *US News & World Report*, July 18, 1977; William H. Bruvold, "Residential Water Conservation: Policy Lessons from the California Drought," *Public Affairs Report* 19, no. 6 (December 1978): 1–7.

21. Reisner, *Cadillac Desert*, 307.

22. For works on Carter's presidency, see Peter G. Borne, *Jimmy Carter: A Comprehensive Biography from Plains to Post-Presidency* (New York: Scribner, 1997); W. Carl Biven, *Jimmy Carter's Economy: Policy in the Age of Limits* (Chapel Hill: University of North Carolina Press, 2002).

23. S. J. Hadeed, "Several Water Resources Projects Survive Carter Cutback," *Water Pollution Federation* 49, no. 5 (May 1977): 727–30.

24. Reisner, *Cadillac Desert*, 317–19.

25. For more information on the controversy surrounding the Tellico Dam exemption, see William Bruce Wheeler and Michael J. McDonald, *TVA and the Tellico Dam, 1936–1979* (Knoxville: University of Tennessee Press, 1986).

26. Reisner, *Cadillac Desert*, 329.

27. Frisch and Kelly, *Jimmy Carter and the Water Wars*, 90.

28. For more on Carter's environmental attitudes and water policies, see ibid.; Dennis L. Soden, *The Environmental Presidency* (Albany: SUNY Press, 1999); Biven, *Jimmy Carter's Economy*.

29. R. Keith Higginson, Oral History Interview with Brit Allan Storey, March 22, 1995, Bureau of Reclamation Library and Archives, Denver, CO, 89.

30. US Government Accounting Office, Colorado River Basin Water Problems: How to Reduce Their Impact, May 4, 1979, Bureau of Reclamation Library and Archives, Denver, CO (GAO Report 1979).

31. Ibid., 54.

32. Ibid., 23.

33. Bureau of Reclamation, Colorado River System Consumptive Uses and Losses Report, 1976–1980, Bureau of Reclamation Library and Archives, Denver, CO.

34. Bureau of Reclamation, Colorado River System Consumptive Uses and Losses Report, 1981–1985, Bureau of Reclamation Library and Archives, Denver, CO.

35. Public Law 97-293, Reclamation Reform Act, 1982, http://www.usbr.gov/rra/Law_Rules/public%20law%2097-293.pdf (accessed September 9, 2012). See also E. Phillip LeVeen, "A Political Economic Analysis of the 1982 Reclamation Reform Act," *Western Journal of Agricultural Economics* 8, no. 2 (1983): 255–65.

36. "Colorado River Still Delivering Goods, but Could Change in 10 Years," *Los Angeles Times*, October 8, 1980; Philip L. Fradkin, "Time to Turn off the West's Water Spat," *Los Angeles Times*, October 25, 1981, E5; "Capturing Our Precious Water," *Los Angeles Times*, July 25, 1982.

37. Fradkin, "Time to Turn off the West's Water Spat," E5.

38. "Capturing Our Precious Water."

39. Raul H. Castro, governor, Arizona State Water Plan, Phase II, State of Arizona, February 1977, Bureau of Reclamation Library and Archives, Denver, CO, 14, 53.

40. Arizona Department of Water Resources, Arizona Groundwater Management Act, 1980, http://www.azwater.gov/AzDWR/WaterManagement/documents/Groundwater_Code.pdf (accessed July 28, 2010). See also Robert Glennon, *Water Follies: Groundwater Pumping and the Fate of America's Fresh Waters* (Washington, DC: Island, 2002); Larry Mays, *Integrated Urban Water Management: Arid and Semi-arid Regions* (New York: Taylor and

Francis US, 2009); Donald E. Agthe, R. Bruce Billings, and Nathan Buras, *Managing Urban Water Supply* (London: Springer, 2003).

41. Penelope McMillan and Richard West, "Peril to Agriculture along River Seen in Arizona, Mexico," *Los Angeles Times,* June 24, 1983; Times Wire Service, "5,000 Flee Floodwaters in Baja California Fields," *Los Angeles Times,* June 26, 1983.

42. David Smollar, "Critics Claim Water Proposal Is All Wet," *Los Angeles Times,* September 11, 1984.

43. Bill Boyarsky, "Sharing the Colorado River's Water: West Braces for a Change," *Los Angeles Times,* September 23, 1985.

44. Total cost figure from Central Arizona Project, http://www.cap-az.com/AboutUs /FAQ.aspx (accessed January 12, 2012). See also Thomas E. Sheridan, "The Big Canal: The Political Ecology of the Central Arizona Project," in John M. Donahue and Barbara Rose Johnston, eds., *Water, Culture, and Power: Local Struggles in a Global Context* (Washington, DC: Island, 1998), 163–86.

45. "Busted Water Project," *Los Angeles Times,* March 25, 1989, 6.

46. See Robert W. Adler, Jessica C. Landman, and Diane M. Cameron, *The Clean Water Act 20 Years Later* (Washington, DC: Island, 1993); Claudia Copeland, *The Clean Water Act: Current Issues and Guide to Books* (Hauppauge, NY: Nova Science, 2003).

47. Patrick Porgans, "The Great Drought Hoax," *San Francisco Bay Guardian,* September 7, 1988, Box 1, Folder 35, Malca Chall Collection, Water Resource Center Archives, University of California, Berkeley.

48. Richard L. Gardner and Robert A. Young, "Assessing Strategies for Control of Irrigation-Induced Salinity in the Upper Colorado River Basin," *American Journal of Agricultural Economics* 70, no. 1 (February 1988): 37–49.

49. Larry D. Morton, Oral Interview by Brit Allen Storey, May 24, 1996, vol. 1, 412–14, Bureau of Reclamation Library and Archives, Denver, CO.

50. C. Dale Duvall, Oral Interview by Brit Allen Storey, January 26, 1993, vol. 1, 40–41, Department of Veterans Affairs, Washington, DC, printed and filed at the Bureau of Reclamation Library and Archives, Denver, CO.

51. Roland Robison, Oral Interview by Brit Allen Storey, September 27, 1993, vol. 1, 52–53, Bureau of Reclamation Library and Archives, Denver, CO.

52. Grand Canyon Protection Act, 1992, http://www.gcmrc.gov/library/reports /LawoftheRiver/GCPA1992.pdf (accessed July 29, 2010).

53. Elliot Diringer, "Chronic Shortages of Water Predicted," *Sacramento Chronicle,* June 2, 1995.

54. Mohammad Karamouz, ed., "Water Resources Planning and Management: Saving a Threatened Resource—In Search of Solutions," *Proceedings of the Water Resources Sessions at Water Forum '92,* Bureau of Reclamation Library and Archives, Denver, CO.

55. Evan Ward, *Border Oasis: Water and the Political Ecology of the Colorado River Delta, 1940–1975* (Tucson: University of Arizona Press, 2003), 113. See also Robert Glennon, *Unquenchable: America's Water Crisis and What to Do about It* (Washington, DC: Island, 2009), 2.

56. "Arizona Seeking to 'Bank' Its Share of the Colorado River," *US Water News Online*, December 1995, http://www.uswaternews.com/archives/arcrights/5azbank.html (accessed July 29, 2010).

57. Linda L. Nash and Peter H. Gleick, The Colorado River Basin and Climate Change, Report prepared for the United States Environmental Protection Agency, December 1993, Bureau of Reclamation Library and Archives, Denver, CO.

58. Bureau of Reclamation, Colorado River System Consumptive Uses and Losses Report, 1991–1995, September 2002, Bureau of Reclamation Library and Archives, Denver, CO.

59. Robert S. Gooch, Paul A. Cherrington, and Yvonne Reinink, "Salt River Project Experience in Conversion from Agriculture to Urban Water Use," *Irrigation and Drainage Systems* 21, no. 2 (May 2007): 145–57.

60. Alvin Sokolow et al., "California Communities Deal with Conflict and Adjustment at the Urban-Agricultural Edge," *California Agriculture* 64, no. 3 (July-September 2010): 121–28.

61. Clean Water Action Council, Land Use and Urban Sprawl, Green Bay, Wisconsin, http://www.cwac.net/landuse/index.html (accessed January 28, 2012).

62. Jana Prewitt and Victoria Voytko, *A History of the US Department of the Interior during the Clinton Administration, 1993–2001* (Washington, DC: Clinton Administration History Project, 2000), xxii, Box 10, Folder OA 24113, Department of the Interior Papers, William J. Clinton Presidential Library, Little Rock, AK, xxii.

63. Ibid., xv.

64. US Department of the Interior, Data from Grand Canyon Flood Positive, More Than 50 Large Beaches Created, May 22, 1996, Box 12, Folder 24115.2, Department of the Interior Papers, William J. Clinton Presidential Library, Little Rock, AK.

65. Clinton Administration History Project, xxiv. See also the USGS Report, http://walrus.wr.usgs.gov/grandcan/flood.html (accessed July 30, 2010).

66. John Keys, "The Future of Western Water Developments," paper presented at the Natural Resource Law Center Symposium, Boulder, CO, June 2–4, 1997, Box 5, Folder A-2-067, Dividing the Waters Collection, Water Resource Center Archives, University of California, Berkeley.

67. US Bureau of Reclamation, Draft Environmental Assessment Development and Implementation of a Protocol for High-Flow Experimental Releases from Glen Canyon Dam, Arizona, 2011–2020, http://www.usbr.gov/uc/envdocs/ea/gc/HFEProtocol/ch3 .pdf (accessed March 1, 2011), 44. See also David Frey, "Artificial Flooding May Help Grand

Canyon," *New West Development*, February 9, 2011, http://www.newwest.net/topic/article /artificial_flooding_may_help_grand_canyon/C35/L35/ (accessed March 1, 2011).

68. US Department of the Interior, News, October 18, 1999, Box 12, Folder 24115.2, Department of the Interior Collection, William J. Clinton Presidential Library, Little Rock, AK.

69. Ibid.

70. Ibid.

71. Court of Appeal Reinstates Historic Quantification Settlement Agreements Cases, Decided December 7, 2011, http://www.bhfs.com/portalresource/lookup/wosid/content pilot-core-2301-33102/pdfCopy.name=/Court%20of%20Appeal%20Reinstates%20 Historic%20Quantification%20Settlement%20Agreements.Client%20Alert.pdf (accessed January 28, 2012).

72. US Department of the Interior, Quality of Water Colorado River Basin Progress Report no. 23, *Reclamation: Managing Water in the West*, Bureau of Reclamation, 2011, http:// www.usbr.gov/uc/progact/salinity/pdfs/PR23final.pdf (accessed January 28, 2012), 24.

73. A. Dan Tarlock, "River Management in the Twenty-First Century: The Vision Thing," paper presented at the Natural Resources Law Center Symposium, University of Colorado, Boulder, June 2–4, 1997, Box 5, Folder A-2-067, Dividing the Waters Collection, Water Resource Center Archives, University of California, Berkeley.

74. Daniel F. Luccke, "Dams, Their Costs and Benefits," paper presented at the Natural Resources Law Center Symposium, University of Colorado, Boulder, June 2–4, 1997, Box 5, Folder A-2-067, Dividing the Waters Collection, Water Resource Center Archives, University of California, Berkeley.

75. Grand Canyon Trust, http://www.grandcanyontrust.org/about.php (accessed July 30, 2010).

76. Grand Canyon Trust, Colorado River Basin Management Study, April 1997, Box 21, Folder E-3-003, Dividing the Waters Collection, Water Resource Center Archives, University of California, Berkeley.

77. Colorado River Basin Science and Resource Management Symposium, November 18–20, 2008, Scottsdale, AZ, http://www.watereducation.org/userfiles/Draft%20Science %20Conference%20Program.pdf (accessed September 15, 2010).

78. US Fish and Wildlife Service, Region 6 Environmental Contaminants, Selenium /Irrigation Drain Water, http://www.fws.gov/mountain-prairie/contaminants /contaminants4.html (accessed March 8, 2011).

79. USGS, National Irrigation Water Quality Program, http://toxics.usgs.gov /interagency/irrig_prog.html (accessed March 8, 2011).

80. USGS, Project Sites, http://www.usbr.gov/niwqp/niwqpprojects/index.html (accessed March 8, 2011).

81. US Department of the Interior, Reclamation: Managing Water in the West: Evaluation of Options to Offset Selenium Impacts in Tributaries/Drains, Grand Valley, Western Colorado, January 2007, http://www.seleniumtaskforce.org/images/GV_Trib_Offset _report_Final.pdf (accessed March 8, 2011), 4.

82. USGS Colorado Water Science Center, Characterization of Selenium Concentrations and Loads in Select Tributaries to the Colorado River in the Grand Valley, Western Colorado, http://co.water.usgs.gov/projects/BSD00/index.html (accessed March 8, 2011).

83. US Bureau of Reclamation, Draft Environmental Assessment Development and Implementation of a Protocol for High-Flow Experimental Releases from Glen Canyon Dam, Arizona, 2011–2020, http://www.usbr.gov/uc/envdocs/ea/gc/HFEProtocol/ch3.pdf (accessed March 1, 2011), 122.

84. Arizona Department of Environmental Quality, 2004, http://www.azdeq.gov /environ/water/ccra.html (accessed March 8, 2011).

85. Clean Colorado River Alliance, April 15, 2005, Bullhead City, AZ, http://www.azdeq .gov/environ/water/download/0415m.pdf (accessed March 8, 2011); Recommendations Report, January 2006, http://www.azdeq.gov/environ/water/download/ccra06.pdf (accessed March 8, 2011).

86. American Rivers, America's Most Endangered Rivers Report, 2010 edition, http:// www.americanrivers.org/assets/pdfs/mer-2010/uppercolorado_factsheet_2010.pdf (accessed March 8, 2011).

87. Upper Colorado River Endangered Fish Recovery Program, 2012, http://www .coloradoriverrecovery.org/general-information/about.html (accessed January 28, 2012).

88. Bureau of Reclamation, Biological Assessment for Non-native Fish Control Downstream from Glen Canyon, January 28, 2011, http://www.usbr.gov/uc/envdocs/ea/gc /nnfc/Appdx-C-BA.pdf (accessed January 28, 2012), 6–8.

89. Citizen's Environmental Assessment Report, Glen Canyon Institute, 2000, http:// www.glencanyon.org/pdfs/CEA_Report.pdf (accessed January 28, 2012).

90. Glen Canyon Institute, Restoration of Glen Canyon, http://www.glencanyon.org /about/faq (accessed January 28, 2012).

Part 2

Currents of Today

5

The Metropolis and the Desert

Growing Cities in the West

NONE OF THE EARLIEST SETTLERS in the American Southwest would have dreamed of the twentieth-century metropolitan growth the region experienced. Some of its major cities grew out of Spanish missions or military outposts, but the gold rush in the mid-nineteenth century quickly expanded them. Many urban areas flourished because of a favorable climate, others because of their locations at important railroad junctions. One common trait they all shared, however, was an insatiable need for water.

While the primary purpose of reclamation and dam building in the West was to provide irrigation to farmers, growing towns and cities also benefited and eventually became some of the largest and hungriest users of the Colorado River. The growth of San Diego, Los Angeles, Tucson, Denver, Phoenix, Las Vegas, Mexicali, and Tijuana has been possible because these urban centers tap into the river. Although Tucson has continued to rely on groundwater, it now receives important augmentation from Colorado River water flowing through the Central

Arizona Project (CAP). Of these eight cities, Phoenix is perhaps the luckiest when it comes to water, sitting as it does at a confluence of several Colorado River tributaries. Yet as its metropolitan area spread to cover the entire "Valley of the Sun" by the late twentieth century, it, too, needed main-stream water from the Colorado. Denver, San Diego, Los Angeles, Mexicali, and especially Las Vegas cannot survive without that water.

San Diego (est. 1769)

The oldest metropolis that depends heavily on the Colorado River is San Diego, whose recent history is tightly connected with that of the younger Los Angeles. A very mild climate makes the San Diego area highly desirable, although its water supply has long been problematic. As with its large neighbor to the north, the first European settlers in San Diego were Spanish. Named after a Catholic saint, the town became the site of an important military post, the Presidio of San Diego, in 1769. The first Franciscan mission in the region was built here; by the end of the eighteenth century it had become the largest mission in the area.

When California became part of the United States in 1850, San Diego was a tiny town near the old presidio, and it had fewer than 1,000 inhabitants at the outbreak of the Civil War. While some of its neighbors to the north swelled with the influx of 49ers hoping to strike it rich, San Diego remained small and provincial. This status would begin to change in the second half of the nineteenth century. In 1867 a businessman named Alonzo Horton bought up a large amount of coastline to extend the town down to the waterfront. With bay frontage and a wharf, Horton hoped San Diego would grow as a result of trade. A population boom did begin, which increased after 1885 when the transcontinental railroad reached the town. By 1890, San Diego's population had reached more than 16,000.[1]

For most of its early history, the town of San Diego obtained most of its water supply from wells. Although some water came from the San Diego River, the river was most often dry, and wells were sunk down through the dry riverbed to the water table below. In 1816, Franciscan monks built a small dam to try to store and divert some of the water from the river, but it was never enough to create a large, stable supply.[2] Later, in 1873, the San Diego Water Company was created to try to develop a more coordinated approach to water for the growing city. By the end of the nineteenth century, more than seven major water corporations and projects existed. Some of these ventures involved more dams and a thirty-seven-mile flume that carried water from the San Diego River down to the town in the valley below the mountains.[3]

While San Diego supplied itself with water from its own watershed for many years, drought in the late nineteenth and early twentieth centuries led city leaders to think about other sources. For a time, Charles Mallory Hatfield made a career for himself in Southern California as a "rainmaker." Although he worked as a salesman for a sewing machine company, Hatfield became fascinated with the idea of altering the atmosphere to make rainfall, and he experimented with various chemicals. By 1902 he felt he had the magic formula and advertised his skills. After two or three successful attempts, word spread and he successfully produced rain for the city of Los Angeles.[4]

In 1915, San Diego leaders asked Hatfield to make rain to fill the Morena Dam reservoir. After building a tower beside the lake, he prepared to work his magic. When rain began to fall on January 5, 1915, everyone praised Hatfield's skills. The rain kept falling, however, and caused flooding—destroying roads, farms, and bridges. On January 27 the Lower Otay Dam failed, creating even more havoc and causing twenty deaths. The town council refused to pay Hatfield for making the rain unless he agreed to take responsibility for the damage, likely to run into the millions. Although he sued for payment, he never received it, but he did go on to make rain for other towns. When he died in 1958, he claimed he had succeeded more than 500 times but told no one his formula. Some people think Hatfield simply had a good sense for weather patterns that would naturally bring rain, for which he claimed responsibility.[5]

San Diego was an ideal port city, and Theodore Roosevelt's Great White Fleet made it an important naval station around the turn of the twentieth century. In July 1921, city officials began discussing the possibility of using Colorado River water.[6] Los Angeles had already demonstrated how to build aqueducts to pipe water a long distance, and the city was asking for government help to tap the Colorado. To create its own solution, San Diego sought federal assistance to bring water from the All-American Canal out beyond Imperial Valley to the coast. The Bureau of Reclamation supported the idea and made plans in 1933 for San Diego to receive 112,000 acre-feet of water from the Colorado through the canal, then under construction. The plan sounded good, but there was no infrastructure to bring the water from Imperial Valley to the city, and no one offered to build it. By the 1940s San Diego's population was booming, and by the end of World War II the city's water resources were overused by war industries in the area.

In 1944 the San Diego County Water Authority was formed and continues to hold responsibility for the city's water resources. Its original purpose was to secure and deliver alternative water supplies, specifically from the Colorado River. After much negotiation with the city of Los Angeles and the Bureau of Reclamation, in 1945 construction finally began on an aqueduct for San Diego that would tap into the Colorado River Aqueduct near the San Jacinto Tunnel. Over seventy-one miles long, the San Diego Aqueduct was completed in 1947. A second barrel was added in 1952, and another aqueduct was completed in 1958. Five major aqueducts provide water to the city today through water purchases from the Metropolitan Water District of Los Angeles and other agencies.[7]

During the Cold War years, San Diego continued to be an important military location and boasted US Navy, Marine Corps, and Coast Guard military bases. Although the city experienced economic decline in the post–Cold War years, its naval base is still very important and houses the largest naval fleet in the world. Other important industries today include cellular manufacturing, tourism, and environmental industries such as recycling, solar energy, and environmental engineering and consulting. In recent years the city's population, which is now more than 3 million, has outstripped its water resources once again. In early 2002, San Diego looked to the Imperial Irrigation District for more water and purchased almost 200,000 acre-feet per year through 2073, the longest and largest water transfer in US history. Approved in 2003, this water transfer became part of a larger Quantification Settlement Agreement that would enable several very large agriculture-to-urban water transfers in Southern California. Imperial County was continuing legal efforts to block the Imperial Irrigation District from transferring the water to San Diego as late as December 2012.[8]

In the last few decades, San Diego has had to be creative in its search for water. With a growing population and a static water supply, the city has been actively encouraging conservation efforts through a heavy public relations campaign, including educational programs and incentives. The city has also become a leader in recycling water and currently processes and reuses approximately 13,000 acre-feet of water each year. Water board officials hope to increase that amount to more than 53,000 acre-feet by 2020.[9] In 2005 and 2006 the city also developed a very complex drought management plan in case water reductions became necessary. Although conditions have eased, San Diego continues to feel that its water supply is tenuous. The city will probably seek more water transfer purchases from agriculture.

Los Angeles (est. 1781)

The story of water in Los Angeles is long and complicated, filled with tales of intrigue, political deal making, and the determination of city leaders. The second-largest city in the United States, Los Angeles boasts a population of almost 4 million—nearly 13 million in the entire metropolitan area. Consumers in its sprawling suburbs continue to seek new supplies of water and have long depended on steady streams from the Colorado River.[10]

Established by the Spanish in 1781, Los Angeles ("the Angels" in Spanish) was initially little more than an outpost for Franciscan missions for the indigenous population. By the time the Spanish gave up Mexico in 1821, the missions and rancheros, who mostly raised cattle on the arid lands, dominated the region. American settlers began moving to California in the early 1800s, but most of the settlement was centered in the port of San Francisco. When California became part of the United States in the 1848 Treaty of Guadalupe Hidalgo that ended the Mexican-American War, many settlers streamed into the new state. When California applied for statehood in 1850, San Francisco boasted a population of around 21,000.[11] Los Angeles had only a little more than 1,600 people.[12] No one expected that the little dusty outpost would grow into an enormous metropolis.

As the Gold Rush began to swell California's population farther to the north, Los Angeles became the supplier of beef to the miners. Southern slave-holding states wanted a transcontinental railroad that would run along a southern route to connect the region to valuable markets and raw materials in the West. The 1853 Gadsden Purchase added the last slice of territory to the continental United States for the purpose of a transcontinental railroad. Even though the southern route was not chosen, the region soon had rail lines connecting the southern part of the state with the north and with the hub for the new transcontinental railroad, completed in 1869. Land speculators began to see the benefits of the region and helped swell the population. Travelers on the transcontinental railroad helped contribute to a real estate boom in the 1880s that resulted in a population of around 50,000 in Los Angeles by 1890.[13]

As the city grew, officials searched for water sources that would both feed its growing population and water the rich agricultural lands on its outskirts. The story of how William Mulholland, an Irish immigrant, brought water to the city has been told many times before. By the time he started working for the city's water department, it had already dammed and channeled all the water the small Los Angeles River could provide. Although pumping up groundwater gave Los Angeles all the water it needed for quite some time, Mulholland convinced the city that there would be an impending shortage if something was not done. He looked north to the Owens Valley and began his long effort to bring its water to Los Angeles or at least to the San Fernando Valley outside of Los Angeles, where he and some of his financial and political backers—Harry Chandler, Henry Huntington, and General Harrison Gray Otis—bought up valuable agricultural land. They would see to it that most of the water Mulholland brought south would water their lands.[14]

Mulholland, like many other early-twentieth-century westerners, believed natural resources needed to be used, a Theodore Roosevelt type of conservation ideology. He also believed that if Los Angeles did not purchase and immediately use as many water rights as possible, the city would one day lose its chance. For the sake of achieving security in the desert, watering investment property worth millions, and ensuring a water supply for unlimited growth in Los Angeles, Mulholland went after all the water he could find. His "use it or lose it" mentality was typical for the times, and his interest in benefiting his own assets and city came before any interest in the natural environment or anyone else's rights to water.

The doctrine of prior appropriation that drove Mulholland and other American westerners was similar to Hispanic water law in the region prior to the Mexican-American War. Both the Hispanic and American versions based water rights on use of the water, but while the Hispanic version had also used concepts of equity and fairness to direct government decisions on water, the American version based rights exclusively on who used the water first, regardless of the impact on others.[15]

In 1904 Mulholland and Fred Eaton, a former Los Angeles mayor, began looking at Owens Valley, over 200 miles from the city. Reclamation engineer Joseph Lippincott, who was supposed to be planning a government irrigation project there, helped Eaton and Mulholland identify and purchase valuable land with water rights in the valley. Lippincott had long had a lucrative engineering business in Los Angeles that had started when he was hired by the Bureau of Reclamation to run the California project in the Owens Valley region. Because Lippincott's divided loyalties were not seen clearly, he was able to help Mulholland and Eaton procure land and water rights that effectively derailed the very reclamation project he was supposed to construct and oversee.[16]

By the end of 1905, Los Angeles had all the water rights it needed to build an aqueduct. Construction began in 1907 and ended in 1913 with a 233-mile aqueduct that drained the Owens Valley of water. By 1924 the lake was almost dry, and most of the farmers had sold out to Los Angeles. Because of drought and increased pumping that turned the Owens Valley into a desert, farmers who still owned land there declared war. In a number of episodes, men from Owens Valley destroyed or damaged parts of the aqueduct, and in 1924 they dynamited a large section. Later that same year, a group seized control of a series of diversion weirs and turned the water out of the aqueduct and back into the lake.[17]

With violence escalating, Mulholland was ready to stop the damage with bloodshed if necessary. The failure of Owens Valley's major banks at the end of 1927, however, ended the sabotage, and Mulholland was victorious. This famous water grab would long haunt Los Angeles when it struggled to find other sources of water in later years. In 1941 the city built an extension to the aqueduct up to Mono Lake. More recently, environmentalists' efforts to revitalize the Owens Valley and restore the lake led to litigation in 1979 and again in 1984. By a final adjudication in 1994,

Los Angeles was required to release water for Mono Lake, and in 2005 water was returned to the lower part of the Owens River.[18]

Mulholland's final chapter with Los Angeles water, however, was a disaster. When Fred Eaton demanded too high a price for the perfect dam site near Owens Valley, Mulholland built his dam in the San Francisquito Canyon just north of Los Angeles in the San Fernando Valley. He made it too large, and when it filled in March 1928, the dam started leaking. It failed on March 12, sending approximately 38,000 acre-feet, or 2 billion gallons, of water rushing down the canyon in a wall about 125 feet high. It is still unclear how many people died in the disaster, but the estimate is around 600. Mulholland resigned from the Los Angeles Department of Water and Power soon thereafter. He died seven years later, in 1935.[19]

Meanwhile, a plan for which Mulholland had lobbied slowly began to move forward, and in 1928, Los Angeles formed the Metropolitan Water District, which soon raised enough money to begin work on the Colorado River Aqueduct. This canal, built between 1932 and 1941, brought water from the Colorado River over 242 miles to Los Angeles. The Bureau of Reclamation built Parker Dam from 1934 to 1938 to create a reservoir (Lake Havasu) that would supply the aqueduct.[20] Over time, the aqueduct was extended to San Diego and was enlarged in 1961 to carry more than 1.2 million acre-feet (MAF).[21]

In the 1950s, California began moving toward plans for what would be called the State Water Project, approved in 1960. As Arizona began fighting for its share of the Colorado River, lawyers filed suit against California. Settled in 1963, the lawsuit's conclusion in favor of Arizona ended a twenty-five-year conflict between the two states over how much water from the Colorado River each was entitled to. Arizona was able to move forward with building the Central Arizona Project canal to bring its share of river water to the central part of the state. As the project finally began to deliver water in 1985, Los Angeles sought alternative sources. Eventually, more water from the Central Valley Project helped, but most additional water to Los Angeles now comes from water transfers from agriculture in the Imperial Valley or through conservation, canal lining, and recycling efforts.[22]

Tucson (est. 1775)

The area around the present city of Tucson was long populated with indigenous peoples who dug canals and channeled water from the Santa Cruz River to their fields of corn and beans. The Spanish first arrived in the region in the 1600s, and Jesuit missionaries established the Mission San Xavier del Bac nearby in 1692. Years later, in 1775, the Spanish established a fortress called Presidio San Agustin del Tucson, from which the town took its name. Following the revolution against Spain in 1821, Tucson remained in Mexican territory until the Mexican-American War. During this conflict, the Mormon Battalion, recruited from Mormon migrants on their way west, captured Tucson from Mexico, but the town remained in Mexican territory at the conclusion of the war. Finally added to the United States by the 1853 Gadsden Purchase, Tucson was part of New Mexico Territory until 1867, when it became the capital of Arizona Territory.[23]

For a number of years, Tucson grew slowly and found enough water in wells and old *acequias* (Spanish canals) from the Santa Cruz River to meet its needs. Water problems began when the US military established Fort Lowell in 1873. Most of the time, windmills helped pump deep groundwater out of wells into storage tanks that were high enough to allow for gravity flow to most buildings in the city and the fort. Some inhabitants, however, complained about the hot, stagnant water coming out of the tanks. The city eventually solved the problem by installing new steam pumps and storage tanks and sinking new wells, but most of this water supply went only to the fort.

As the city's population began to grow after the Civil War, residents found it increasingly challenging to find water. People hired water carriers to gather water from the springs and wells and haul it to their homes on the backs of burros. Some entrepreneurs began to cart water from house to house for a small charge. One such businessman, Adam Saunders, built a bathhouse where men could purchase an occasional bath for twenty-five cents. Although some attempts were made to build dams on the Santa Cruz River, they were damaged and destroyed in strong floods during the 1890s and were never rebuilt.[24]

When the Tucson Water Company was formed in the late nineteenth century, it sought to coordinate the delivery of fresh, safe water to the city through carefully constructed and maintained systems. The company created a pipeline that was more than four miles long and transported water from the river to the center of town, bypassing problems with nearby agricultural runoff and animal waste. In 1889 the company also dug wells and installed a steam pump to lift the water out of its nearly forty-foot depths. In spite of the increased and improved water supply, the city of Tucson sometimes had to limit water usage because of a swelling population and drought in the 1890s. By the early twentieth century the city had ordinances that limited the amount of water one could use for outside irrigation and specified time periods for such use, with fines imposed for violations.

Ongoing water shortages during the early twentieth century led to new wells, a new City of Tucson Water and Sewage Department, and water meters. By the 1920s the water level in area wells was already dropping, and by the 1940s the Santa Cruz River flowed only when it carried floodwater. In the 1950s, Tucson had grown beyond the water supply's capacity and was forced to dig more wells; by the 1960s the city was purchasing water from surrounding farmers' wells and making plans to bring water in through a pipeline from the San Pedro River. The pipeline was never constructed, but Tucson continued to purchase water rights from farms in the surrounding region to safeguard its supplies.[25]

In the 1970s city leaders once again faced a shaky water situation. The Central Arizona Project had finally been approved and legislated by Congress, and construction would bring Colorado River water to the city if it so desired. City leaders, however, were not sure they wanted to sign contracts for CAP water. They were skeptical of its quality, knowing it would be much saltier than the well water they were used to using. They also worried about its cost, fearing it would be much too expensive to constitute a practical water supply. The only other alternative to CAP would be to do a better job of conserving water and limiting its agricultural use in the basin area around Tucson. In 1976 the city tried to recoup the real expenses of the water system and general funds for increasing the network by substantially raising water rates, but the major change in rates resulted in a public revolt and a recall of the majority of the members of the city council.[26]

Although strong voices in Tucson advocated for urban infill development instead of suburban sprawl, they eventually lost out to those supporting unlimited growth.[27] To allow such growth, the city was forced to agree to accept CAP water. By the 1980s Tucson had also developed water conservation programs that put it far ahead of other municipalities in the Southwest.[28] When the state of Arizona was under threat of losing CAP funding if it did not conserve water, the state legislature passed the Groundwater Management Act in 1980. As they learned that Tucson planned to use less groundwater and more CAP water, people in the area protested. Colorado River water was much saltier and more unpleasant tasting than well water, and some were worried about its health impacts. Fears were calmed after much public campaigning, and CAP water was delivered, but complaints about taste and corrosive damage to pipes and appliances immediately plagued the city. By 1994, CAP water was no longer being sent to urban users but was used only to recharge the groundwater and by farmers for irrigation. Today, CAP water is considered a major part of the Tucson water supply, but it is still mostly recharged into the aquifer to mix with and augment groundwater.[29]

Denver (est. 1858)

As with many western urban areas, the city of Denver had its beginnings in early mining. This semiarid region on the western edge of the High Plains, with a view of the Rocky Mountains, was originally part of Spanish colonial territory. The Denver area was along the disputed border of the Louisiana Purchase, sold by Napoleon to the United States in 1803. Explored by Zebulon Pike in 1806, the region remained at the edge of the American frontier until the mid-nineteenth century. Except for mountain men trapping furs and passing through for an occasional trading rendezvous, this area remained the exclusive domain of Native Americans, particularly the Arapaho.[30]

The discovery of gold in California in 1848, however, began to change the region permanently. As 49ers headed across the plains to the goldfields, the US government took action to ensure safe passage for wagon trains through Indian lands. Accordingly, federal commissioners negotiated and oversaw the signing of the 1851 Fort Laramie Treaty that established territorial divisions among the Arapaho, Cheyenne, Sioux, Crow, Shoshone, Assiniboine, Mandan, Hidatsa, and Arikara Nations. The treaty also ensured safe passage for wagon trains on the Oregon Trail.

In 1849–50, a group of prospectors discovered gold in the South Platte River Valley at the foot of the Rocky Mountains. As disappointed gold seekers returned home from California across the plains over the next decade, rumors of gold in the region led to a major discovery along Cherry Creek in 1859.[31] Here, the mining town of Denver was born and grew, serving settlers with food, supplies, and saloons. The Pikes Peak gold rush went on for several years, boosting the area's population until it was incorporated in 1861.

It seems somewhat ironic that Denver, within sight of large amounts of snowpack in the Rocky Mountains, struggles to find a stable water supply. The mountain range serves as the Continental Divide, where watersheds drain either east to the Atlantic Ocean or west to the Pacific. The mountains also serve as a "rain shadow," blocking rain-producing weather systems from moving from the west over to the eastern slope. As prevailing winds pull moist air up the western slope of the Rockies, the moisture condenses and falls on the top of the range, leaving dry air to move east. Positioned on the eastern slope of the mountain range in what is referred to as the Colorado Front Range, Denver receives very little moisture, making water sources scarce.[32]

In the early years, Denver City (as it was then called) relied on water from the South Platte River and its tributary, Cherry Creek. Soon, irrigation ditches were dug to channel water from the South Platte directly into the city. In 1867 the first major irrigation ditch was dug, eventually managed by the Denver City Water Company (DCWC), formed in 1870. Major concerns had already arisen about water quality in Denver, not just in the streams but also in the ground. The DCWC quickly began to develop a municipal water system with pumps, distribution mains, taps, and fire hydrants. By 1878 the DCWC was probably supplying 1,000 water taps.[33]

Silver deposits were discovered in the 1880s, and the resulting rush gave the city another boost in population, from about 5,000 at the start of the decade to over 100,000 by 1890. As settlers headed west for mining, Denver became a major railway hub as well as the capital of the new state of Colorado in 1876. Although the mining bubble did eventually bust and the national economic collapse in 1893 brought troubles to Denver, enough agriculture and ranching were established in the region for it to remain an important supply center as long as it could find enough water.[34]

Over the coming years, numerous water companies were formed and either failed or merged with rival companies—all seeking to supply the growing city of Denver with a safe and steady water supply. By the mid-1880s, as the population passed 50,000, the city's water supply was supplemented by artesian wells. A coal miner struck water while drilling in the area, and others were soon digging wells that tapped the same artesian water source. Still others worked to create water storage facilities for Denver, drawing the attention and investments of various bankers and other businessmen who would battle over water stocks during the next decades.[35]

In an effort to provide better water sources for a growing population, a water corporation called the Laramie County Ditch Company devised a plan in 1890 to bring more water to the Denver area and surrounding agricultural lands. Using primarily Japanese immigrant crews, the company dug what would become known as the "Grand Ditch" to direct snowmelt flowing down from the Never Summer Mountains to the eastern slope.[36] At first only a mile long, the canal was slowly extended sixteen miles over the next forty years to capture much of the mountain runoff that normally flowed into the Colorado River. It ran along a twenty-foot-wide shelf across the mountains at Poudre Pass and brought western slope water to the Front Range.

Eventually creating a water monopoly in Denver, a merged company, the Denver Union Water Company, formed in 1894 with Walter S. Cheesman and David Moffat as its heads. The two men had finally consolidated water control in Denver and sought to stabilize the supply and make sure it was free of contaminants. By 1906 the company was chlorinating its water supply to prevent typhoid and other diseases. In 1879 at least forty residents had died of typhoid fever as a result of drinking contaminated water.[37]

Soon, Cheesman and the Denver Union Water Company hired an engineer to build a large dam and reservoir on the South Fork of the South Platte River. It was a gravity-arch masonry dam; when completed in 1905, it was the highest dam in the world, at 221 feet. City leaders such as Cheesman wanted Denver to continue its growth and to become a major center between Chicago and the West Coast, but it needed a better supply of water if it were going to grow much larger. Although the 1902 Reclamation Act had already started to implement a string of reclamation projects in the West, Denver continued to take care of its own water issues through private and local funds. The Cheesman Dam became not only a feat of engineering but also an example of a private water enterprise.[38]

During the early decades of the twentieth century, Denver's demand for water continued to grow, as the population increased to 256,000 by 1920.[39] While the Cheesman Dam and reservoir became the key components of Denver's water system, residents soon needed more irrigation canals and reservoirs. During the boom years for the cattle business right after the Civil War, Denver residents had thought about and studied possible importation of water from the western slope of the Continental Divide. Nothing was done along these lines, however, until the city faced serious water shortages in the early 1930s. Several studies were conducted, and the Colorado–Big Thompson Project was authorized by Congress; construction began in 1937. Although the final pieces of the project were not completed until 1957, the system's lakes, dams, tunnels, siphons, and aqueducts brought water to the dry eastern slope of Colorado, including Denver.

In the years before World War II, most of Denver's economy was based on the transportation and processing of meats and minerals. During and after the war, the population of Denver and of eastern Colorado in general grew significantly. Since Colorado was centrally located, far from vulnerable coastlines, borders, and government centers, many saw it as the perfect location for military installations. After Lowry Air Force Base was constructed in the 1930s, the area around Denver soon saw the addition of the Federal Center in the 1940s, the North American Air Defense Command center in the early 1950s, and the US Air Force Academy in the late 1950s.

This strong military presence created jobs and another increase in population, requiring even more water.[40] Thus more reservoirs and dams were built along the South Platte River during the 1950s and 1960s, supplemented by water from the Colorado River. The Cold War years brought a different kind of attention to the state of Colorado and the city of Denver in particular. An oil boom in the 1950s that lasted through the 1970s helped swell the population again, and the issue of water became increasingly important.

The last major water storage facility, Strontia Springs Dam, was constructed in 1983. As the twenty-first century dawned, Denver realized that the only new source of water it would find for a growing population would be from conserving and recycling water. A state-of-the-art water recycling plant was constructed in 2004 and supplements Denver water by filtering and purifying wastewater.

Denver's population continued to grow and in January 2012 was estimated at around 2.83 million. Since Denver was ranked sixth in the nation among the "Best Cities for Business" in 2011, the population will likely keep increasing.[41] Although water resources are limited, ongoing conservation efforts have actually led to a decrease in overall household water usage in Denver since 2000.[42]

Phoenix (est. 1881)

The first settlers of the large metropolis now called Phoenix were the Hohokam. Over a period of approximately 1,000 years, these people dug about 135 miles of irrigation canals and watered their fields, growing a mixture of crops that included cotton, maize, tobacco, beans, and squash. Somewhere between 1300 and 1450 the Hohokam left the area, abandoning their canals and fields. Archaeologists and other scholars debate the reasons for their disappearance. While war or disease could have caused a dramatic departure, the most likely reason is a combination of long-term drought and high levels of salinity that accumulated over time from irrigation. It appears that the Akimel O'odham (Pima) and Tohono O'odham are likely descendants of a remnant of the Hohokam who stayed in the area.[43]

As the Spanish conquistadors and, later, Jesuit missions brought Europeans to the region, they stayed mostly in the southern part of Arizona, around present-day Tucson.[44] In the early nineteenth century, Anglo-American explorers and mountain men hunted in the area, finding the Salt River an excellent place to camp on the long desert trail west. After the Mexican-American War the region passed into American hands, and the area that was eventually Phoenix became part of the New Mexico Territory. In the 1850s a number of settlers filtered into the region along the Salt River. Some came to the valley lured by the discovery of gold in California. One of those prospectors was Jack Swilling, who arrived in 1856. After prospecting for gold and fighting Apaches with the Gila Rangers militia, Swilling joined the Confederate Army in Arizona during the Civil War. After the war, the US military established Fort McDowell on the Verde River.[45]

Swilling formed the first irrigation company in the area of Wickenburg and later another in the Salt River Valley after seeing military farming near Fort McDowell. He used the old Hohokam canal channels. A small town formed, called Pumpkinville after the primary local crop. Later, the town called itself Swilling's Mill in honor of its founder, but a friend of Swilling's, Darrell Duppa, suggested the name "Phoenix" because a civilization was rising up out of the ashes of a previous one long gone. The name stuck, and the city of Phoenix was incorporated in 1881, with a population of 1,708.[46]

As the population expanded in the 1880s and experienced drought from 1898 to 1904, city officials looked for additional water supplies to supplement the wells they had drilled previously. Phoenix is fortunate to be near the confluence of three rivers—the Salt, Gila, and Agua Fria—with a fourth river, the Verde, joining the Salt River to the north. Since water from the Salt River was considered far too salty to drink, city officials made plans to bring Verde River water into the city through a redwood pipeline that stretched for twenty-eight miles. Begun in 1920 and completed in 1922, the conduit served the city until leaking made it necessary to replace it with a concrete lining in 1930.

Although the Salt River was indeed very salty, area farmers had long used it for irrigation. Droughts and cyclical flooding did a great deal of damage, and when the Reclamation Act was signed in 1902, a group calling itself the Salt River Users Association formed and lobbied for an irrigation project. In 1904, construction began on the Roosevelt Dam north of Phoenix; when completed in 1911, it was the tallest masonry dam in the world, at 280 feet high. Additional dams were built on the Salt River in the following years: Mormon Flat Dam in 1925, Horse Mesa Dam in 1927, and Stewart Mountain Dam in 1930. To extend the Salt River Project's services, dams were also built on the Verde River: Bartlett Dam in 1939, Horseshoe Dam in 1946, and Blue Ridge Dam in 1965.[47]

The Agua Fria River flows only part of the year and empties into Lake Pleasant, a reservoir created by the Waddell Dam, built in 1928. At that time, Lake Pleasant was intended for agricultural irrigation but was later used to store water from CAP. Conceived very early in the twentieth century but delayed by decades of controversy and water wars, the Central Arizona Project pipeline finally became feasible after Arizona won its court case against California in 1963. Authorized in 1968, construction began in 1973, and the pipeline began delivering water in 1985. A new Waddell Dam was constructed at the older dam site, enlarging the storage capacity of Lake Pleasant. The lake now receives most of its water from CAP and stores it for usage in Phoenix and the surrounding area.[48]

From 1989 to 1996, Roosevelt Dam on the Salt River was raised seventy-seven feet to increase the lake's storage capacity by about 20 percent. In 1994, Phoenix built a storage and recharge project on the Salt River just northeast of the city. An old diversion dam built in 1908, Granite Reef, was later used as part of the new Granite Reef Underground Storage Project, completed in 1993. Recharge basins allow excess river and CAP water to be absorbed and stored in the underground aquifer. Also in 1994, the Tres Rios Constructed Wetlands Demonstration Project was authorized. Part riparian restoration, part research project, it is a series of three major sites that experiment with using various wetlands as water purifiers for wastewater. In 2000 the Army Corps of Engineers received congressional approval to expand the demonstration project into a full-scale restoration wetlands project that will eventually cover 480 acres.[49]

Perhaps one of the most interesting aspects of current water negotiations involving the city of Phoenix is the transfer of agricultural land to urban uses, along with the corresponding water rights. According to recent statistics, a substantial amount of farmland has been converted to residential and commercial uses. In 1955 irrigated farmland within Salt River Project boundaries totaled more than 200,000 acres. By 2005 the number had dropped to about 20,000 acres.[50] However, because of legal restrictions and the ability of third parties to protest water transfers, very few outright water purchases from agriculture have been made. In an exchange agreement, the city now delivers reclaimed water to farmers and in return receives credits for pumping Salt River canal water owned by the Roosevelt Irrigation District.[51]

Tijuana (est. 1889)

Two cities in Mexico use the Colorado River as a major part of their water supply. Tijuana, right across the border from San Diego, is the busiest border town between the two countries, with an estimated 40 million people crossing every year. It is the largest city in the Mexican state of Baja California and the sixth largest in the country. Population estimates for the city in 2010 vary widely, from 1.5 million to 3.5 million. The most commonly cited number from the 2010 census is just under 1.6 million.[52]

The very early history of the Tijuana region is similar to the histories of other cities along the border: the area consisted of a mix of indigenous communities, sometimes loosely united through trade.[53] In the area of San Diego and Tijuana, the Kumeyaay people (sometimes spelled Kumiai) claimed the area as home. Two distinct groups within the Kumeyaay, the Ipai and the Tipai, lived on their respective sides of the San Diego River.[54] First contact with Spanish conquistadors occurred in 1540, when Hernando de Alarcón led part of Coronado's expedition up the Gila River.

After the Mexican-American War ended in 1848, gold strikes in California stimulated similar prospecting in Baja California, where both gold and silver deposits were discovered. With various ebbs and flows, a mining boom that extended from 1850 to 1890 led to much travel through the border region and stimulated the growth of both San Diego and Tijuana. At first only a location for a border customs house, the small settlement slowly grew into a service community for those traveling south into Baja California looking for gold.[55]

One of the biggest challenges to mining in Baja California was the lack of water. The peninsula is narrow, with a mountain range in the middle, so most of the water from rainfall quickly sheds away to the ocean on both sides. Because of this water shortage, Mexican miners used wind to separate ore from dust and dirt by tossing the deposits into the air. This technique was called "dry-washing" and was much less effective than using water, but it was still a workable solution.[56] Although gold mining shifted away from Baja for a time, new strikes in the 1860s and 1870s again drew a large number of prospectors and miners to the Tijuana area. During this time, more permanent settlers moved into the Tijuana Valley to farm and to profit from trading with and supplying miners.[57]

Soon, the people of Tijuana established a stagecoach service from San Diego to Santo Tomás, 120 miles to the south. So much traffic crossed the border heading for goldfields to the south that Mexico established a customs house at Tijuana in 1874. In 1885, rail lines connecting San Francisco to Los Angeles helped spark a land speculation boom in the region, bringing more permanent populations to San Diego and Tijuana. As the gold played out, Tijuana developed a brisk tourist business for visitors to San Diego who wanted a taste of something foreign. Soon, regular stagecoach excursions took people to and from the border several times a day. Tijuana began hosting horse races, bullfights, dances, and other festivals to entertain American visitors.[58]

The town of Tijuana was officially organized in 1889 when another gold deposit discovery near Santo Tomás brought a new flood of prospectors. Although the town's permanent population remained very small, it increased in the 1920s because of its legalized drinking and gambling establishments, which Americans happily frequented. Until the Mexican president ended legal gambling in Baja California in 1935, many of the rich and famous from Hollywood and elsewhere vacationed in Tijuana. By the 1950s, Tijuana's population was around 63,000, and the city worked to orient its tourist attractions less to gangsters and more to vacationing families.[59]

As Tijuana's population grew during the first half of the twentieth century, so did its economy, based mostly on tourism. In the 1960s, however, industrial development programs helped diversify the economy. A number of companies began building assembly plants (called *maquiladoras*) to take advantage of cheap labor. Companies such as Sanyo, Ford, General Electric, and others set up factories there. Over time, these companies and their factories competed for water with a growing agriculture industry in surrounding areas. Both industry and agriculture contaminated the water supply with everything from pesticides and salt to chemical toxins.[60] By the late 1970s and early 1980s, environmental degradation had led to a drop in agricultural exports and to multiple water-quality issues.[61] However, in 1975 an aqueduct was built to the city from the Colorado River. In the twenty-first century, this aqueduct is providing 90 percent of Tijuana's water supply.[62]

Because Tijuana's wastewater began contaminating the Tijuana River, which flows along the border just south of San Diego, the United States became interested in sewage treatment plants inside Mexico. As the border population expanded in the late twentieth century, sewage treatment plants could not meet the challenge. San Diego's wastewater system helped treat some of its Mexican sister city's water as early as 1965, but more plants were desperately needed. In 1983 the International Boundary and Water Commission built new sewage facilities to protect the Tijuana River from further contamination. Population growth, however, meant the facility was at full capacity and insufficient as soon as it began operating. When the North American Free Trade Agreement (NAFTA) became law in 1994, growth in industry in Tijuana followed, increasing both sewage and industrial wastewater contamination. In 1994, talks between the United States and Mexico about Tijuana sewage problems resulted in an International Treatment Plant that could process 25 million gallons of sewage a day.[63]

Recent water challenges for Tijuana involve quantity as well as quality issues. Population increases and industrial demands have made the 1975 aqueduct completely inadequate, so the city began to negotiate with San Diego over the construction of a bi-national aqueduct. This pipeline would be a shared resource, pumping Colorado River water to both cities. San Diego was negotiating for water transfers from farmers in the Imperial Valley and would use the aqueduct to transport that water. Tijuana would receive an increased measure of the 1.5 MAF of river water sent across the border every year by international treaty. San Diego was in no hurry to implement the plan, however. In 2003, talks about a shared aqueduct foundered when Tijuana decided to try expanding its own pipeline without San Diego's help. Water managers in Tijuana predicted a major shortage for the city by 2007 without any help, but San Diego could afford to drag its feet—it had a guaranteed water supply that would likely last until 2011.[64]

The local environment is receiving more attention as the consequences of unmanaged growth have become evident in the Tijuana area. In 2009, engineers from the University of California–Davis proposed a unified water plan for Tijuana and the entire state of Baja California that included large construction projects for water transport. Although water markets could sell Tijuana the extra water its expanding population needed, it lacked the infrastructure through which to transport that water.[65] The city recommended expansion of the aqueduct along with desalination and wastewater treatment to enhance the water supply.

In 2010 the La Morita Wastewater Treatment Plant was completed and began processing what will eventually be 5.8 million gallons of waste each day.[66] In early 2011 the North American Development Bank approved a $4.8 million loan to Tijuana to complete the Tecolote–La Gloria treatment plant, which will begin treating 2 million gallons of waste per day.[67] In August 2011 a bi-national working group consisting of Southern California and Nevada water authorities and government agencies inside Mexico discussed plans to create a turnout on the All-American Canal that could divert some of Mexico's share of Colorado River water to Tijuana in times of crisis, such as the 2010 earthquake in Baja California. If these plans come to fruition, the turnout could be completed in early 2014.

Mexicali (est. 1903)

The history of Mexicali is very similar to that of its neighbor, Tijuana. Much smaller in population, Mexicali is the capital of the state of Baja California, with an estimated 600,000 to 700,000 inhabitants. Like Tijuana, the city boasts a mixed economy, but historically the largest sector has been agriculture. Prior to Spanish settlement, the region was home to the Cucapá people, who supported themselves with fish from the Colorado River and subsistence farming. Aside from this small indigenous population, the region remained largely unsettled until the late 1880s. With no counterpart similar to San Diego across the border, there was little reason to settle there until someone decided to develop irrigated agriculture nearby.

In 1901, land developer Charles Rockwood began to irrigate and sell farms in the United States around the Salton Sink. Rockwood called the area the Imperial Valley, and he collaborated with Mexican landowner Guillermo Andrade, who owned most of the land bordering the Colorado River near the US border. The two men reached an agreement to build a canal from the Colorado River that would run partly within the Mexican border up to the Imperial Valley. During the first few years of the plan, the population on the American side rose to 7,000, and a small settlement of tiny buildings formed a town that straddled both sides of the border. The US side was called Calexico, and within Mexico it was called Mexicali.[68]

The story of what happened to Rockwood's canal and the flood that created the Salton Sea is well-known (see chapter 1). Following the disaster, farmers returned to the rich soil and continued to irrigate their land using the canal that ran partly through Mexico. Seeing an opportunity, land developer and *Los Angeles Times* owner Harry Chandler and his partner, Harrison Gray Otis, had already purchased large tracts of land on the Mexican side of the border, south of the Imperial Valley. Through their Colorado River Land Company, Chandler and Otis used water from the canal Rockwood had built. The company leased land to tenant farmers, both Mexican and Asian, to grow cotton and other crops in the rich soil.[69]

For several decades, Mexicali sat among large farm tracts owned by Americans. In 1936–37, Mexican president Lázaro Cárdenas pressured the Colorado River Land Company to sell its holdings to Mexican farmers. He sought to "Mexicanize" the region along the border by encouraging migration of Mexican farmers to Mexicali to farm. Cárdenas believed Mexico should appropriate as much Colorado River water as possible on the Mexico side of the border prior to any future negotiation of a formal treaty with the United States.[70] By the 1950s, farmers around Mexicali had become the most important cotton producers in Mexico and were growing many vegetables for export, including asparagus, broccoli, carrots, lettuce, tomatoes, peas, and peppers. For much of the first two decades of the twentieth century, Mexicali's water came from the old canal that ran to the Imperial Valley. When the Boulder Canyon Project Act authorized the construction of Hoover Dam in December 1928, it also contained provisions for an "All-American Canal" that would bring Colorado River water to the Imperial Valley without cutting through Mexico.[71]

For some time, farmers around Mexicali had depended on seepage from the All-American Canal and groundwater. When Mexico and the United States signed a water-sharing treaty in 1944, Mexicali found itself entitled to the lion's share of 1.5 MAF of Colorado River water that would cross the border each year. The Morelos Dam was completed in 1950, and diversion canals were soon bringing the river water to farms and the city of Mexicali. In 1982 an aqueduct was completed to bring the river water to Tijuana.[72] Both cities now boast manufacturing plants, or maquiladoras, mostly of American companies that want to benefit from cheap labor and low taxes.

Water quality in Mexicali has some of the same problems as that in Tijuana. Industrial waste and sewage from a growing population pollute the water, as do large amounts of pesticides and salts from irrigation runoff. All of the waste from surrounding farmland and the city of Mexicali flows out and north across the US border through the New River, an ancient streambed that became the channel for Rockwood's famous mistake in 1904. The river no longer has a natural source of water other than irrigation and waste.[73] As the population grew and pollution in the New River increased, people on both sides of the border became frustrated with the odor and unsafe conditions. In the 1980s a wastewater treatment plant was built to help clean water that entered the river. In 1992 both Mexico and the United States adopted Minute 288, a treaty that created a long-term plan for cleaning up the river.[74] Part of the cooperation also involved cost sharing; in 1997 and 1998 several such projects were funded, including new sewer pipes, lining replacements, and better pumping equipment.[75]

In 2002–2003, California authorized a controversial project meant to conserve water in view of impending shortages. The plan was to line the All-American and Coachella Canals with concrete to prevent the seepage of almost 68,000 acre-feet of water per year. The Coachella lining project was finished in 2006, and the All-American Canal lining was completed in 2010.[76] The problem the canal lining has created for Mexicali and the surrounding agricultural region is that the yearly seepage has contributed an important amount of water to local wells. Mexican farms along the border fear that while California will save and enjoy more water, Mexican farms will wither. Although the seepage comprises only 12 percent of the groundwater supply, that amount is important because it helps dilute the much saltier water already in the ground.[77] It is estimated that more than 1,000 families living in the Mexicali Valley depend on this seepage for all their water needs. If the groundwater dries up because it is not being replenished, then these families would likely move into the urban area of Mexicali, placing serious pressure on its municipal water system.[78]

On April 4, 2010, a 7.2-magnitude earthquake shook the Baja Peninsula and tore up the region's irrigation canals and the aqueduct that brings Colorado River water to Tijuana and Mexicali. In December 2010 the United States and Mexico signed Minute 318, an agreement allowing Mexico to bank much of its Colorado River water allocation for several years while it repairs the aqueducts and canals. Included in the agreement are stipulations for ensuring a water supply to the Ciénega de Santa Clara wetlands in the river's delta, along with other protections for border wildlife.[79] In the coming years, Mexicali and Tijuana will need the cooperation of the United States to maintain a steady supply of water from the Colorado River.

Las Vegas (est. 1905)

The history of Las Vegas begins with the discovery of an oasis in the desert. In 1829, Rafael Rivera became the first Euro-American to gaze upon the valley oasis; he named it "Las Vegas" (The Meadows).[80] Rivera was part of the Mexican trading company led by Antonio Armijo, and he was scouting for water when the group camped after leaving the main Spanish Trail route to Los Angeles. The artesian well he discovered there made Las Vegas an important stop for traders heading to Los Angeles from Mexico and other areas farther north. In 1844, John C. Fremont camped at the springs and wrote about them in his journals. In the years that followed, many others decided on a similar route west to take advantage of the natural supply of water.[81]

A few years later, Mormons began to use Las Vegas as an important stop on the Salt Lake City–to–Los Angeles mail and trading route. In 1855 they built a fort at the springs to help protect the outpost, but they abandoned it in 1858 because of Indian attacks and raids. It remained an important halfway stop between the two cities, however, and the site attracted the attention of railroad developers looking to join East and West with a transcontinental railroad. Settlers headed to Nevada in larger numbers, and in 1864 it became a state. In 1890 the railroad scouts arrived, and in 1904 the first grading began, soon followed by all the usual trappings of a small railroad town. The natural supply of water made the location an important stopping point for steam-powered engines winding their way between Salt Lake City and Los Angeles.

Because it was a railroad stop, Las Vegas became a twenty-four-hour town, catering to railroad workers and the many travelers who began coming through the region. Saloons and gambling halls became primary segments of the town's operations. The mining industry also began to spawn important businesses in the area as first gold and then silver, zinc, and lead became important parts of Nevada's economy. Still, the history of Las Vegas might have been quite different without two important developments: legalized gambling, instituted statewide in 1931, and the building of the Hoover Dam, begun in 1932.[82] Jobs were plentiful because of the dam project, and workers poured in from across the country to obtain those jobs. As important as the workers were the tourists who began arriving in Las Vegas to see the dam project, heralded as one of the wonders of the world. The town also seemed attractive because of the legalized gambling, prostitution, and abundant drinking establishments. Thus Las Vegas experienced its first boom during the 1930s while the rest of the country was struggling with the Great Depression.[83]

In 1935, Nevada formed the Colorado River Commission, a group that would oversee Nevada's rights to water from that river. It replaced the older Colorado River Development Commission, formed by Nevada in the 1920s for much the same reason. The new commission worked both to safeguard Nevada's claim to water and to guarantee a fair supply of hydroelectricity from the Hoover Dam and other generating plants to be built in the future. It would become the major body responsible for contracting for the delivery of Lake Mead water to Las Vegas and for land and power purchases from the Colorado River Storage Project.

In the 1940s, World War II brought major growth to Las Vegas. The development of magnesium for military uses became the impetus for the establishment of Basic Magnesium, Incorporated (BMI), outside Las Vegas in 1941. Using hydroelectricity generated by the dam and Colorado River water obtained for its use by the Colorado River Commission, BMI brought many jobs to the area. The army also established a gunnery school outside Las Vegas. During World War II the population had doubled, to around 15,000 permanent residents. Adding to this number, thousands of weekend visitors crammed into the city, mostly defense workers from Los Angeles eager for entertainment and gambling. The main route to and from Los Angeles (Highway 91) had recently been paved, making it easy for people to get away to Las Vegas for the weekend.[84]

As the town grew, its underground water supplies shrank. Some developers issued early warnings about drawing down the aquifer in the 1930s, but most people ignored them. The population boom in the 1940s, however, began to bring sharp attention to a possible impending water crisis. Water consumption in the city was far above the national average, and it had doubled in just one year (1942 to 1943). It was time to make use of Nevada's annual Colorado River allocation of 300,000 acre-feet, so city leaders entered into negotiations with the Bureau of Reclamation to begin tapping into the allocation. In 1942, Nevada signed an agreement for the delivery of 100,000 acre-feet of water per year, and in 1944 the agreement was changed to reflect the full 300,000-acre-foot annual allocation allotted to Nevada in the 1922 Colorado River Compact.[85]

In 1947, Las Vegas city officials formed the Las Vegas Valley Water District, directed to formulate and carry out plans to pipe water directly from Lake Mead to the city. To accomplish this task, the water district, with the help of the Colorado River Commission, began negotiating and eventually, in 1952, purchasing water rights from the Union Pacific Railroad and the older Land and Water Company. More challenges delayed construction of a pipeline until the next decade, however. Since the city of Las Vegas did not have the financial resources it would take to build the pipeline itself, it relied on the Bureau of Reclamation.[86] In 1960 the Colorado River Commission and the bureau formulated a plan they called the Southern Nevada Water Project (later renamed the Walter B. Griffith Project). In 1967 the Colorado River Commission signed an agreement with the bureau for the delivery of 138,000 acre-feet of water per year. Construction of the pipeline began in 1968, and it became operational in 1971.

In 1977 the second phase of the Southern Nevada Water Project began and was completed in 1982. In this part of the project, engineers constructed a second pipeline from Lake Mead, parallel to the first one, and included lateral lines to bring water out to the suburbs in the valley. They also built the Alfred Merritt Smith Water Treatment Plant in 1971 to process the river water and bring it up to proper quality standards. The Southern Nevada Water Authority, formed in 1991, bought complete title to, and interests in, the water project from the bureau for $121.2 million in 2001.[87]

Even though pipelines were built and water was running into Las Vegas from Lake Mead by the early 1980s, the city still faced an impending water crisis. The city's exponential growth is best exemplified by business magnate Steve Wynn's empire. Beginning in the late 1970s, Wynn renovated older gambling halls such as the Golden Nugget and the Frontier Hotel and Casino. Then he built new ones: the Mirage in 1989, Treasure Island in 1993, Bellagio in 1998, Wynn in 2005, and Encore in 2008. These new casinos stimulated a period of fast-paced growth along the Las Vegas Strip and brought nearly 5,000 new residents per month to the city. Between 1985 and 2000, water consumption in Las Vegas doubled.[88]

Patricia (Pat) Mulroy, the general manager of the Las Vegas Water District during this period, looked for answers to the city's water needs. Soon after she took the position in 1989, she began what would become known as the great "water grab" and stirred up controversy far and wide.[89] No other municipal water agency had ever proposed anything so bold. Her proposal was to purchase the water rights under almost half of the state of Nevada, claiming that rural areas were not making the best use of water desperately needed in Las Vegas. Those outlying parts of the state saw her proposal as just one more example of greed coming out of "Sin City." Many residents compared the plans to the Los Angeles water grab in the Owens Valley more than sixty years earlier.

How the water would get from the far-flung desert to the city was another matter entirely. Mulroy's plan called for a pipeline more than 1,000 miles long, and many saw immediately that this idea was not feasible for anyone. Eventually, Mulroy's true intentions became clear as she began talking about alternatives. The real plan proposed by the Southern Nevada Water Authority and Mulroy was to build a dam and reservoir on the Virgin River, a major tributary of the Colorado River that ran through the barren desert in southwestern Nevada. As various groups geared up to protest the damage such a dam, reservoir, and pipeline would cause, Mulroy surprised everyone with what sounded like a reasonable alternative. Instead of building the dam and pipeline, she indicated that Las Vegas would prefer to take the same amount of water out of Lake Mead. Through skillful negotiation, Mulroy eventually won over environmentalists, rural water users, and even the Department of the Interior. By the mid-1990s, Mulroy had the support of most interest groups in Nevada for a larger share of Lake Mead water for Las Vegas.[90]

To succeed, however, Mulroy would have to convince the lower basin states of the Colorado River to give Las Vegas more than the 300,000 acre-feet allocated by the 1922 compact. Banking and leasing water was the answer, but a new pipeline would also be required to bring more water to the city. Amid much doubt and fears among upper basin states, Nevada began banking water in underground aquifers in Arizona, which recharged excess water into the aquifer. Nevada paid to use some of this water and take it out of Lake Mead, while Arizona agreed to take the same amount out of the aquifer instead of Lake Mead in times of drought. The Southern Nevada Water Authority completed a second pipeline to Lake Mead in the late 1990s, and in 2009, plans began in earnest for a third intake line. From 1999 to 2004, severe drought challenged the entire Colorado River basin, especially Las Vegas, a city already scrambling to find more water. To meet growing demand, the Southern Nevada Water Authority completed a second pipeline to Lake Mead in the late 1990s. When the severe drought from 1999 to 2004 dropped levels in the lake, plans for a third intake tunnel began. A contract for the project was awarded in 2008, and it is scheduled for completion in 2014.[91] Thus, in the early twenty-first century, Las Vegas finds its well-being tied directly to the future health and viability of the Colorado River.

Urban Water Users and the Colorado River

By the mid-twentieth century, these eight major cities that depend on the Colorado River were growing quickly, with few signs of slowing down. The histories of all these southwestern cities were fueled by the gold rush and the concept of Manifest Destiny that drove Americans west to take and tame the continent. When wealth proved elusive for gold prospectors, each city found other important avenues for commerce and economic development. All of them benefited from locations either at important junctions or at the terminus of railways west. Several became important military locations because they were either on the coast or far inland, safe from enemy attackers. Gold prospecting gave way to other types of mining, ranching, and military production industries that fed each city's economy. Some benefited most from a favorable climate; all of them except Denver are located in climates that are warm for much of the year.

A permanent condition for each of these cities, however, is the reality of aridity and water scarcity. Officials in most of these metropolitan areas were concerned about water shortages long before they became a real threat, but no one was willing to make tough decisions to limit growth. Even Tucson, as dependent as it was and still is on overdrawn groundwater, could not persuade citizens to halt expansion. To most people, slowing the growth meant economic stagnation or even decline. Thus in the early twenty-first century these cities and their large populations find themselves trying to find solutions to future shortages while weathering cyclical droughts in the present.

Throughout the Colorado River basin, both agricultural and urban water users have never seriously limited themselves; they continue to use water as though it were not a scarce resource. None of the states has curtailed agriculture because they do not want to limit an important sector of state economies. Municipal water districts have also been reluctant to raise water prices or discuss rationing. Throughout the stories of these cities that rely on the river, conservation is never a starting point but instead is always a last-resort response to crisis.

Even though no strong policies to limit growth ever took hold, Tucson, Los Angeles, and San Diego began pursuing water market transfers and conservation alternatives to meet potential water shortfalls. Las Vegas, like the other cities, is paying citizens to remove grass from lawns in favor of Xeriscape landscaping.[92] In

Phoenix, some concerned citizens have been worried about the apparent disregard for limits to water in the Southwest. In 2006 a nonprofit public development association called Valley Forward conducted a survey of residents in the greater Phoenix area about water issues. The results revealed that 60 percent of valley residents believed there was not enough water for Maricopa County to continue to support sustained growth. Around 49 percent also recognized that the rest of the state faced similar water shortages.[93]

However, while 58 percent of those surveyed believed the government was not doing enough about water issues, only about 6 percent knew anything about Colorado River rights and water allocations. Although 66 percent said they would follow mandated conservation laws if implemented, the lack of awareness about Colorado River water was of serious concern to the organization. In response, the group planned public awareness campaigns vital to achieving full public support for water conservation, whether mandatory or voluntary.[94]

Since the 1980s, Arizona State University (ASU) has conducted environmental research involving water. Evolving out of the Center for Environmental Studies, established in 1997, ASU now boasts a School of Sustainability, which opened in 2007. One of its research projects is called PASS—the Phoenix Area Social Survey. The pilot survey was conducted in 2003 and a second survey was done in 2006, with plans for similar surveys every five years. The data collected will provide the foundation for many important research projects that seek to answer a multitude of questions.

An interesting component of the PASS surveys involves questions about water. In both 2003 and 2006, people seemed to understand that problems existed and that water conservation was an important issue. However, their desires as community members of a large metropolis frequently conflicted with conservation policies. In the 2003 survey, respondents were asked to express the amount of concern they had about water supplies in the valley on a four-point scale, from "very concerned" to "not concerned at all." Approximately 53 percent stated that they were "very concerned" about the overall water supply in the valley, indicating an awareness of a limited supply. However, only 21 percent of these same people were concerned about shortages stemming from water usage in their own neighborhoods.[95] Another interesting finding is that six of every ten respondents stated that they were conserving water by using low–water-usage showerheads and toilets. Apparently, some efforts to conserve seemed reasonable to make while others, such as those focused on reducing lawn watering, did not.

Another interesting but perhaps not surprising result was the correspondence between income level and conservation efforts. According to the measurements established by the survey, high-income households reported water conservation efforts most often, while lower-income households reported much less effort. The survey's summary concluded that the higher the income level, the better able people are to make conservation efforts.[96] If urban areas are going to be serious about water conservation, they will have to address the issue of cost for water-saving devices.

One set of questions in the 2003 survey focused on perceptions about landscaping. Even though lawn watering accounts for approximately 60 percent of domestic water use, most survey respondents felt grass and trees make neighborhoods much more appealing. Most lower-income–level housing had little greenery, but residents still found it highly desirable. While desert landscaping has become popular in upper-income–level neighborhoods, the survey showed that most of this desert landscaping is used only in the fronts of homes, while the backs are most often an oasis filled with grass, trees, and shrubs.[97] Recent statistics show that a nationwide average of more than 50 percent of domestic water is used outdoors, but behaviors highlighted in the PASS report illustrate people's cultural disconnect between their water source and the Colorado River specifically.[98]

Three years later, in 2006, a new PASS report showed similar concerns and discrepancies. Although the survey posed different questions, respondents were again asked about water supply. A much higher percentage, 85 percent, said they were "very concerned" about both the amount of water people were using in the Phoenix Valley and the ongoing regional drought.[99] Interestingly, however, nearly half of the respondents believed they would be unable to reduce their water consumption. While people agreed that there was a problem, many could not see any way they could participate in a possible solution.[100] Although most people recognized the significance of drought for the water supply, many also were aware that people were using too much water. When asked what specifically was at fault, the answers were interesting. Most respondents blamed lawn watering and swimming pools as the biggest culprits in the over-consumption of water in the valley. Other causes with almost as many votes included the expanding population and the creation of manufactured lakes and ponds.[101] Yet while individuals might agree about causes, they seldom felt willing to specifically include themselves and were unwilling to change their current practice unless forced to do so.

Many environmentalists and scholars of water issues, using Newtonian economic theory, argue that the best way to make urban dwellers or anyone else conserve water is to raise the price. Most economists who support water marketing also argue that pricing water according to laws of supply and demand would make it much more expensive and thus reduce consumption. In 2010 the Sociology Department at the University of Nevada at Las Vegas conducted a social survey very similar to the

2003 and 2006 PASS surveys in Phoenix. In the 2010 survey, residents of Las Vegas demonstrated that they were much more aware of water shortages than residents of Phoenix had been. All respondents said they supported water conservation efforts such as recycling and finding ways to operate with less. They were not willing to pay higher prices for water, however, no matter how dire the situation.[102]

Although urban dwellers are resistant to changes that might cost more, city officials and water districts have a variety of incentive programs designed to save money. All six US cities have rebate programs for water-saving landscaping. Most of them also advertise significant rebates for business water use, including landscaping, low-flow indoor devices, and rinse nozzles for restaurant dishwashing. As metropolitan water districts work to keep the water flowing through conservation efforts, many users are gradually making changes in their usage. If average citizens and urban business owners make insufficient changes to reach a sustainable balance between use and resources, however, significant price increases might be the only answer. For now, this action is one most cities are trying to avoid.

Notes

1. For a comprehensive history of San Diego, see Iris Wilson Engstrand, *San Diego: California's Cornerstone* (San Diego: Sunbelt, 2005).

2. For an early history of San Diego water systems, see Theodore Andrew Strathman, "Land, Water, and Real Estate: Ed Fletcher and the Cuyamaca Water Company, 1910–1926," *San Diego History* 50, no. 3 (Summer-Fall 2004): 124–44.

3. A very early account of San Diego's history and construction of the flume is William Ellsworth Smythe, *History of San Diego, 1542–1908: The Modern City,* vol. 2 (San Diego: The History Company, 1908), 443–52.

4. Dan Glaister, "Cloudbusting," *The Guardian*, January 2, 2007, http://www.guardian.co .uk/world/2007/jan/02/worlddispatch.usa (accessed March 11, 2011).

5. Garry Jenkins, *The Wizard of Sun City: The Strange True Story of Charles Hatfield, the Rainmaker Who Drowned a City's Dreams* (New York: Basic Books, 2005).

6. Mike Sholders, "Water Supply Development in San Diego and a Review of Related Outstanding Projects," *Journal of San Diego History* 48, no. 1 (Winter 2002): 61–70.

7. To receive Colorado River water, San Diego had to join the Metropolitan Water District, giving up the dreams some held for an aqueduct separate from the river. See Theodore Andrew Strathman, "Dream of a Big City: Water Politics and San Diego County Growth, 1910–1947," PhD diss., University of California, San Diego, 2005.

8. For more information on recent legal challenges, see Alejandro Davila, "County of Imperial to Continue Legal Challenge to Water Transfer," *Imperial Valley Press* [El Centro,

CA], January 12, 2012. On the Quantification Settlement Agreement, see Donald A. Wilhite, *Drought and Water Crisis: Science, Technology, and Management Issues* (Boca Raton, FL: CRC, Taylor and Francis, 2005), 249–86.

9. San Diego County Water Authority, Recycled Water, http://www.sdcwa.org /recycled-water (accessed September 18, 2012).

10. For a history of Los Angeles, see Marc Reisner, *Cadillac Desert: The American West and Its Disappearing Water* (New York: Penguin Books, 1993 [1986]); Norris Hundley Jr., *The Great Thirst: Californians and Water, a History* (Berkeley: University of California Press, 2001); Richard White, *"It's Your Misfortune and None of My Own": A New History of the American West* (Norman: University of Oklahoma Press, 1991). See also Robert M. Fogelson, *The Fragmented Metropolis: Los Angeles, 1850–1930* (Berkeley: University of California Press, 1993).

11. San Francisco History, Population, http://www.sfgenealogy.com/sf/history/hgpop .htm (accessed January 29, 2012).

12. Los Angeles Almanac, General Population by City 1850–1900, http://www.laalmanac .com/population/po25.htm (accessed January 28, 2012).

13. One of the most comprehensive works on Los Angeles and water is Hundley, *The Great Thirst*, 124. See also William L. Kahrl, *Water and Power: The Conflict over Los Angeles Water Supply in the Owens Valley* (Berkeley: University of California Press, 1982).

14. See Reisner, *Cadillac Desert*; Donald Worster, *Rivers of Empire: Water, Aridity, and the Growth of the American West* (New York: Oxford University Press, 1985). See also a fairly recent work by Mulholland's granddaughter, Catherine Mulholland, *William Mulholland and the Rise of Los Angeles* (Berkeley: University of California Press, 2000).

15. For more information on Hispanic water appropriation doctrine and law, see Hundley, *The Great Thirst*, 74; Michael C. Meyer, *Water in the Hispanic Southwest: A Social and Legal History, 1550–1850* (Tucson: University of Arizona Press, 1984), 159–62.

16. Abraham Hoffman, "Origins of a Controversy: The U.S. Reclamation Service and the Owens Valley–Los Angeles Water Dispute," *Journal of the Southwest* 19, no. 4 (Winter 1977): 333–46.

17. In addition to Reisner and Hundley, see John Walton, *Western Times and Water Wars: State, Culture, and Rebellion in California* (Berkeley: University of California Press, 1992); Robert A. Sauder, *The Lost Frontier: Water Diversion in the Growth and Destruction of Owens Valley Agriculture* (Tucson: University of Arizona Press, 1994); Gary D. Libecap, *Owens Valley Revisited: A Reassessment of the West's First Great Water Transfer* (Stanford: Stanford Economics and Finance, 2007).

18. Daniel Kelly, "Owens River Restoration—Los Angeles Takes First Steps toward Restoring Portions of the Lower Owens River," *Somach Simmons and Dunn* (December 2006), http://www.somachlaw.com/pubs/OwensRiverRestoration.pdf (accessed March 8,

2011); Randal C. Archibold, "A Long-Dry California River Gets, and Gives, New Life," *New York Times,* January 12, 2008.

19. Donald C. Jackson and Norris Hundley Jr., "Privilege and Responsibility: William Mulholland and the St. Francis Dam Disaster," *California History* 82, no. 3 (2004): 8–47.

20. David Jason Zetland, "Conflict and Cooperation within an Organization: A Case Study of the Metropolitan Water District of Southern California," PhD diss., University of California, Davis, 2008.

21. See Reisner, *Cadillac Desert*; Desert USA, "Parker Dam," http://www.desertusa.com /colorado/parker_dam/du_parkerdam.html (accessed January 28, 2012). See also information on the dam on the Bureau of Reclamation website, http://www.usbr.gov /projects/Project.jsp?proj_Name=Parker-Davis%20Project (accessed January 28, 2012).

22. Gary D. Libecap, "'Chinatown': Owens Valley and Western Water Reallocation— Getting the Record Straight and What It Means for Water Markets," *Texas Law Review* 83, no. 2055 (2005): 2055–89.

23. One of the best recent works on Tucson is Michael F. Logan, *Desert Cities: The Environmental History of Phoenix and Tucson* (Pittsburgh: University of Pittsburgh Press, 2006). See also two earlier works, C. L. Sonnichsen, *Tucson: The Life and Times of an American City* [Norman: University of Oklahoma Press, 1987 [1982]), and Bradford Luckingham, *The Urban Southwest: A Profile History of Albuquerque, El Paso, Phoenix, and Tucson* (El Paso: Texas Western Press, 1982).

24. William D. Kalt, *Tucson Was a Railroad Town: The Days of Steam in the Big Burg on the Main Line* (Mountlake Terrace, WA: VTD Rail Publications, 2007). See also Kenneth Seasholes et al., "Water in the Tucson Area: Seeking Sustainability," Water Resources Research Center, no. 20, July 31, 1999, 1–55, http://wrrc.arizona.edu//publications /water-tucson-area-seeking-sustainability (accessed September 18, 2012).

25. Lisa Song, "Thirsty Cities: Water Management in a Changing Environment," *Earth: The Science behind the Headlines* (December 31, 2009), http://www.earthmagazine.org /earth/article/2de-7d9-c-1f (accessed January 28, 2012).

26. Logan, *Desert Cities*, 179.

27. Urban infill is a growth strategy that utilizes land within a built-up area through reuse and renovation of old, blighted, or abandoned urban areas.

28. Karen Dotson and Timothy Francis, "Urban Infill—It's Not Just for Vacant Land: Tucson, Arizona Identifies Low Volume Reclaimed Water Customers," Water Reuse Association, Tucson, Arizona, December 8, 2011, http://ebookbrowse.com/urbaninfill-pdf -d244576358 (accessed September 18, 2012).

29. See information on Tucson's Clearwater Program for recharging groundwater with CAP water in the 2008 update of the "Water Plan: 2000–2050," section 4, http://cms3 .tucsonaz.gov/water/waterplan-2008.htm (accessed January 28, 2012).

30. For an extensive examination of the Denver area prior to white settlement, see Sarah M. Nelson, *Denver: An Archeological History* (Boulder: University Press of Colorado, 2008). See also Frederic P. Miller, Agnes F. Vandome, and John McBrewster, *History of Denver* (Saarbrücken, Germany: VDM, 2009).

31. See Robert Leaman Brown, *The Great Pikes Peak Gold Rush* (Caldwell, ID: Caxton, 1985).

32. For more information on rain shadow, see David C. Whiteman, *Mountain Meteorology: Fundamentals and Applications* (New York: Oxford University Press, 2000).

33. Charles C. Fisk, "The Metro Denver Water Story: A Memoir," unpublished manuscript, Papers of Charles C. Fisk, Water Resources Archive, Colorado State University, Fort Collins, 73; digitized by Water Resources Archive, http://digitool.library.colostate.edu/exlibris/dtl /d3_1/apache_media/L2V4bGlicmlzL2RobC9kM18xL2FwYWNoZV9tZWRpYS8iNjQ5 .pdf (accessed January 29, 2012).

34. See Duane A. Smith, *The Trail of Gold and Silver: Mining in Colorado, 1859–2009* (Boulder: University Press of Colorado, 2009).

35. Denver Water, History of Denver Water, http://www.denverwater.org/AboutUs /History/ (accessed January 29, 2012).

36. Philip L. Fradkin, *A River No More: The Colorado River and the West* (Berkeley: University of California Press, 1995 [1968]), 42–44. See also Carl Abbot, Stephen J. Leonard, and Thomas J. Noel, *Colorado: A History of the Centennial State,* 4th ed. (Boulder: University Press of Colorado, 2005).

37. Fisk, "Metro Denver Water Story," 73.

38. Denver Water, History of Denver Water.

39. For a recent social history of Denver, see Phil H. Goodstein, *Denver from the Bottom Up: A People's History of Early Denver* (Denver: New Social Publications, 2004).

40. George Radosevich, *Evolution and Administration of Colorado Water Law, 1876–1976* (Fort Collins: Water Resources Publications, 1976); Kay Collins, "The Transmountain Diversion of Water from the Colorado River: A Legal-Historical Study," MA thesis, University of New Mexico, Albuquerque, 1965; Andrew J. Prelog, "Water Scarcity and Rapid Complex Change in Colorado: An Evaluation of Historical Patterns of Water Appropriation and Socio-Demographic Growth," MS thesis, Colorado State University, Fort Collins, 2007.

41. Metro Denver, Economic Forecast, January 18, 2012, http://www.metrodenver.org /metro-denver-economy/forecasts (accessed January 28, 2012).

42. Denver Water, Supply and Planning, 2012, http://www.denverwater.org/SupplyPlanning (accessed September 18, 2012).

43. Stephen Plog, *Ancient Peoples of the American Southwest* (New York: Thames and Hudson, 1996); George J. Gumerman, *Exploring the Hohokam: Prehistoric Desert Peoples of the American Southwest* (Albuquerque: University of New Mexico Press, 1997).

44. James E. Officer, *Hispanic Arizona, 1536–1856* (Tucson: University of Arizona Press, 1987).

45. Patricia Gober, *Metropolitan Phoenix: Place Making and Community Building in the Desert* (Philadelphia: University of Pennsylvania Press, 2006). See also Logan, *Desert Cities*; Douglas E. Kupel, *Fuel for Growth: Water and Arizona's Urban Environment* (Tucson: University of Arizona Press, 2003).

46. City of Phoenix, Phoenix Water Service Celebrates Its 100th Anniversary, 2007, http://www.phoenix.gov/webcms/groups/internet/@inter/@dept/@wsd/documents /web_content/d_038000.pdf (accessed September 18, 2012).

47. For details on these dams, see the Bureau of Reclamation Arizona website, http://www .usbr.gov/lc/yuma/ (accessed January 29, 2012). For further discussion of water resources and conservation policies in Phoenix, see Kelli L. Larson, Annie Gustafson, and Paul Hirt, "Insatiable Thirst and a Finite Supply: An Assessment of Municipal Water-Conservation Policy in Greater Phoenix, Arizona, 1980–2007," *Journal of Policy History* 21, no. 2 (2009): 107–37.

48. Bradford Luckingham, *Phoenix: The History of a Southwestern Metropolis* (Tucson: University of Arizona Press, 1989).

49. City of Phoenix, Tres Rios Wetlands, http://phoenix.gov/waterservices/tresrios /index.html (accessed January 29, 2012).

50. University of Arizona, The Water Report: Agricultural Water to Municipal Use, December 15, 2008, http://wrrc.arizona.edu/sites/wrrc.azwater.edu/files/ag_to_muni _article.pdf (accessed September 18, 2012).

51. City of Phoenix, Water Services Department, 2011 Water Resource Plan, http:// phoenix.gov/WATER/wsd2011wrp.pdf (accessed January 12, 2012).

52. Instituto Nacional de Estradística y Geografía (INEGI) Información Nacional, por Entidad Federativa y Municipios, Tijuana, Baja California, 2011, http://www.inegi.org.mx /sistemas/mexicocifras/default.aspx?ent=02 (accessed January 29, 2012); Estrada y Asociados, Total Population 2010 Census, Tijuana, http://www.estradayasociados.com.mx /area-information/tijuana/demographic-data (accessed January 29, 2012).

53. See Marco Antonio Samaniego López, ed., *Breve Historia de Baja California* (Tijuana: Universidad Autónoma de Baja California, 2006); David Piñera Ramírez and Jesús Ortiz Figeroa, eds., *Historia de Tijuana*, 2nd ed. (Tijuana: Universidad Autónoma de Baja California, 1989).

54. Kumeyaay.Com, Pre-Contact History, 2011, http://www.kumeyaay.com/kumeyaay -history/34-articles.html (accessed January 29, 2012). See also M. Steven Shackley and Steven Lucas-Pfingst, *The Early Ethnography of the Kumeyaay* (Berkeley: University of California Press, 2007).

55. Lawrence D. Taylor, "The Mining Boom in Baja California from 1850 to 1890 and the Emergence of Tijuana as a Border Community," *Journal of the Southwest* 43, no. 4 (Winter 2001): 463–92.

56. Ibid., 474.

57. Ibid., 477.

58. Ibid., 480. See also Thurber Dennis Proffitt, *Tijuana: The History of a Mexican Metropolis* (San Diego: San Diego State University Press, 1994).

59. See Paul J. Vanderwood, *Satan's Playground: Mobsters and Movie Stars at America's Greatest Gaming Resort* (Durham, NC: Duke University Press, 2010); Daniel D. Arreola and James R. Curtis, *The Mexican Border Cities: Landscape Anatomy and Place Personality* (Tucson: University of Arizona Press, 1994).

60. David A. Sonnenfeld, "Mexico's 'Green Revolution,' 1940–1980: Towards an Environmental History," *Environmental History Review* 16, no. 4 (Winter 1992): 28–52.

61. Roberto A. Sanchez, "Water Quality Problems in Nogales, Sonora," *Environmental Health Perspectives* 103 (February 1995): 93–97.

62. Lana K. Mowers and Eduardo T. Riuz, "Go Ahead, Drink the Water," *Motor Management* (July 2004), http://www.pump-zone.com/articles/58.pdf (accessed March 11, 2011).

63. History of San Diego and Tijuana Sewage Systems, February 22, 2008, http://www.sewagehistory.com/index.htm (accessed September 20, 2012); "Sewage History of Tijuana," *San Diego Magazine* (December 2007).

64. "Bi-national Aqueduct for Tijuana Pitched, but Pipeline May Not Come in Time," *US Water News* (September 2002), http://www.uswaternews.com/archives/arcglobal/2binaqu9.html (accessed March 11, 2011); Stephanie Jackter, "More on the Liquid Gold of the Future," *Tienda Baja Nomad,* August 25, 2003, http://forums.bajanomad.com/viewthread.php?tid=762 (accessed March 11, 2011).

65. Josue Medellin-Azuara, Leopoldo Mendoza-Espinosa, Jay R. Lund, Cynthia Watter-Barrera, and Richard E. Howitt, "Water Supply for Baja California: Economic Engineering Analysis for Agricultural, Environmental and Urban Demands," January 2009, http://cee.engr.ucdavis.edu/bajacalvin/BAJA_CALVIN_FinalReport.pdf (accessed March 11, 2011).

66. Sandra Dibble, "$4.8-Million Loan to Expand Tijuana Sewage Treatment," *Sign on San Diego,* February 9, 2011, http://www.utsandiego.com/news/2011/feb/09/48-million-loan-expand-tijuana-sewage-treatment/ (accessed September 15, 2011).

67. Ibid.

68. George Kennan, *The Salton Sea: An Account of Harriman's Fight with the Colorado River* (Ithaca, NY: Cornell University Press, 2009 [1917]); Pat Laflin, *The Salton Sea: California's Overlooked Treasure* (Indio, CA: Coachella Valley Historical Society, 1999 [1995]), 17–30; Worster, *Rivers of Empire,* 196; Reisner, *Cadillac Desert,* 122–23.

69. Casey Walsh, " 'To Come of Age in a Dry Place': Infrastructures of Irrigated Agriculture in the Mexico-U.S. Borderlands," *Southern Rural Sociology* 24, no. 1 (2009): 21–43.

70. Evan Ward, *Border Oasis: Water and the Political Ecology of the Colorado River Delta, 1940–1975* (Tucson: University of Arizona Press, 2003), 14–17. See also Samaniego López,

Breve Historia de Baja California, 124–30; John Joseph Dwyer, *The Agrarian Dispute: The Expropriation of American-Owned Rural Land in Postrevolutionary Mexico* (Durham, NC: Duke University Press, 2008).

71. Boulder Canyon Project Act, 1928, http://www.usbr.gov/lc/region/g1000/pdfiles /bcpact.pdf (accessed March 11, 2011).

72. San Diego Dialogue Report, Providing a Reliable Water Supply in the San Diego /Imperial Valley/Baja California, September 2001, http://www.sandiegodialogue.org/pdfs /Water_Paper_Sept01.pdf (accessed March 11, 2011).

73. Sonnenfeld, "Mexico's 'Green Revolution,'" 40.

74. Introduction to the New River/Mexicali Sanitation Program, Colorado River Basin Regional Water Quality Control Board, June 1, 2009, http://www.swrcb.ca.gov/rwqcb7 /water_issues/programs/new_river/nr_intro.shtml (accessed March 11, 2011).

75. See Alfonso A. Cortez Lara, "Irrigation and Transboundary Water Management in the Lower Colorado River: The Changing Role of Agriculturists in the Mexicali Valley, Mexico," PhD diss., Michigan State University, Lansing, 2010.

76. See Vicente Sánchez Munguía, *El Revestimiento del Canal Todo Americano: ¿Competencia o Cooperación por el Agua en la Frontera México–Estados Unidos?* (Tijuana: El Colegio de la Frontera Norte, 2004); San Diego County Water Authority, Fact Sheet, Canal Lining Projects, January 2010, http://www.sdcwa.org/canal-lining-projects (accessed September 20, 2012).

77. Haley Nolde, "Mexicali: Living on Borrowed Water," http://journalism.berkeley.edu /projects/border/mexicali.html (accessed March 11, 2011).

78. Jose Luis Castro-Ruiz and Vencente Sanchez-Munguia, "The Lining of the All-American Canal: Effects on Mexico," *Southwest Hydrology* (September-October 2005), http://journalism.berkeley.edu/projects/border/mexicali.html (accessed March 11, 2011).

79. Sandra Dibble, "U.S., Mexico Boost Collaboration on Colorado River," *Sign On San Diego,* February 6, 2011, http://www.signonsandiego.com/news/2011/feb/06/us-mexico -boosting-collaboration-colorado-river/ (accessed March 11, 2011).

80. For early Las Vegas history, see Eugene P. Moehring and Michael S. Green, *Las Vegas: A Centennial History* (Reno: University of Nevada Press, 2005); Barbara Land and Myrick Land, *A Short History of Las Vegas* (Reno: University of Nevada Press, 1999).

81. For an entertaining yet thorough history of Las Vegas, see Hal K. Rothman and Mike Davis, eds., *The Grit beneath the Glitter: Tales from the Real Las Vegas* (Berkeley: University of California Press, 2002), esp. pp. 115–25.

82. Kevin Wehr, *America's Fight over Water: The Environmental and Political Effects of Large-Scale Water Systems* (New York: Routledge, 2004); Michael Hiltzik, *Colossus: Hoover Dam and the Making of the American Century* (New York: Free Press, 2010).

83. Michael Duchemin, "Water, Power, and Tourism: Hoover Dam and the Making of the New West," *California History* 86, no. 4 (2009): 60–78.

84. See Thomas Ainlay and Judy Dixon Gabaldon, *Las Vegas: The Fabulous First Century* (Mount Pleasant, SC: Acadia, 2003), 54–102.

85. James W. Hulse, *Nevada's Environmental Legacy: Progress or Plunder* (Reno: University of Nevada Press, 2009), chapters 3 and 4.

86. "A Brief History of Water in Southern Nevada," Southern Nevada Water Authority Resource Plan 2009, http://www.snwa.com/assets/pdf/wr_plan_exec_summary.pdf (accessed January 29, 2012).

87. Jedediah Rogers, *Robert B. Griffith Water Project (formerly Southern Nevada Water Project)* (Denver: Bureau of Reclamation, 2001), 12.

88. G. Tracy Mehan III, "Energy, Climate Change, and Sustainable Water Management," *Environment Reporter* 38, no. 48 (2007): 5.

89. Jon Christensen, "Build It and the Water Will Come," in Hal K. Rothman and Mike Davis, eds., *The Grit beneath the Glitter: Tales from the Real Las Vegas* (Berkeley: University of California Press, 2002), 115–16.

90. Ibid., 119. See also Michael Seissenstein, "The Water Empress of Vegas: How Patricia Mulroy Quenched Sin City's Thirst," *High Country News* [Paonia, CO], April 9, 2001; Charles Fishman, *The Big Thirst: The Secret Life and Turbulent Future of Water* (New York: Simon and Schuster, 2011), 51–87.

91. Southern Nevada Water Authority Resource Plan 2009; Henry Brean, "OSHA Clears Work to Resume on Lake Mead Tunnel," *Las Vegas Review-Journal,* June 26, 2012, http://www.lvrj.com/news/osha-clears-work-to-resume-on-lake-mead-tunnel-160458325.html (accessed September 9, 2012).

92. For more information on this low-water landscape technique, see Eartheasy: Solutions for Sustainable Living, http://eartheasy.com/grow_xeriscape.htm (accessed January 29, 2012).

93. Valley Forward Phoenix Water Survey of Valley Residents, March 2006, http://www.valleyforward.org/issues/44/ (accessed January 29, 2012).

94. Ibid.

95. Julie Russ, ed., The Phoenix Area Social Survey (PASS), March 2003, published in association with the Center for Environmental Studies, Arizona State University, Tempe, http://caplter.asu.edu/docs/contributions/pass.pdf (accessed January 29, 2012).

96. Ibid., 29.

97. Ibid., 19–23.

98. "Outdoor Water Use in the U.S.," EPA Water Sense, an EPA Partnership Program, February 8, 2012, http://www.epa.gov/WaterSense/pubs/outdoor.html (accessed September 18, 2012).

99. PASS Report 2006, http://caplter.asu.edu/docs/contributions/2007_PASS2.pdf (accessed January 29, 2012).

100. Ibid., 17.

101. Ibid., 27.

102. Las Vegas Metropolitan Area Social Survey, 2010, http://strata.unlv.edu/docs/LVMASS.pdf (accessed January 29, 2012).

6

Owning the River

Indian Water Rights and Settlements

INDIGENOUS PEOPLE AROUND THE WORLD have always struggled to preserve their language, their culture, even their very existence. They must also battle for access to natural resources, such as water. While told largely today from the Anglo-American perspective, there is a long and complicated history of Indian rights to the Colorado River. By the end of the nineteenth century, many of these original water users had lost access to its waters. As southwestern towns became vibrant urban centers in the twentieth century, native peoples of the river basin were subjugated, placed on otherwise unwanted land, and left to farm without necessary water supplies.

Since the early 1980s the story has changed as Indian nations have asserted their sovereignty and sought access to natural resources. This part of the story is not yet finished, as lawmakers and politicians continue to address unsettled Indian water claims in the twenty-first century. However, the power relationship in the West has certainly changed. At first, Indian communities gained only promises, or "paper water," which existed legally but with no actual water or pipes through which it could flow. Eventually, Indian tribal negotiators and state officials hammered out agreements that did bring water to the Indian communities, along with the necessary infrastructure to use it. Some of these agreements still look much like the earlier empty successes, but Indian communities in the Colorado River basin are increasingly seeing "wet water" that actually flows to the reservations. As these claims are settled, non-Indian water users find themselves newly dependent on Indian compromises or water sales and leases. While not all indigenous people have been thus empowered, many have become important players in the contest for water in the Colorado River watershed.

The Dry Reservation

Long before Euro-American westward expansion in the nineteenth century, the Colorado River was the lifeblood of many native communities. As white settlers became interested in the area referred to as the "Great American Desert," the government built military forts on Indian lands to regulate the movements of Indian communities to smaller and smaller areas. Eventually, large groups of Navajo and Mescalero Apache were at first provoked to resist or starved into submission, then rounded up for removal from their lands in present-day Arizona to the Bosque Redondo Indian Reservation. This 1 million–acre piece of dry and barren land surrounded Fort Sumner, approximately seventy-five miles north of present-day Roswell, New Mexico. Many died along the forced march in 1863–64, called "the Long Walk" by the Navajo, and many more died after they arrived at the reservation. It soon became obvious that farming the barren land was impossible and that assimilation plans were failing. In 1865 the Apache left Fort Sumner without permission, and in 1868 the government reluctantly allowed the Navajo to return to their homelands.[1]

Other groups, including the Pima-Maricopa, Mohave, Quechan (Yuma), and Cocopah, had their ancestral lands reduced and split into separate pieces by the international border or forced relocations. Some of the resulting reservations directly border the Colorado River, while others retain access to its various tributaries. The result is a significantly large Indian claim to the river's water, one that is only partially quantified.[2] Already over-allocated, future quantifications of Indian shares guarantee that someone will experience a shortage. The efforts to settle competing water claims have thus been very emotional and, at times, confrontational.

The story of Indian diversion of Colorado River water in the modern era begins in the years just following the Civil War. While most of the famous "Indian Wars" on the Great Plains took place after that conflict, the Mohave had already been defeated by federal forces and confined to reservation land. In 1858 the Mohave attacked a wagon train heading to California and killed eighteen people.[3] The next year they attacked a federal cavalry unit, igniting a fierce battle that resulted in a Mohave defeat. In the process of making peace, one leader, Chief Ireteba, agreed to move his people to a reservation south of their current homeland, along the banks of the Colorado River.[4] Thus split into two groups, Ireteba's band of around 800 agreed to the reserved lands (called the Colorado River Indian Reservation) because the government promised them it would build an irrigation system from the river so they could farm.

In 1865 the Office of Indian Affairs (later the Bureau of Indian Affairs) put its plan before the US Congress, which approved it and allocated $100,000 for the project. In 1867, work began on what would eventually be an eight-mile canal, dug completely with shovels mostly by the Mohave. This plan was the first federally sponsored irrigation project in US history, thirty-five years before the 1902 Reclamation Act created what would become the Bureau of Reclamation.[5] The problems, however, were many. Although warned by Ireteba and other engineers who evaluated the region, the federal government ignored evidence that the land was too alkaline for agriculture. Drawing water from the Colorado River would also be difficult from that location because of the depth of its canyon. Regardless, the project began, and by 1874 a nine-mile stretch had been completed.

Almost immediately, problems with collapsing tunnels and silting made the canal almost unusable. About two decades later, Charles Rockwood, the famous developer of Imperial Valley, California, would discover some of the same challenges with the river. Even though Rockwood's canal created the first major disaster in western irrigation history (flooding and the resultant creation of the Salton Sea; see chapter 1), it was also the first successful irrigation of arid land in the region.[6] The Mohave project was the first attempt and the first failure. Between the 1870s and the early twentieth century, Congress reluctantly spent nearly $1 million on a project that simply would not work. Even though Congress approved a long-overdue expansion of the project in 1910 to serve Indian farms, problems with drainage and alkalinity continued. In the long run, Indian farmers could not afford the expense of leveling and properly watering their land. The promised canal was never extended beyond the original nine miles, and by 1926 two-thirds of Mohave land on the reservation was being leased to non-Indian farmers who had more resources for startup costs.[7]

Another early, and perhaps the longest, Indian struggle for water involved the Gila River, a major tributary of the Colorado River. A large group of Pima (Akimel O'odham) farmed the river's banks for generations and watched the first permanent white settlement in the region. Unlike the Colorado River Mohave, who were moved south of their traditional lands, the Pima were given reservation land where they traditionally resided, although a greatly reduced amount. This land allowed the Pima to develop a self-sufficient lifestyle based on agriculture, irrigated with Gila River water. However, like the Mohave, the Pima were forced to request help from the federal government when their fields began to dry up during a severe drought in the 1870s. To compound the problem, the completion of the Florence Canal in 1886 to water the farms of non-Indians upstream reduced the flow to the Pima Reservation even further.[8]

Even though President Rutherford B. Hayes extended the boundaries of the reservation in 1887 to help the Pima by incorporating the confluence of the Gila and Salt Rivers, crops continued to fail, and the people sank into poverty and dependence on government aid. Meanwhile, Arizona's white population more than doubled between 1880 and 1890 (from 40,440 to 88,243), putting pressure on already limited water supplies.[9] Those above-stream from the Pima allocated more and more of its water until the flow was nearly completely dried up by the time it reached the reservation.

By 1900, the plight of the Pima had reached the major newspapers, which published descriptions of desperation and poverty. The stories decried the fact that even though the Pima had asked for help as early as the 1870s, nearly three decades had passed with no help forthcoming.[10] Although the drought finally ended in 1905, the rains came too late to save the Pima. After eleven years of crop failures and the destruction of their sacred mesquite forest to drought and for cutting trees to sell, many Pima left the reservation for work elsewhere. For those who remained, the 1902 Reclamation Act seemed to offer the long-awaited help. Those in Congress who lobbied for the act held up the Pima and their suffering as a primary argument for reclamation. Because of this fact, many people expected the first major reclamation project to occur along the Gila River to help the Pima.[11]

When private land developers along the Salt River won the first project, the Pima were astonished. Even though the US Geological Survey had identified a site along the Gila for a dam and a reservoir and Congress had appropriated nearly $1 million to study the site, nothing ever happened. While it seems clear that Congress intended that the Pima would receive the first reclamation project, the newly created Reclamation Service listened to those who lobbied for the Salt River Project instead. The long-promised Gila Project did not begin until after 1924, when plans for the Coolidge Dam and the connected San Carlos Irrigation Project were finally approved. The Pima fight for water rights was far from over, however, and would not end completely until 2004.[12]

These first indigenous struggles for Colorado River water resulted mostly in benefits for non-Indians. This story was repeated throughout the first half of the twentieth century as the new Reclamation Service, formed by the 1902 Reclamation Act, constructed dams and reservoirs up and down the length of the Colorado River.[13] The western water law of prior appropriation made water secure for most western settlers in the region. Ironically, however, this same basis of law later came to support Indian claims to water.[14] In 1908, in what became a landmark legal case, native communities obtained at least verbal rights to water no matter how long white landholders had been using it.[15] Along the Milk River in Montana, white farmers, cattle ranchers, and Indians all diverted water from its streams. The Fort Belknap Native Reservation, established in 1888 for the Gros Ventre and Assiniboine Indians, had the Milk River as its northern boundary, and Indians used the water to irrigate a small portion of their land.

In 1898, Fort Belknap Indians began diverting a much larger amount of water onto about 30,000 acres in an effort to boost the reservation's economy. Shortly before this diversion began, white ranchers purchased land nearby and announced plans to appropriate a large share of the Milk River's water. When the Fort Belknap Reservation began its diversion, the ranchers, represented by Henry Winters, quickly protested. They claimed a prior water right according to western law because they had started building dams and diversion channels to use the water before the Indians started their own diversion. Faced with an impasse, the federal government filed suit on behalf of the reservation for water rights.[16]

After making it through two lower courts, the case reached the US Supreme Court in 1908, where the judges ruled in an 8 to 1 decision that the Fort Belknap Reservation had priority claims because the establishment of the reservation in 1888 contained "implied" water rights to support its inhabitants' livelihoods.[17] The resulting Winters Doctrine gave Indians reserved water rights that had priority over the rights of most white settlers in the American West. The Winters Doctrine became the primary basis of all future Indian water claims and legal challenges, and it remains so to the present day. Yet although this landmark decision seemed to settle the matter, Indians found it hard to actually obtain this water to which they had legal claim. For the next fifty years, it seemed all the Indians had won was "paper water" instead of any actual "wet water."[18]

One of the issues that complicated Indian water claims was the contradiction between the western water doctrine of prior appropriation, which required beneficial use, and the Winters Doctrine, which granted water rights based on present and possible future uses. This conflict continues to be part of ongoing negotiations between states and Indian nations seeking to quantify their water rights. No one agreed on how to measure future water needs to accommodate population growth. The other point of contention was the way water rights for Indian reservations were used. Some argued that since the reservations' main purpose was to teach Indians to become self-sufficient farmers, water rights were only implied for that purpose. Others disagreed and argued that water was an absolute right and could be leased, even sold, and certainly used for pursuits other than agriculture.[19]

In the years following the *Winters* case, the Reclamation Service and the Bureau of Indian Affairs became, in the words of historian Donald Pisani, "uneasy allies."[20] The Bureau of Reclamation's task was the development of western lands for agriculture. By providing irrigation projects, the bureau was to make it possible for small farmers to populate the arid region. The Bureau of Indian Affairs also aimed to construct irrigation projects on western Indian lands. Another goal was to promote agriculture—in this case, to encourage Indians to stay on their reservations and to eventually help them assimilate into US society by becoming independent farmers.

The organizations' aims diverged in some very important ways, however. The Reclamation Service was mostly interested in providing irrigated farmland for white Americans and was very interested in buying and selling "surplus" Indian lands to non-Indians. The Bureau of Indian Affairs usually deferred to the Reclamation Service to build irrigation projects, but it did not always believe agriculture was the only way Indians could be assimilated into US culture. It was just one solution, and even then, only if conditions were right.[21] Thus the development of Indian water projects was plagued by conflicting aims and the lack of clearly quantified water allocations.

Only "Paper Water"

As the population of the southwestern states continued to expand in the early twentieth century, it became clear that some agreement on Colorado River usage was necessary. Thus in 1922 the seven basin states gathered to divide the waters. Arizona made itself the spoiler by refusing to sign the resulting Colorado River Compact because the agreement did not specify how much water would be allocated to each state. Yet while the seven states argued over water shares, no one was really thinking about the Indian reservations, which had perhaps the largest claim to the river. Only Secretary of Commerce Herbert Hoover demanded the insertion of a statement that read: "Nothing in this compact shall be construed as affecting the obligations of the United States of America to Indian tribes."[22] This small statement was largely ignored until Indian communities began seeking water rights through litigation. In the first few decades after the 1908 *Winters* decision, several dozen lawsuits were filed for Indian water rights reserved under the Winters Doctrine. Most of these cases, however, resulted in paper promises, with no financial means to actually develop the rights granted by the courts. One of the first groups to see water development in the

Colorado River basin was the Pima. In 1924, when federal money was allocated for construction of the Coolidge Dam on the Gila River and related irrigation works, the Pima believed water would finally flow to their farms.[23]

Although the Bureau of Indian Affairs completed Coolidge Dam in 1930, there was still no clear allocation of Gila River water. Finally, in 1935, the state of Arizona issued an allocation granting the Pima priority rights to enough water to irrigate 35,000 acres. This small promise of water did little to help their economy in the long run. Pisani argues that irrigated agriculture actually severely damaged what had been a diversified economy that included fishing and hunting. Limited to small allotments that could be irrigated, the Pima lost much of their land to whites.[24]

For forty-plus years following the *Winters* decision, Indian nations made very little progress in obtaining water rights. Various lawsuits did result in victories on paper, but again there was little or no means to develop the projects. As the federal government began filing suits on behalf of Indian nations, some members of Congress worried about the implications. The Winters Doctrine rested on the concept of "reserved rights," which applied to all federal land, not just Indian reservations. State governments worried that federal power to sue on behalf of reservations might lead to government appropriation of other state water rights.[25] Since states could not sue the federal government, members of Congress began working to give states some control over water litigation.

In 1951 Senator Patrick McCarran, a Democrat from Nevada, proposed an amendment rider to the Department of Justice's 1952 appropriation bill. The rider made it possible to waive the federal government's immunity and allow it to be sued in state courts over any water acquired under its law.[26] The scope of the amendment was not really tested until 1969, when Colorado brought a suit against the federal government over control of the waters of the Eagle and Colorado Rivers. The resolution of this case, and later federal appeals to higher courts, affirmed that the federal government could be named a party in a suit over water rights and that state courts had the authority to adjudicate water claims that affected state water rights.[27] The eventual effect of moving water litigation into state courts was to convince Indian tribes that their best chance of obtaining "wet water" was through negotiated settlements rather than lawsuits. Bringing suit in state courts seemed much more risky when state water holders could become parties to the suit.

During the 1950s, ongoing conflicts between states over the Colorado River became the most important factors affecting tribal water claims. In 1948 the upper basin states of the Colorado River watershed gathered to decide on the amount of water to which each state would be entitled. Representatives from these states worried that unless they apportioned their own water rights and began developing them soon, the lower basin would lay claim through prior appropriation to all of the river's water. By this time the lower basin, mostly California, was enjoying the results of Hoover and Parker Dams and the Colorado River Aqueduct to Los Angeles and was planning extensions of the All-American Canal that brought water to the Imperial Valley. The upper basin was eager to begin its own projects, and the 1948 Upper Colorado River Basin Compact paved the way.[28]

After deciding on their respective allotments, the upper basin states proposed various development projects that were eventually combined into a plan for dams and reservoirs. As Secretary of the Interior Stewart Udall argued for a broader plan that would also include projects for the lower basin, Indian water interests became involved. Along with plans for a large dam in Glen Canyon, Udall incorporated ideas for one or possibly two dams in the Grand Canyon that would generate electricity and serve as "cash machines" to fund other proposed projects, such as the long-sought Central Arizona Project (CAP) canal. Arizona wanted to claim its water before California took it all and argued that the only way it could do so was by constructing a very expensive canal. California effectively blocked any legislation that would fund such a project, thus initiating a longstanding conflict over the amount of water to which Arizona and California were entitled.[29]

In an effort to finally bring about a resolution, Arizona filed suit against California in federal court in 1952 (see chapter 2). In one of the longest US Supreme Court cases in history, both sides debated their positions with great emotion and legal expense. Finally, after eleven years of litigation, the Supreme Court ruled in 1963 (published in 1964) in favor of Arizona. What had been at stake was whether water from the Gila River would count as part of Arizona's annual Colorado River allotment of 2.8 million acre-feet (MAF).[30] The ruling eventually allowed Arizona to take the full amount from the main stream of the Colorado, paving the way for legislation to fund the Central Arizona Project.

The most important part of the case for Indian water use was a clause that placed the burden of meeting these water rights clearly on the states in which a tribe resided. The case specifically quantified water allocations from the Colorado River for five tribes: the Chemehuevi, Cocopah, Quechan (Yuma), Colorado River Indian Community, and Fort Mojave Indian Reservations. The total amount of water allocated to these tribes amounted to 761,000 acre-feet from the Colorado River.[31] The Court's findings also stated that when reservations were established, they came with water rights "intended to satisfy the future as well as the present needs" of the Indians.[32] The only way to measure how much water a reservation might need was by determining the amount of acreage that could be irrigated. This concept became known as Practically Irrigable Acreage (PIA). This method for quantifying water rights assumed that most water would be used for agriculture, but tribes later challenged this notion by expanding water rights to industrial uses.[33]

Once *Arizona v. California* was decided, Arizona again sought authorization for CAP to bring its share of Colorado River water into the state's interior. California continued to lobby against CAP, but plans for dams in the Grand Canyon threatened to derail it completely. First proposed in the late 1940s, these dams met with general support among politicians in Arizona and within the Department of the Interior. Carl Hayden and other supporters argued that a dam at Bridge Canyon, the lower end of the Grand Canyon, would only flood a part of the park that visitors never saw from any of the rim trails and would provide added beauty with a navigable lake.[34]

No one seemed to protest plans for Marble Canyon at the upper end of the Grand Canyon except for the superintendent of Grand Canyon National Park, George F. Baggley. As engineers surveyed the locations for both dams, no one worried about the reaction of the Indians on whose reservations the dams would reside. Marble Canyon Dam would be partially on Navajo land, and Bridge Canyon Dam on the Hualapai Reservation.[35] One reason no one asked the Indian communities for their opinion was the changing relationship between the government and Indian tribes. In 1953, Congress passed House Resolution 108, which terminated the special trustee relationship between Indians and the federal government.[36] Called the Termination Policy, it removed Indian sovereignty over reservation lands and was meant to ensure assimilation into mainstream culture.

The Termination Bill and Public Law 280 (passed the same year, giving states jurisdiction over Indian reservations) dealt a heavy blow to native control over any issue, including water. Bureau of Reclamation engineers did not think Indian tribes had any say in the subject of dam building in the Grand Canyon. Although the Termination Policy began to die in the late 1950s and early 1960s, none of the Arizona politicians supporting Grand Canyon dams was worried. As the Supreme Court began to reaffirm tribal sovereignty in the early 1960s, however, Arizona faced tribal demands and even outright opposition. In 1958, Arizona found itself negotiating with the Hualapai, who retained lawyers to fight for their demands. The Hualapai had questioned plans for Bridge Canyon Dam as early as 1949, but then it was mostly because they had not been consulted.[37] The Hualapai wanted the dam built, but only if they benefited from it. They saw the dam as a source of jobs and revenues from hydroelectric generation. When negotiations ended in 1960, the Hualapai held a contract for annual payments of between $550,000 and $620,000 over the life of the project (forty to fifty years).[38]

Indian resistance to Bridge Canyon Dam had ended, but the issue of Marble Canyon Dam on Navajo land remained. Arizona did not think it needed to negotiate with the Navajo because the federal government had granted rights to power sites along that part of the Colorado River long before extending the boundaries of the Navajo Reservation. The Navajo believed they had clear rights and argued strongly against Marble Canyon Dam. However, in 1962, Arizona residents were eagerly anticipating what appeared to be a victory in the Supreme Court case against California. Some in the state, including Carl Hayden and Morris Udall, believed they should apply again for federal funding for CAP.

No one believed Arizona could actually build the project without federal help, so state politicians began pushing another bill through Congress after the US Supreme Court ruling in 1963. However, now that the original state plan for CAP had been abandoned, the contract with the Hualapai was null and void. If CAP were approved with the two Grand Canyon dams still in the plan, the Hualapai and Navajo would lose around 60,000 acres to flooding and the Havasupai nearly a mile of canyon land in which they currently resided.[39] Just when things looked bleak, plans for the dams hit the public press and enormous protests began, orchestrated by David Brower and the Sierra Club. Other environmentalists quickly joined the publicity campaign, making it look as though Arizona's dams would completely destroy a national landmark and one of the most popular tourist destinations in the United States (see chapter 3).[40]

A logical approach for the tribes was to ally themselves with the environmental groups. Anticipating such an alliance and the devastating press it might bring, Arizona finally agreed to negotiate with the tribes. Talks began in earnest in 1965, and the Hualapai demanded compensation or threatened that they would find funding to build the dam on their own. Clearly, members of the tribe believed a dam would benefit them, but only if they received a substantial portion of annual revenue generated in exchange for flooded lands. Ultimately, the Hualapai demanded payments equal to what Arizona had promised earlier, the naming of the dam Hualapai Dam, control over the tourist revenue generated by the lake, a road to the reservoir, and the ability to purchase power generated by the dam at the lowest market rates.[41]

Sensing a possible advantage, supporters of CAP began publicizing Hualapai support for a dam, arguing that environmentalists who opposed it were racist and anti-Indian. However, the tables soon turned when the press learned what Barry Goldwater had blatantly stated at an environmental symposium on the Grand Canyon rim in March 1966. He talked about how much the Hualapai needed the Bridge Canyon Dam. When a young geography professor in the audience asked what the Navajo thought of the plan, Goldwater stated that the Navajo did not have any rights in the matter of the Marble Canyon Dam and thus did not need to be consulted.[42] Environmentalists thus gained something for their side. They began to paint Goldwater's words as indicative of a blatant disregard for the Navajo people and perhaps of the racist attitude of Arizona as a whole.

Although the Grand Canyon dams had the support of the Hualapai as well as the White Mountain Apache, San Carlos Apache, Pima, and Ute Nations, opponents could say that the Navajo represented a much larger number than all these nations combined. Because CAP supporters refused to negotiate with the Navajo as they had with the Hualapai, the Navajo protested the dam projects, and environmental groups such as the Sierra Club widely publicized their opposition. Instead, Navajo leaders supported the development of coal-fired electric plants that would be constructed on the Navajo Reservation and would use coal procured from reservation land.[43]

By 1967, politicians were forced to take the Grand Canyon dams out of CAP bills in an effort to pass them. Pressure from environmental groups and the general public made such dams politically impossible. When Arizona politicians and Secretary of the Interior Stewart Udall decided to remove the dams from the legislation, they ended their problems with the Navajo Nation but abandoned the Hualapai in the process. Left without supporters or alternative resources for a dam, the Hualapai had nothing to raise their standard of living. CAP was funded without the Grand Canyon dams, and while the Hualapai continued to experience economic hardships, the environmentalist support for coal-fired plants in the Navajo Nation led to widespread ecological damage—perhaps the greatest irony of the entire story.

Quantification and "Wet Water"

Although the Hualapai and Navajo Nations did not really benefit from the Grand Canyon dams controversy, plans for CAP eventually brought water to native communities as they sought quantification of their water rights. Beginning in the 1970s, Indian nations throughout the West sat down with state representatives at negotiating tables to hammer out water agreements. The impact of the McCarran Amendment (which moved water litigation to state courts) convinced many tribes that they had much more to gain through negotiation, and states eagerly agreed to settlements, assuming they would save money and perhaps even water. Although the settlements often cost as much as litigation would have, Indian communities gained faster access to water and were able to trade some water for infrastructure.

The context of the 1970s was also important for Indian water rights. As tribal representatives began to negotiate, a growing number of nonprofit organizations were evolving to help fight legal battles on behalf of Indians and other minority groups. Stimulated by Lyndon B. Johnson's "war on poverty," these groups served as legal advocates for the underrepresented. One example is the Native American Rights Fund (NARF), founded in 1970.[44] Sponsored by the Ford Foundation and the

TABLE 6.1. Water settlements along the Colorado River

Chemehuevi, Cocopah, Colorado River Indian Community, Quechan (Yuma), and Fort Mojave Tribes	1963 (*Arizona v. California*)
Ak-Chin	1978
Tohono O'odham	1982
Salt River Pima–Maricopa	1988
Ute Mountain Ute, Southern Ute	1988
Fort McDowell–Yavapai Nation	1990
Northern Ute	1992
San Carlos Apache	1992
Jicarilla Apache	1992
Yavapai-Prescott	1994
Zuni	2003
Gila River Indian Community	2004
White Mountain Apache	2009
Navajo	2009 and ongoing
Hualapai	Ongoing

Carnegie Corporation, this group provided assistance in a growing number of water rights adjudications.[45]

Eventually, ten nations within the Colorado River basin received or negotiated for priority rights to Colorado River water: the Chemehuevi, Cocopah, Colorado River Indian Community, Fort Mojave, Jicarilla Apache, Navajo Nation, Northern Ute, Quechan (Yuma), Southern Ute, and Ute Mountain Ute communities. The Navajo continue to negotiate with Arizona for what they consider their full water rights. The Gila River Indian Community should also be included if one considers all major tributaries of the Colorado. One could also include the Tohono O'odham, Ak-Chin, Salt River Pima–Maricopa, Fort McDowell, San Carlos Apache, Hualapai, Yavapai-Prescott, White Mountain Apache, and Zuni communities, since most or part of their water claims are met either by tributaries of the Colorado or by main-stream water transported through CAP and other channels.[46]

While some of the settlements specified shares of Colorado River water out of its main stream, others, such as the Gila River Indian Community and Tohono O'odham (Papago), received shares of water from CAP before it was even complete. Passed in 1980, the Arizona Groundwater Management Act included quantifications of CAP water for various Arizona tribes. Although the state was reluctant to bargain away so much CAP water, such agreements were deemed necessary to obtain funding for completion of the project pipeline all the way to Tucson. The Tohono O'odham settlement in 1982 furthered this aim, granting the tribe at least 66,000 acre-feet of water annually from the yet-to-be-finished project. Throughout the 1990s, many other tribes sat down at the negotiating table and settled on water rights. The Ute Nation finally saw completion of most of the Animas–La Plata Project in Colorado by mid-2011. The state of Colorado needs to pay the Bureau of Reclamation approximately $36 million for its 10,460-acre-foot share of water from the project. The Ute community suggested that the state allow the water to revert back to the tribe, and the tribe would then sell the water back to the state at a much lower cost than the $36 million debt repayment. In late January 2012, however, the Colorado Water Conservation Board ended the nearly two-year discussion with the tribe and began talks with the bureau. Too many legislators feared future complications if they purchased water from the Ute community.[47]

The largest water settlement in US history occurred in 2004, when the Gila River Indian Community finally received confirmed water rights after more than a century of struggle. The settlement granted the Gila and Tohono O'odham more than 650,000 acre-feet of water annually from CAP, almost twice the amount reserved for the state of Nevada. This water represents 47 percent of CAP's annual flow. The Gila Indians also obtained rights to water from the Gila, Salt, and Verde Rivers—which, when combined with their share of CAP, totaled more than 650,000 acre-feet annually. Earlier, in 1995, plans were laid for the Pima-Maricopa Irrigation Project, meant to irrigate 146,330 acres throughout the 2,400-mile system.[48] Now, with an assured supply of water, the reservation is able to grow produce to sell as well as provide a better variety of fresh foods for the community. A clinical study conducted from 1987 to 2000 diagnosed the Gila River Indian Community with one of the highest rates of type II diabetes in the world.[49] They are hoping better access to good foods will help.[50]

Although not fully implemented until 2008, the resulting Arizona Water Rights Settlement Act finally settled water claims for the Gila River Indian Community and the Tohono O'odham Nation.[51] This largest water settlement in US history received harsh criticism from non-Indians in Arizona. Retired journalist Earl Zarbin voiced the opinions of many in an interview with the *Arizona Republic* in November 2004:

> Why 1 percent of our population should be able to control that much water is beyond reason or comprehension. It sets up a mechanism for these reservations to control the future for Arizona's population growth. These Indian tribes are under no compulsion to lease water to the non-Indians. They can either lease or not lease.[52]

In spite of complaints, the water settlement put to rest the longstanding uncertainty over CAP water and Indian claims to it. In 2009 the White Mountain Apache also signed a settlement agreement for 99,000 acre-feet annually. Control of water has shifted to new players.

The largest claim, however, remains only partially resolved—the Navajo (Diné) water claim. Starting in the 1970s, Navajo tribal chairman Peter MacDonald stirred up controversy both inside and outside the nation by asserting legal claim to all water in the Colorado River basin. In fact, the Navajo Reservation constitutes 25,000 square miles (more than 18 million acres) and rests entirely within the Colorado River watershed. Two tributaries also pass through the reservation: the Little Colorado and the San Juan River. When quantifying what the Navajo should be legally entitled to under the Winters Doctrine, Stanley Pollack (assistant attorney general for the tribe) estimated that it would not be less than 5 MAF of water per year.[53] This amount is more Colorado River water than the entire state of Arizona receives (2.8 MAF) and more than even California's share (4.4 MAF). According to a stipulation in *Arizona v. California* that makes a state responsible for meeting the water rights of Indian communities residing in that state, New Mexico, Utah, and Arizona will have to apportion water to the Navajo Nation from their shares.

When MacDonald first hired water engineers in the late 1970s to assess how much water the Navajo might legally claim, he directed them to measure water in the ancient homeland between the four sacred mountains, nearly doubling the actual size of the reservation. That measurement led MacDonald to conclude that the Navajo had claim to "every drop of the water . . . used by New Mexico, Arizona, California, Nevada, Utah, Colorado, and Wyoming" out of the Colorado River watershed.[54] Of course, no basin state would ever accept a claim that size. MacDonald was prepared to sue for Navajo water rights in court, but in late 1992 he was sent to prison for having committed federal crimes, including extortion and racketeering. Although his sentence was commuted in 1995, MacDonald's days in power were over. Since then, Stanley Pollack and tribal president Joe Shirley have led the Navajo water fight.

In 1958, as part of the Colorado River Storage Project, the Bureau of Reclamation began construction of Navajo Dam on the San Juan River in New Mexico. Completed in 1962, this dam and future pipelines and other structures were intended to water the Navajo Indian Irrigation Project. During the early 1970s, some of the necessary structures were built, but huge delays plagued the project and some vital parts were not completed until 2003.

Part of the delay was caused by the settlement of water claims with the state of New Mexico. MacDonald continued to urge the tribe to hold firm on water rights, in spite of his legal troubles and the end of his political career. In 2004 he argued to the press that the federal government and the nation's legal advisers were simply trying to cheat the Navajo out of water. He further argued that the promise of federal dollars to build infrastructure was a clever way to force the Navajo to give up their rights under the Winters Doctrine:

> That is still the law of the land. Obviously it is not favoring states at all. The Winters Doctrine says Indians have senior and prior rights regarding water. Indians have the right to as much water as they can put to beneficial use . . . The reason the federal government and states do not want Indian tribes to take advantage of the Winters Doctrine is because it gives tribes first cut of the water they need for tribal purposes. Whatever is left is to be given to the settlers and latecomers. According to the Winters Doctrine, the states are the latecomers.[55]

In spite of MacDonald's words of warning, the tribe voted to sign the agreement with New Mexico on April 19, 2005.[56] According to apportionment agreements in the upper basin of the Colorado River, New Mexico has rights to approximately 700,000 acre-feet annually, or 11.25 percent of the upper basin's 7.5 MAF allocated by the 1922 Colorado River Compact. The 2005 agreement allocated the Navajo 325,670 acre-feet of water annually from the San Juan River. Without federal funding, however, no infrastructure construction was possible to carry the water to where it was needed.

Before any congressional support could be gained, those opposed to the agreement publicized their anger and outrage. One Navajo grassroots leader, Ron Milford of Concerned Citizens for Diné Water, argued that Pollack, Lena Fowler, and others had negotiated away far too much of the San Juan River. He also accused Pollack of misleading the tribe—by first telling them they would receive half of the water in the river (approximately 750,000 acre-feet) but then signing an agreement for only half that amount.[57] Both Milford and MacDonald hoped the water agreements would bring much more—enough water to sell on the market and help reverse the terrible poverty and unemployment on the reservation.

Part of the settlement was a provision to construct a Navajo-Gallup Water Supply Project. This stipulation allocated 7,500 acre-feet of water per year to the city of Gallup, long the site where Navajo people spend hours in line with trucks of empty drums to fill with water and haul home. The significance of this water settlement cannot be overstated. As of 2009, the poverty rate in the Navajo Nation was greater than 50 percent. Even more surprising in the United States, between 30 and 40 percent of houses on the reservation do not have running water. Once the project has been completed, two pipelines will carry water from the river south. One will go to Gallup, with a spur west to the Navajo capital, Window Rock, and the second pipeline will carry water farther to the east to the Jicarilla Apache Nation.[58]

Although the agreement was made in 2005, it did not receive necessary funding until congressional maneuvering resulted in a US Senate bill that allocated the needed money over a fourteen-year period. Even though there were critics, the Bureau of Reclamation conducted a study in 2007 to see if there was enough water for these allocations so water users in the rest of the upper basin would not be threatened. In its Hydrologic Determination, the bureau asserted that it was "likely that sufficient water [would] be available from the Navajo Reservoir water supply through at least 2060."[59] In 2009, President Barack Obama signed the bill and appropriated $3 million for the first phase. It is expected that the completed projects will cost more than $800 million.[60]

In December 2010, Navajo Nation president Joe Shirley and Secretary of the Interior Ken Salazar signed a final agreement, but detractors almost immediately began gathering forces for a challenge. The New Mexico cities of Aztec, Farmington, and Bloomfield hired an attorney and protested the settlement because they feared losing water for their own irrigation needs. Some of these suits continued in early 2012. In late 2011 the Navajo were close to arriving at a settlement with Utah over 81,500 acre-feet of water, with the possibility of another 314,851 acre-feet per year if there is no shortage.[61] Some people are also upset because they think the Navajo will sell the water outside the state. Although there are no specific plans for such a sale, lawyers argue that they have the right to sell the water to anyone at any time.[62]

It is unlikely, however, that much of this water will be sold in the near future. The need for water is too great on the Navajo Reservation alone. Even those in rural areas who have wells are forced to haul water because of unsafe levels of uranium and arsenic. In 2011 the University of Arizona's Institute of the Environment conducted a study that discovered the unsafe levels in tribal groundwater. Both contaminants are known to pose a high risk of cancer. There are treatment methods to remove the contaminants, but they are expensive and require large amounts of electricity to operate. To solve this problem, project participants planned a system that would run on solar power alone and remove contaminants.[63] The need for water purification systems is very strong. According to an ongoing study conducted by Northwestern University in Evanston, Illinois, Uranium 234 was found in many of the samples at a much higher level than is deemed safe. Other contaminants found in unsafe amounts include arsenic, lead, Thorium 230, and other radioactive metals.[64]

The rest of the Navajo water claim—Colorado River water from Arizona's apportionment—remains unsettled. In 2003 the Navajo Nation filed suit against the states and the US secretary of the interior to stop allocations of water from CAP.[65] The Gila settlement was finalized the next year, in spite of Navajo protests. After the nation signed agreements with New Mexico, Arizona began to realize it was only a matter of time before the state would be forced back to the bargaining table. In late 2010 a proposal for a settlement was sent to the Navajo Tribal Council, which tabled the bill until it could hold public hearings. The proposal gave the nation 31,000 acre-feet of water per year from the Colorado River and any unappropriated amounts from the Little Colorado River, which runs through the reservation. The proposal would also end a lawsuit the Navajo filed against the Department of the Interior in 2003 for mismanaging tribal resources.[66]

Critics of the proposed settlement argued that such an agreement was giving away water to which the Navajo had rights under the Winters Doctrine. But Lena Fowler, a member of the community's Water Rights Commission, supported the proposal. A few years earlier, in 2008, Fowler had voiced her frustration in a newspaper article when she clearly stated what at least some of her fellow tribal members were thinking: that the grandiose claim for the amounts of water MacDonald had originally sued for was foolish. "Let's say we claim all of that 100 percent," she said. "Now where are we going to get the money to put our water to use? That's what a settlement does. When you negotiate, that's what you are negotiating for." In the same article in the *High Country News*, journalist Matt Jenkins wrote, "The Navajos would rather have 100 percent of nothing than 50 percent of something."[67]

Indeed, this "something" is what most negotiated water settlements have procured: infrastructure such as dams, reservoirs, and pipelines. Some community members are still pushing for an amount much closer to the 5 MAF MacDonald proposed back in the 1970s, but in November 2010 the Navajo Tribal Council agreed to the much smaller package offered by Arizona. In addition to the 31,000 acre-feet from the Colorado and approximately 161,000 acre-feet from the Little Colorado each year, the tribe would also receive $693 million for construction of pipelines.[68] One stipulation, however, is that the tribe would not be allowed to lease or sell groundwater or water from the Little Colorado. This limit prevents the Navajo Nation from profiting from the water settlement in a way those such as Milford and MacDonald had hoped. The tribe was less than enthusiastic about the deal.[69]

In August 2011 the previously agreed-to settlement encountered more roadblocks. Arizona senator Jon Kyl feared the expensive deal would not make it through the US Congress, so he asked for removal of the $515,000 pipeline that would deliver water to Navajo and Hopi homes. The focus then moved from the main Colorado River and pipeline to a water settlement from the Little Colorado River. The proposed deal gives the Navajo any unclaimed water in the tributary and "nearly unlimited access" to groundwater under the reservation.[70] Senator Kyl hoped to complete the Arizona settlement with the Navajo by the end of 2012, since he planned to retire and not run for reelection. On February 14, 2012, Kyl and fellow Arizona senator John McCain introduced the refined proposal called the Navajo-Hopi Little Colorado River Water Rights Settlement Act.[71] The bill waved any Navajo and Hopi claims to water in the Little Colorado River in exchange for water delivery projects that would bring potable groundwater to the two reservations. The bill also authorized water deliveries of 6,411 acre-feet annually to the eastern part of the Navajo Nation through the Navajo-Gallup pipeline. Most significant for Arizonans, it also dismissed other claims against the United States "regarding management of the Lower Colorado River."[72] Although many Navajo and Hopi supported the bill, others called it blackmail because it tied the water from CAP to the extension of land and coal leases for the Navajo Generating Station.[73] By August, the bill had collapsed and Senators Kyl and McCain publicly lamented the failure.[74] In late 2012, Interior Secretary Salazar invited tribal members to continue talking about a water settlement.[75] Amounts of Navajo and Hopi water rights to both the Colorado River and the Little Colorado River remained unquantified by late 2012.

Indian Water Marketing

One of the most controversial issues surrounding Indian water rights in recent decades is water marketing. Ever since the *Winters* case in 1908, states and courts have been trying to decide whether Indians can be restricted from marketing their water. As early as 1955, the US Congress agreed that when tribes leased land to non-Indians, they included water rights.[76] Such leases accounted for an estimated 400,000 acre-feet of water in the Colorado River basin in 1988. When a portion of *Arizona v. California* was decided in 1963, the principle of PIA was meant only to help quantify water rights for Indian reservations. To prevent misinterpretation, the Supplemental Decree for Arizona (1979) specified that PIA "shall not constitute a restriction of usage of them [water rights] to irrigation or other agricultural application."[77]

Since the late 1980s, tribes have been looking at the possibilities of selling water and profiting from their water rights while scholars debate the legalities of their doing so.[78] Legal scholar Christine Lichtenfeld argued in 1989 that states had no jurisdiction over how Indian communities used their water rights, calling any attempt to do so "paternalism."[79] Carefully laying out each related court interpretation since 1908, she asserted that allowing Indian nations to sell or lease their water rights was legally protected because any kind of use of the water enabled the group to develop self-sufficiency for its "permanent homeland."[80] She further argued that if a state was worried about water leaving the state, it should purchase the water at fair market value. Thus the water would be valuable to both the state and the tribe; if Indian nations were able to place a value on their water, they might be more likely to conserve it as well.

Other scholars argued similar points, asserting that preventing Indian nations from marketing their water was an infringement on their sovereignty.[81] During the 1990s, some scholars argued that federal law enabled Indian water marketing and that states could not prohibit it, in spite of the McCarran Amendment.[82] In 2001, scholars Laura Kirwan and Daniel McCool argued that water marketing without land attachments was difficult because of state laws against such marketing. They asserted, however, that allowing such marketing would be very beneficial to both Indians and non-Indians. Communities would gain much-needed revenue to improve the economies of their people, nearby cities would find another important source of water without building new dams or reservoirs, and environmentalists would be happy to avoid having any new projects built.[83]

In almost all Indian water rights adjudications, however, the states have managed to place at least some restrictions on water marketing. In Arizona, the Ak-Chin, Fort McDowell, Salt River Pima–Maricopa, San Carlos Apache, and Tohono O'odham communities have all been allowed to market water, but only within certain specified areas.[84] In the Colorado Ute Water Settlement in 1988, the right to sell or lease water is specifically mentioned and protected, but it is limited to sales in the upper basin only.[85] Although the right to market water was originally part of the act and was fairly unrestricted, lower basin states protested, fearing that water sold out of the upper basin would impact shares in the lower states.[86]

Although many questions have been raised about whether tribes have the right to market their water out of state, they have been selling water within states to interested buyers. As yet, there are no firm data on how much water has been marketed this way or how much money has been earned, but it is undoubtedly going to be a significant issue in the coming years. By March 2011, various tribes had signed either long- or short-term leases for more than 1.3 MAF in the Colorado River watershed alone.[87] There have also been many water transfers within states. San Diego, Los Angeles, and Phoenix regularly purchase water rights from area farmers. States in the lower basin of the Colorado are also making agreements to exchange and store water for each other.

It is a natural step, then, for Indian nations to begin to profit from their water rights. An organization with offices in Maryland and Phoenix, Native American Water Management, LLC, concentrates solely on marketing Indian water. This company helps negotiate water agreements, measures water and analyzes potential uses, and conducts sales for tribes, charging the end user a fee for services. At essentially no cost, Indian tribes can finally profit from their huge water rights negotiations, regardless of whether they have managed to build water projects and pipelines.[88] As more marketing of water occurs and urban areas seek new sources, it is inevitable that city leaders will look to Indian water to fill the gaps in supply.

An unexpected challenge to Indian water rights, however, has been the impact of the environmental movement in the United States. The passage of the National Environmental Policy Act (NEPA) in 1969 and the establishment of the Environmental Protection Agency (EPA) in 1970 did not at first seem problematic for Indian nations along the Colorado. These acts, followed by clean air and water acts and the Endangered Species Act in 1973, seemed to promise better protection of natural resources. Most believed better protection of nature would benefit Indian nations, and, in fact, these laws did help some groups preserve wilderness areas against development. The great irony, however, is that the environmental movement has also meant that water development projects have often slowed down just as Indian communities managed to obtain the rights to develop water resources.[89]

During the dam-building and irrigation project era, non-Indian groups throughout the western states benefited from the generation of cheap electric power, access to irrigation for agriculture, and assured access to water for industry and urban growth. Indian reservations waited and waited for their turn as they sought redress through the courts, but they received no financial help for development. Once they began to negotiate with state governments, the infrastructure they bargained for ran into the formidable roadblocks put up by the EPA. Indian communities were told that dams posed serious threats to endangered fish and that irrigation projects spelled doom for protected river and marshland wildlife.[90]

In many ways, this new attention to the environment placed what some have called an "unfair burden" on Indian tribes and their lands.[91] Although most of the main stream of the Colorado River was thoroughly developed before the passage of NEPA in 1969, many of the Indians' water developments involve undeveloped tributaries. Thus non-Indian waters in the Colorado River basin do not have to conform to the act's standards, while new developments—mostly Indian—do. These regulations have slowed and even altered and reduced water development projects for the Colorado Ute, the Jicarilla Apache, and the Navajo Nation, among others.[92]

One of the best recent examples of the stresses environmental issues have placed on indigenous water projects is the Animas–La Plata Project in southwestern Colorado. First authorized in 1968, the project languished as environmental and cost concerns were voiced. Its construction was necessary to fulfill the terms of the 1988 Colorado Ute Water Settlement, and area farmers needed irrigation water. However, even those who would benefit the most from the project were divided over the issue. When the final environmental impact statement was finished in 1980, the project was declared safe, even though various considerations for fish passage needed to be added. Ute members who supported the project were frustrated by the twelve-year delay.[93]

In spite of EPA approval, getting construction of the dam and reservoir under way proved difficult. The Animas River is one of the last free-flowing rivers in the Rocky Mountains, and many environmental groups have protested the plan. The Sierra Club worked to stop the project in 1992, and in 1993 a group of Ute members formed the Southern Ute Grassroots Organization to fight it. Sage Remmington, an important spokesman for the organization, argued that the project would mostly provide water to non-Indian land developers, who would cause even more environmental damage in the region. The reservoir would also flood a wildlife preserve. Remmington argued that the "arrogance that exists in Western water law . . . has to stop."[94] Other tribal members supported the project, including Leonard Burch, who was tribal chair during most of the period 1966 to 2002. He and others on the tribal council believed the water would help bring prosperity to the tribe. By 2005, approvals were in place, and blasting at the reservoir site had started.[95]

Although indigenous voices are faint and usually ignored across the border in the Colorado River delta (see chapter 7), Indian communities within the river basin inside the United States have gained power. Indian leaders finally achieved a partial victory in 1908, yet many decades passed before real, or "wet," water actually flowed onto the reservations. As these new players began settling their claims, power over Colorado River water shifted from white settlers to state governments and finally to Indian reservations. As climate change and ongoing drought create challenges for the river basin, native water may possibly bring unexpected power to historically marginalized people.

NOTES

1. See Lynn R. Bailey, *Bosque Redondo: An American Concentration Camp* (Pasadena: Socio-Technical Books, 1970); Frank McNitt, *Navajo Wars* (Albuquerque: University of New Mexico Press, 1972); Raymond Bial, *Great Journeys: The Long Walk—The Story of Navajo Captivity* (New York: Benchmark Books, 2003).

2. Monroe E. Price and Gary D. Weatherford, "Indian Water Rights in Theory and Practice: Navajo Experience in the Colorado River Basin," *Law and Contemporary Problems* 40, no. 1 (Winter 1976): 97–131; Robert H. Abrams, "Reserved Water Rights, Indian Rights and the Narrowing Scope of Federal Jurisdiction: The Colorado River Decision," *Stanford Law Review* 30, no. 6 (July 1978): 1111–48; Ann Caylor, "'A Promise Long Deferred': Federal Reclamation on the Colorado River Indian Reservation," *Pacific Historical Review* 69, no. 2 (May 2000): 193–215.

3. See Charles W. Baley, *Disaster at the Colorado: Beale's Wagon Road and the First Emigrant Party* (Logan: Utah State University Press, 2002).

4. For an early analysis of Mohave migration and homeland, see Kenneth M. Stewart, "The Aboriginal Territory of the Mohave Indians," *Ethnohistory* [Columbus, OH] 16, no. 3 (Summer 1969): 257–76. See also Alfred Louis Kroeber and Clifton B. Kroeber, *A Mohave War Reminiscence, 1854–1880* (New York: Dover, 1994).

5. Caylor, "Promise Long Deferred," 198.

6. For early accounts of Rockwood's disaster, see George Wharton James, *The Wonders of the Colorado Desert* (Boston: Little, Brown, 1911); Frank Waters, *Colorado* (New York: Rinehart, 1946).

7. Caylor, "Promise Long Deferred," 211.

8. See David H. Dejong, "Forced to Abandon Their Farms: Water Deprivation and Starvation among the Gila River Pima, 1892–1904," *American Indian Culture and Research Journal* 28, no. 3 (2004): 29–56. Ranchers who thought the banks of the Gila River were inviting settled in the Florence area. The discovery of silver nearby and the development of the Silver King Mine created a boomtown during the 1880s. See Jack San Felice, *When Silver Was King: Arizona's Famous 1880's Silver King Mine* (Mesa, AZ: Millsite Canyon, 2006).

9. US Census Bureau, Arizona, 2000, http://www.census.gov/dmd/www/resapport /states/arizona.pdf (accessed January 28, 2012).

10. Dejong, "Forced to Abandon Their Farms," 40.

11. Report in the Matter of the Investigation of the Salt and Gila Rivers—Reservations and Reclamation Service, 62nd Cong., House of Representatives, 3rd sess., 1912–13 (Washington, DC: Government Printing Office, 1913), 5–7, http://water.library.arizona.edu /body.1_div.39.html (accessed January 28, 2012).

12. See David H. Dejong, "'Abandoned Little by Little': The 1914 Pima Adjudication Survey, Water Deprivation and Farming on the Pima Reservation," *Agricultural History* 81, no. 1 (Winter 2007): 36–69.

13. The Reclamation Service was renamed the Bureau of Reclamation in 1923.

14. For the most thorough history of the early years of the Bureau of Reclamation, see Donald J. Pisani, *Water and American Government: The Reclamation Bureau, National Water Policy, and the West, 1902–1935* (Berkeley: University of California Press, 2002).

15. Daniel McCool, "Precedent for the Winters Doctrine: Seven Legal Principles," *Journal of the Southwest* 29, no. 2 (Summer 1987): 164–78; McCool, "Winters Comes Home to Roost," in Char Miller, ed., *Fluid Arguments: Five Centuries of Western Water Conflict* (Tucson: University of Arizona Press, 2001), 120–38. See also Norris Hundley Jr., "The Dark and Bloody Ground of Indian Water Rights: Confusion Elevated to Principle," *Western Historical Quarterly* 9, no. 4 (October 1978): 454–82.

16. Pisani, *Water and American Government,* 164–66.

17. *Winters v. United States,* 207 US 546 (1908), http://caselaw.lp.findlaw.com/cgi-bin /getcase.pl?court=us&vol=207&invol=564 (accessed January 28, 2012). See also John

Shurts, *Indian Reserved Water Rights: The Winters Doctrine in Its Social and Legal Context, 1880s–1930s* (Norman: University of Oklahoma Press, 2000).

18. McCool, "Winters Comes Home to Roost," 122.

19. Gary Weatherford, Mary Wallace, and Lee Herold Storey, *Leasing Indian Water: Choices in the Colorado River Basin* (Washington, DC: Conservation Foundation and John Muir Institute, 1988), 19–20; David Getches, "Defending Indigenous Water Rights with the Laws of a Dominant Culture," in Dik Roth, Rutgerd Boelens, and Margreet Zwarteveen, eds., *Liquid Relations: Contested Water Rights and Legal Complexity* (Piscataway, NJ: Rutgers University Press, 2005), 44–65.

20. Pisani, *Water and American Government*, 154.

21. Ibid., 155.

22. Colorado River Compact, 1922, Article VII, Colorado River/Central Arizona Project Collection, Hayden Library, Arizona State University, Tempe.

23. San Carlos Irrigation Project, Bill S. 966, May 1, 1924, House of Representatives, 68th Cong., 1st sess., Report no. 618 (Washington, DC: Government Printing Office, 1924).

24. Pisani, *Water and American Government*, 168.

25. For a recent discussion of reserved water rights, see Nathan Brooks, "Indian Reserved Water Rights: An Overview," CRS Report for Congress (Washington, DC: Congressional Research Service, January 24, 2005), http://www.policyarchive.org/handle/10207/bitstreams /1917.pdf (accessed January 28, 2012).

26. Department of Justice, Appropriation Act of July 1952, 66 Stat. 560, http://www.uspto. gov/web/trademarks/PL107_273.pdf (accessed January 28, 2012).

27. For a full description of the intent of the McCarran Amendment, see Senate Report no. 755, 82nd Cong., 1st sess. (Washington, DC: Government Printing Office, 1951), 5–6.

28. Upper Colorado River Basin Compact, 1948, http://www.usbr.gov/lc/region/pao /pdfiles/ucbsnact.pdf (accessed January 28, 2012).

29. Russell Martin, *A Story That Stands Like a Dam: Glen Canyon and the Struggle for the Soul of the West* (Salt Lake City: University of Utah Press, 1999 [1989]), 251–52.

30. *Arizona v. California,* 376 US 340 (1964).

31. Dale Pontius, principal investigator, *Colorado River Basin Study for the Western States Water Policy Review Advisory Commission* (Tucson: SWCA, Inc., Environmental Consultants, March 1997), http://wwa.colorado.edu/colorado_river/docs/pontius%20 colorado.pdf (accessed February 28, 2012).

32. Ibid., 596–601.

33. Jack L. August Jr., *Dividing Western Waters: Mark Wilmer and* Arizona v. California (Fort Worth: Texas Christian University Press, 2007), 111–12; Elizabeth Weldon, "Practically Irrigable Acreage Standard: A Poor Partner for the West's Water Future," *William and Mary Environmental Law and Policy Review* 25, no. 1 (2000): 203–31.

34. Marc Reisner, *Cadillac Desert: The American West and Its Disappearing Water* (New York: Penguin Books, 1993 [1986]), 287.

35. Byron E. Pearson, "We Have Almost Forgotten How to Hope: The Hualapai, the Navajo, and the Fight for the Central Arizona Project, 1944–1968," *Western Historical Quarterly* 31, no. 3 (Autumn 2000): 297–316.

36. Paul Stuart, "United States Indian Policy: From the Dawes Act to the American Indian Policy Review Commission," *Social Service Review* 51, no. 3 (September 1977): 451–63. See also Donald L. Fixico, *Termination and Relocation: Federal Indian Policy, 1945–1960* (Albuquerque: University of New Mexico Press, 1986).

37. Statement of Hualapai Indians of Arizona on Bridge Canyon Bill (S. 75), June 2, 1949, Box 17, Folder 8, CM MSS-87, Colorado River/Central Arizona Project Collection, Hayden Library, Arizona State University, Tempe.

38. Hualapai Contract, August 30, 1960, Box 8, Folder 4, Carl T. Hayden Papers, Special Collections, Hayden Library, Arizona State University, Tempe.

39. Pearson, "We Have Almost Forgotten How to Hope," 305.

40. Robert Dean, "Dam Building Still Had Some Magic Then: Stewart Udall, the Central Arizona Project, and the Evolution of the Pacific Southwest Water Plan, 1963–1968," *Pacific Historical Review* 66, no. 1 (February 1997): 92.

41. Letter from Acting Commissioner of Reclamation to Morris Udall, October 12, 1965, Box 477, Folder 5, MS 325, Morris K. Udall Papers, University of Arizona Library, Special Collections, University of Arizona, Tucson. See also Pearson, "We Have Almost Forgotten How to Hope," 307.

42. Pearson, "We Have Almost Forgotten How to Hope," 308.

43. Apparently, Stewart Udall was angry with Navajo chief legal counsel Normal Littell and the advice he was giving the tribe regarding plans for Marble Canyon Dam and CAP. A story in the *Hartford Courant* reported that Barry Goldwater was telling Udall to leave the Navajos alone, that they could choose whomever they wanted as lead counsel. Udall had taken Littell to court to try to remove him from his post. "Barry Raps Udall on Navajo Move," *Hartford* [CT] *Courant,* October 14, 1966.

44. See the NARF website for information on the organization; http://www.narf.org /about/about_whatwedo.html (accessed January 28, 2012).

45. Jon C. Hare, "Indian Water Rights: An Analysis of Current and Pending Indian Water Rights Settlements," 1996, Dividing the Waters, Box 6, Folder A-3-021, Water Resource Center Archives, University of California, Berkeley.

46. As water rights settlements proceeded throughout the 1980s and 1990s, many argued that such agreements must be concluded without taking water away from non-Indians. See Reid Payton Chambers and John E. Echohawk, "Implementing Winters Doctrine Indian Reserved Water Rights: Producing Indian Water and Economic Development without

Injuring Non-Indian Water Users?" Western Water Policy Project, 1991, Dividing the Waters, Box 6, Folder A-3-014, Water Resource Center Archives, University of California, Berkeley.

47. Catherine Tsai, "Tribes Talk with Colorado on Animas–La Plata Water," *Native American Times* [Tahlequah, OK], January 29, 2012, http://www.nativetimes.com/news /environment/6747-tribes-talk-with-colorado-on-animas-la-plata-water (accessed January 29, 2012). For a complete list of Indian water rights settlements up to 1995, see Joan Specking, Reserved Water Rights Compact Commission, July 1995, Dividing the Waters Collection, Box 5, Folder A-2-071, Water Resource Center Archives, University of California, Berkeley.

48. Gila River Community, Pima-Maricopa Irrigation Project Vision Statement, 2001, http://www.gilariver.com/vstate.htm (accessed January 29, 2012).

49. Andrea M. Kriska et al., "Physical Activity, Obesity, and the Incidence of Type II Diabetes in a High-Risk Population," *American Journal of Epidemiology* 158, no. 7 (2003): 669–75.

50. National Diabetes Information Clearinghouse, the Pima Indians, Genetic Research, May 2002, http://diabetes.niddk.nih.gov/dm/pubs/pima/genetic/genetic.htm (accessed January 29, 2012).

51. Public Law 108-451, December 10, 2004, http://www.gpo.gov/fdsys/pkg/PLAW -108publ451/pdf/PLAW-108publ451.pdf (accessed January 29, 2012).

52. Shaun McKinnon and Billy House, "Congress OK's Water Settlement Empowering Tribes," *Arizona Republic* [Phoenix], November 18, 2004.

53. Joe Gelt, "Indian Water Rights," *Arroyo* 10, no. 1 (August 1997): 6.

54. Matt Jenkins, "The Colorado River's Sleeping Giant Stirs; Navajo Nation Wants Its Long-Overdue Cut of the River," *High Country News* [Paonia, CO], April 28, 2003; Jenkins, "Seeking the Water Jackpot," *High Country News,* March 17, 2008; Bettina Boxall, "Navajo Sue for River Water: Legal Move Asserts Tribal Rights and Seeks to Set Aside Federal Guidelines in Allotting Colorado Surplus to Agencies in the Southwest," *Los Angeles Times*, May 26, 2003.

55. Brenda Norrell, "Former Chairman Urges Navajo to Fight New Mexico Water Rights Settlement," *Indian Country Today* [New York City], December 23, 2004, http://www.enn .com/top_stories/article/649 (accessed February 28, 2012).

56. San Juan River Basin in New Mexico Navajo Water Rights Settlement, April 19, 2005, http://www.ose.state.nm.us/water-info/NavajoSettlement/NavajoExecutiveSummary.pdf (accessed January 29, 2012).

57. Max Goldtooth, Peter MacDonald, and Ron Milford, "Navajo Water Rights: Truths and Betrayals," *High Country News,* July 21, 2008, http://www.hcn.org/issues/40.13 /navajo-water-rights-truths-and-betrayals (accessed January 29, 2012).

58. Navajo-Gallup Water Supply Project, University of New Mexico School of Law, Utton Transboundary Resources Center, November 2009, http://uttoncenter.unm.edu/pdfs/WM _Navajo-Gallup_Project.pdf (accessed January 29, 2012).

59. Bureau of Reclamation, Hydrologic Determination for Navajo-Gallup Water Supply Project, April 2007, http://www.usbr.gov/uc/envdocs/eis/navgallup/FEIS/vol1/attach-N .pdf (accessed March 16, 2011).

60. "Navajos Elated with Water Settlement Bill," *Arizona Republic* [Phoenix], March 25, 2009. See also Bureau of Reclamation, Record of Decision for the Navajo-Gallup Water Supply Project, Final Environmental Impact Statement, September 2009, http://www.usbr .gov/uc/envdocs/eis/navgallup/FEIS/Navajo-GallupROD-Signed.pdf (accessed March 16, 2011).

61. Amy Joi O'Donoghue, "Navajo Water Rights Settlement with Utah Inching Closer," *Deseret News* [Salt Lake City], December 29, 2011, http://www.deseretnews.com/article /705396522/Navajo-water-rights-settlement-with-Utah-inching-closer.html?pg=2&s_cid =s10 (accessed January 29, 2012).

62. Jenny Kane, "Navajo Water Rights Dispute," *Daily Times* [Farmington, NM], February 6, 2011, http://www.krqe.com/dpp/news/navajo-water-rights-dispute (January 29, 2012).

63. Will Ferguson Wick, "UA Project Could Bring Clean Water to Navajos," *Green Valley* [AZ] *News and Sun,* March 12, 2011, http://www.gvnews.com/news/article_de7f4c86-452f -11e0-a4bf-001cc4c03286.html (January 29, 2012).

64. Navajo Nation Water Quality Project, http://navajowater.org/uranium234/ (March 16, 2011).

65. Bill Ibelle, "Attorney for Navajo Nation Sues Federal Government over Water Rights to Colorado River: Liquid Gold," *Lawyers USA*, June 2, 2008; Shaun McKinnon, "Navajos Sue U.S. over Water from the Colorado River," *Arizona Republic*, March 17, 2003.

66. "Navajo Lawmakers Table Proposed Water Settlement," Associated Press, September 30, 2010, http://azcapitoltimes.com/news/2010/09/30/navajo-lawmakers-table-proposed -water-settlement/ (accessed March 16, 2011).

67. Matt Jenkins, "Seeking the Water Jackpot," *High Country News*, March 17, 2008.

68. Matt Jenkins, "In Navajoland, a Contentious Water Deal Divides the Tribe," *High Country News,* February 21, 2011, http://www.hcn.org/issues/43.3/in-navajoland-a -contentious-water-deal-divides-the-tribe (accessed March 16, 2011).

69. Ibid.

70. Felicia Fonseca, "Navajos Focus on Little Colorado River Settlement," *Daily Courier* [Prescott, AZ], August 28, 2011, http://www.dcourier.com/Main.asp?SectionID =1&SubSectionID=1&ArticleID=97438 (accessed January 29, 2012). See also Dan Killoren, "Water Law: Racing an Arizona Senator's Retirement, Dry Navajo Nation Draws Closer to

Securing More Water," *Circle of Blue: Reporting the Global Water Crisis,* November 17, 2011, http://www.circleofblue.org/waternews/2011/world/water-law-racing-an-arizona-senators-retirement-dry-navajo-nation-draws-closer-to-securing-more-water/ (accessed January 29, 2011).

71. Text of S. 2109, Navajo-Hopi Little Colorado River Water Rights Settlement Act, February 14, 2012, http://www.gpo.gov/fdsys/pkg/BILLS-112s2109is/pdf/BILLS-112s2109is.pdf (accessed February 28, 2012).

72. Jon Kyl, United States Senator for Arizona, news release, February 14, 2012, http://kyl.senate.gov/record.cfm?id=335999 (accessed February 28, 2012).

73. Weatherford, Wallace, and Storey, *Leasing Indian Water,* 37.

74. Brett Walton, "Water Rights: Arizona Senators Jon Kyl and John McCain Meet with Navajo Nation Leaders," *Circle of Blue,* April 5, 2012, http://www.circleofblue.org/waternews/2012/world/water-rights-arizona-senators-john-kyle-and-john-mccain-meet-with-navajo-nation-leaders/ (accessed October 22, 2012).

75. Jon Kyl and John McCain, "An Endless Tribal Water Fight: Navajos, Hopis Opted to Let a Long-Sought Settlement Slip Away," *Arizona Republic,* AZ Central Online, August 13, 2012, http://www.azcentral.com/arizonarepublic/viewpoints/articles/2012/08/11/20120811an-endless-tribal-water-fight.html (accessed October 22, 2012).

76. Marley Shebala, "Interior Secretary Invites Tribal Officials D.C. to Renegotiate Water Rights Settlement," *Navajo Times,* October 18, 2012, http://www.navajotimes.com/politics/2012/1012/101812wat.php (accessed October 22, 2012).

77. Supplemental Decree, *Arizona v. California,* January 9, 1979, http://www.usbr.gov/lc/region/g1000/pdfiles/scsuppdc.pdf (accessed January 29, 2012).

78. Weatherford, Wallace, and Story, *Leasing Indian Water,* 51.

79. Christine Lichtenfeld, "Indian Reserved Water Rights: An Argument for the Right to Export and Sell," *Land and Water Review* 131 (1989): 146.

80. Ibid., 140.

81. Lee Herold Storey, "Leasing Indian Water off the Reservation: A Use Consistent with the Reservation's Purpose," *California Law Review* 76, no. 1 (January 1988): 179–220.

82. Chris Seldin, "Interstate Marketing of Indian Water Rights: The Impact of the Commerce Clause," *California Law Review* 87, no. 6 (December 1999): 1545–80.

83. Laura Kirwan and Daniel McCool, "Negotiated Water Settlements: Environmentalists and American Indians," in Richmond Clow and Imre Sutton, eds., *Trusteeship in Change: Toward Tribal Autonomy in Resource Management* (Boulder: University Press of Colorado, 2001), 273.

84. Mary McNally, "Water Marketing: The Case of Indian Reserved Rights," *Water Resources Bulletin* 30, no. 6 (December 1994): 967.

85. Public Law 100-585, Ute Colorado Water Settlement, November 3, 1988, http://www.usbr.gov/uc/progact/animas/pdfs/pl_100585.pdf (accessed January 29, 2012).

86. McNally, "Water Marketing," 968.

87. "Tribal Water Rights: The West's Last Big Bucket of Water," *American Water Intelligence* 2, no. 3 (March 2011), http://www.americanwaterintel.com/archive/2/3/opinion/tribal-water-rights-wests-last-big-bucket-water.html (accessed January 29, 2012).

88. Native American Water Management, LLC, http://nawater.net/ (accessed January 29, 2012).

89. Jerilyn DeCoteau, "The Effects of Development on Indian Water Rights: Obstacles and Disincentives to Development of Indian Water Rights," in *Water and Growth in the West: Natural Resources Law Center Symposium* (Boulder: University of Colorado School of Law, 2000), 1–17.

90. For more discussion, see Michael E. Harkin and David Rich Lewis, eds., *Native Americans and the Environment: Perspectives on the Ecological Indian* (Lincoln: University of Nebraska Press, 2007).

91. Kirwan and McCool, "Negotiated Water Settlements," 265–80.

92. Ibid., 269. See also Mik Moore, "Coalition Building between Native American and Environmental Organizations in Opposition to Development," *Organization and Environment* 11, no. 3 (1998): 287–313; Daniel McCool, "Indian Reservations: Environmental Refuge or Homeland?" *High Country News,* April 10, 2000.

93. Julia Dengal, director, *Cowboys, Indians, and Lawyers* (DVD), Bullfrog Films, 2005.

94. Ibid.

95. Joe Hanel, "Animas–La Plata Project: Colorado Lawmakers Give Initial Approval for State to Buy Project Water," *Cortez* [CO] *Journal,* April 17, 2010.

7

Crossing the Border

US-Mexico Relations and the River

"WHAT ABOUT THE HUMAN BEINGS?" Cucapá chief
Don Madaleno asked government officials and members of the environmental nonprofits who had gathered
in his small shantytown of El Mayor, Mexico. "We are
also endangered."[1] He and his indigenous community of
approximately 300 are the survivors of the Cucapá people
in the Colorado River delta region. Dependent on fishing
in the scarce waters of what was once a vibrant ecosystem,
the Cucapá, the "people of the river," struggle to maintain
their cultural identity amid a radically altered landscape.
Scientists and environmental activists organized a bi-
national partnership in 2002 to address declining wetlands
habitat in the delta in an effort to preserve endangered species of fish and birds. Don Madaleno argues that those who
have come to help "seem to care more about the fish than
the people."[2]

While almost unknown to most users of the Colorado
River, the Cucapá's current dilemma represents an often-
overlooked aspect of a new theory meant to govern

relationships between people and the environment. The concept of environmental justice, born in the early 1980s, is one that recognizes and seeks to address inequities in the distribution of environmental benefits and costs. Theorists and activists seek to address issues such as the degradation of the environment near marginalized people. Pollution and waste contaminants are most often found near the living spaces of ethnic minorities, those without strong political voices. Some of these problems are receiving redress in the twenty-first century through Superfund cleanup projects.[3]

In the case of the Colorado River delta, however, efforts to rejuvenate wetlands and protect wildlife are endangering the very basis of traditional ways of life for the Cucapá people. Environmental justice for these people requires a balance between wildlife preservation and the indigenous fishing culture. The two nations connected to the Colorado River will need to cooperate with each other and with the river itself. In the past, most international interaction regarding the river involved, first, reluctant sharing of resources, then US hegemony over the river to which it claimed full entitlement. Even though Mexico received a guaranteed share of the river through a 1944 treaty, US interests dominated any policies regarding the river's use. In more recent decades, water quality and quantity challenges have required a more cooperative relationship. Instead of a contest between two countries, the natural environment became a third contestant at the border, demonstrating both the importance of ecosystem health and the control the ecosystem places on the humans who interact with it.

Of the nearly 18 million Americans and Mexicans who depend on the Colorado River for water, few ever think about where the water comes from. River geoscientist Ellen Wohl laments that individuals are so removed from water sources that rivers become "virtual rivers."[4] The actual rivers are diverted, siphoned, and pumped through a labyrinth of tunnels and aqueducts, disconnecting them from ecosystems and peoples who once thrived on their waters. Environmentalists and scientists interested in these multiple problems are trying to reconnect people and wildlife to the Colorado River through restoration programs, and they have produced a significant body of research seeking solutions to declining marshes and fish populations.[5] In the midst of that ongoing conversation, the Cucapá Indians are mostly invisible; although Mexican environmental organizations are participating in the conversation, little attention is given to indigenous voices.[6] Madaleno's words ring true.

In the story of the Colorado River borderlands, multiple problems have altered relationships between people and water, as illustrated in this study. The most obvious issue today is quantity, but others, including water quality, continue to challenge river users. The environment itself is fraught with both of these vexing problems: how much water should be used to preserve ecosystems and at whose expense when water is scarce? At the start of the twentieth century, Mexico and the United States reluctantly shared the river's water while simultaneously seeking control. As American users gained power through dams and diversion projects, Mexico sought balance through treaties and other responses that demonstrated an intense nationalism. While bi-national cooperation could have been the most effective way to manage this shared watershed, separate interests and fears regarding the other country's motives have historically worked against such cooperation. However, in spite of national, regional, and local political conflicts, common interests have emerged often enough to maintain a relatively peaceful and, at times, willing partnership.

The Colorado River meanders through seven states in the very arid American West; its long story includes litigation, negotiation, and political battling among those claiming a share of its waters. Beginning as early as the 1850s, Wyoming, Colorado, Utah, Nevada, New Mexico, Arizona, and California fought to use every drop of water in the river's channels. Yet this overused water source is more than an American river; it is an international one, crossing the US-Mexico border between California/ Arizona and the Mexican states of Baja California and Sonora.[7] The river becomes the common life source among these four states, which meet to form the banks of the river at the border. The Colorado River delta—a collection of shallow channels, low wetlands, and broad mudflats—stands at the place where the river once made its final approach to the Gulf of California.

Reluctant Sharing

For a stretch of about twenty-four miles, the Colorado River serves as part of the US-Mexico border between Yuma and San Luis and finishes its last fifty-five miles in Mexico. Although the river has reached the sea only periodically since the early 1900s, its delta remains an important ecosystem of approximately 1,800 square miles. First drawn by the Treaty of Guadalupe Hidalgo in 1848 and later by the Gadsden Purchase in 1853, the demarcation line between the United States and Mexico extends from Texas to the Pacific, bifurcating indigenous peoples such as the Cucapá who lived in the delta region of Sonora, Mexico, and as far north as present-day Yuma, Arizona. These "people of the river" originally settled along this lower end of the Colorado River and met the Spanish when they first explored the region in the 1530s. From the early 1600s to the mid-1800s, the Spanish and later the independent nation of Mexico controlled the Colorado River delta region and the Cucapá. Mostly unsettled, the present-day Mexican states of Sonora to the east of the river and Baja California to the west remained fairly unpopulated until the 1880s. When some settlers began to farm in the region, the Mexican government encouraged farmers to settle near the US border to discourage American encroachment.[8]

Although new borders indicated which part of the region belonged to Mexico and which to the United States, not everyone thought American expansion should halt at these lines. The 1853–54 filibustering campaign of William Walker from Tennessee in Baja California is an example. Although Walker's attempt to take over the peninsula was not successful, the experience created longstanding Mexican mistrust of US intentions in the region. Historian Evan Ward argues that Walker's actions "foreshadowed the dynamics of conquest and intrigue that would characterize US-Mexican contests over land and water resources in the Colorado River Delta for the next century and a half."[9]

The first large-scale farming near the border occurred when American developer Charles Rockwood and the American Land Company brought water to the Imperial Valley in 1901. To introduce water to the rich valley soil, Rockwood collaborated with Mexican landowner Guillermo Andrade, who owned much of the land along the Colorado River and the US border. Once granted right-of-way, Rockwood and his investors constructed a channel that diverted some of the Colorado River into an old riverbed known as the Río Álamo. Flowing for some distance through Mexican territory, the new diversion brought water north into the Imperial Valley, encouraging agricultural investors. By 1905, more than 7,000 settlers were farming the valley, using water from the Colorado River after it passed through Mexico (see chapter 1).[10]

Although Rockwell's company obtained permission for passage through Andrade's private property before the water diversion began, Mexican president Porfirio Díaz feared the arrangement would eventually cause conflict. He was reluctant to force the issue, however, for fear that US retaliation might lead to the loss of Baja California. He lodged a mild protest to the US State Department, which investigated but took no action. Two treaties in 1848 and 1853 defined the borders and required freedom of navigation in all Colorado River boundary waters, but they made no mention of water use for irrigation or urban dwellers. As far as the US government was concerned, no violations of treaties had occurred.[11]

Díaz was, in fact, supportive of American agricultural interests in the region. He believed Mexico's best future required foreign investments so the country could advance. Although he did not wish to sacrifice Mexican sovereignty in the region, he welcomed American capitalists who would open opportunities for Mexican agriculture and business. His political liberalism thus made it possible for American investors Harry Chandler and Harrison Otis and their Colorado River Land Company (CRLC) to thrive in the region after buying out most of the water rights previously owned by Andrade.[12]

As Rockwood's main and bypass intake channels silted closed, in 1904 he asked permission to cut another temporary intake within Mexico until he could clear the others. Again, Mexico agreed, but as a condition it stipulated that up to 50 percent of the water that flowed through the canal should go to farmers on the Mexican side of the border. Díaz worried that Rockwood's diversion would eventually create a water shortage or even cause the river channel to shift. Indeed, such fears proved well founded when the new diversion gates also silted up and the river broke through in 1905, diverting its entire flow into the Imperial Valley (see chapter 1). Southern Pacific Railroad engineers finally managed to stop the flow in 1907.[13]

Another casualty of the flood proved to be the river delta itself. From 1905 to 1907, vegetation grew in the riverbed, preventing it from returning to its old channel. The river spread out over land to the west, not forming another real channel until major flooding in 1909 pushed the river's flow into Volcano Lake. The uncontrollable Colorado River became a meandering river, moving west and then toward the east over time.[14] Over the next few years, American engineers built levees and a dam along the Bee River to push the Colorado back to the east and away from Imperial Valley. In 1911 more flooding caused the dam to break, and the river flowed back naturally to the west into Volcano Lake. Because the land and the lake inside Mexico were higher than Imperial Valley, any excess floodwaters would threaten the US side, as they had in the 1905 break. The US Congress responded with money to help mend the dam and levees, but Mexico refused to help with the cost of repair and often charged high duties to transport necessary equipment across the border.[15]

When the Imperial Irrigation District (IID) was formed in 1911 to make water ownership in the valley public, many residents urged the new entity to build an "All-American Canal" that would flow completely inside the US border. That way, they would not be dependent on Mexico's cooperation. US farmers were angry because of Mexican farmers' refusal to help repair and maintain the canal while receiving half the benefits. If the canal ran completely inside the United States, Americans could monopolize all of its water without any requirement to share.

Ironically, the region containing most of the farmland on the Mexico side of the border actually belonged to Americans. Chandler and his father-in-law, Otis, owned more than 800,000 acres of land watered by the Río Álamo canal that flowed to the Imperial Valley. They had purchased this land from Guillermo Andrade in 1904 and, in turn, leased most of their holdings to Mexican developers.[16] These landowners subleased smaller farms to both Mexican and Asian farmers. Many Chinese and some Japanese arrived as tenant farmers in the Mexicali Valley, paying the IID for water and shipping their cotton to sell to specific US businesses controlled by the CRLC.

Soon, the Southern Pacific Railway built a hub to connect the Mexicali Valley to Los Angeles and the rest of the southern United States.[17]

As the CRLC flourished across the border using water operated by the IID, resentment grew on the US side. Numerous attempts to force Chandler and his company to share in canal maintenance costs failed, and the Mexican government refused to get involved lest it anger American landowners along the border. They employed many Mexican tenant farmers and migrant workers in the cotton fields and maintained at least a modicum of Mexican control of the borderland. Americans on the US side, however, were angered by the behavior of their fellow landholders in Mexico, seeing them as greedy and unpatriotic.[18] However, these men supported President William Howard Taft's "Dollar Diplomacy," which sought regional stability through US investment in foreign countries.[19]

Another crisis along the border developed during the Mexican Revolution in November 1910, and the political unrest continued for more than a decade. For a while, the revolutionary Flores Magón launched his bid for power from Baja California and threatened to gain control of the Colorado River and canals that watered the Mexicali and Imperial Valleys. This radical liberal and the political party he formed, the Partido Liberal Mexicano (PLM), were angry because of the uneven distribution of wealth and power in Mexico and the control foreign investors wielded over Mexican resources.[20] Mexican troops ended his rebellion in 1911, but Americans took notice. US officials threatened Mexico that if it did not protect the waterworks there, the United States would send in its own troops. Mexico did send troops and conditions quieted down, but border relations remained controversial as World War I began. By 1915, Chinese farmers represented 42 percent of the population of the Mexicali Valley, causing significant concern about a "Yellow Peril" across the border.[21]

Eventually of even more concern was the fact that about 9 percent of the population was Japanese. Although that percentage was fairly small, Americans worried about Japan's growing presence. That fear was heightened when Californians learned of a Japanese investor's bid to purchase all of the Colorado River Land Company's holdings from Chandler in 1909.[22] Although the sale did not go through, California newspapers soon reported the potential danger of a Japanese invasion through Baja California. Suddenly, the US-Mexico border along the Colorado River became vital to US security and a possible invasion zone. In many ways, the threat of an Asian invasion and frustrations with Mexico created congressional support in the West for the 1902 Reclamation Act and the 1928 Boulder Canyon Project Act, which authorized Hoover Dam and the construction of the All-American Canal from the Colorado River to the Imperial Valley.[23] In addition, Americans such as Philip David Swing, lawyer for the IID, and Arthur Powell Davis, director of the Bureau of Reclamation, believed complete development of the Colorado River was important for national security as well as for the western economy along the border. In fact, Americans in the Imperial Valley believed the only way to completely secure the necessary water resources for their enterprises was to control all of the Colorado River and perhaps to obtain Baja California from Mexico. The Mexican Revolution, however, left behind a strong legacy of nationalism.[24]

Competing Nationalism

First, however, in order to pave the way for construction of Boulder (Hoover) Dam and the All-American Canal, the seven river basin states had to decide how to share the water. When state officials finally negotiated the Colorado River Compact in 1922, the document made almost no mention of Mexico, and representatives from south of the border were specifically not invited to the negotiating table. The only statement acknowledging that Mexico might have any interest in the river's water was one Secretary of Commerce Herbert Hoover insisted be included. He believed that if the United States were to agree to any sharing of Colorado River water with Mexico in the future, that water would come mostly from surplus water not being used by the states. Hoover's statement did mention that if surplus water was not available to meet any theoretical future agreement, the upper and lower basin American states would equally share the burden.[25]

Most representatives in the Colorado River basin north of the border revealed very specific anti-Mexican attitudes by appropriating as much water as possible, as much to keep Mexico from using it as to fulfill their own perceived needs. California argued that the dam and canal were necessary for the security of water resources in the state, while Mexican and American landholders in Mexico argued that the canal would deprive them of water. They wanted the US State Department to negotiate a treaty with Mexico that would distribute the river's water fairly. American officials and state water lawyers, however, argued that a treaty was unnecessary. Since all of the river's water originated inside the United States, Mexico had no claim to any of its waters under international law. Much earlier, in 1896, US attorney general Judson Harmon had issued a ruling in connection to a US-Mexico dispute over the Rio Grande.[26] Mexico argued that American diversions upstream unfairly deprived its farmers of a rightful share of the river that created the international boundary. Harmon ruled, in what became known as the Harmon Doctrine, that the United States had no obligation to Mexico under international law regarding any water contained within US borders.[27]

Although some would use the Harmon Doctrine to oppose a treaty with Mexico, most scholars of water law repudiated the doctrine. Many people in California did not want a treaty because they saw it as mostly benefiting American landlords in Mexico who refused to pay their share of maintenance costs for the Río Álamo canal. It would serve them right, the argument went, if they lost the canal water after the All-American Canal was constructed. Let Mexico figure out its own water problems and how to deal with real estate magnate Harry Chandler as well. Once again, opponents such as IID's Philip Swing argued that the All-American Canal was necessary and patriotic. If it was not built to support good American farmers, then Mexicans, absentee landlords such as Chandler, and Asian "coolies" would be the ones to benefit.[28]

Throughout the 1920s, tensions continued along the Colorado River and the US-Mexico border. Mexico, in the wake of revolution, sought stronger connections between the nation's periphery and its center, as well as control over natural resources. President Alvaro Obregón sought this integration at the same time US states were negotiating the Colorado River Compact that divided the river's waters among themselves. After the compact was signed in 1922 and the Boulder Dam Project and All-American Canal were approved in 1928, Mexican officials were even more convinced that their access to necessary water for Baja California was in jeopardy. On the American side, there was talk of purchasing the peninsula to avoid future negotiations with Mexico over the river's water. Henry Ashurst, a senator from Arizona, proposed such a purchase in 1931. Mexico politely refused, and the offer stimulated strong anti-American sentiment and fear of US imperialism.[29]

In the meantime, US-Mexico conflicts over the Rio Grande persuaded some politicians that treaty negotiations should begin and that they needed to include the Colorado River. Philip Swing introduced the Tijuana River into the discussion as rumors circulated that Mexico had plans to develop that river and sell its water to San Diego. It made sense to discuss all of the water shared between the United States and Mexico, and both the US Congress and the Mexican government agreed to negotiate on all three rivers simultaneously. Several years of studies, appointments of delegates, and discussions led to a US offer in 1929.[30] Most of the negotiations involved the Colorado River, from which Mexico believed it was entitled to at least 4.5 million acre-feet (MAF) annually. The 4.5 MAF Mexico wanted would be just enough to adequately irrigate the 1.5 million acres of arable land in the river basin on its side of the border and represented approximately 25 percent of the river's estimated annual flow. Mexican officials also pointed to the original agreement in 1904 to allow Charles Rockwood's canal to route water through Mexico up to the Imperial Valley. That agreement had been made with the understanding that Mexico would be entitled to 50 percent of the water.[31] The Americans, however, offered 750,000 acre-feet of water annually, an amount that would maintain current agriculture in the Mexican border states but not increase it. Although disappointed, Mexico reduced its offer to 3.6 MAF. The United States rejected that amount and insisted on its earlier offer, at which point negotiations fell apart.[32]

When Lázaro Cárdenas became president of Mexico in 1934, the Colorado River issue became even more complicated. The memories of US filibustering, heavy foreign investments, offers to buy Baja California, and the failure to include Mexico in negotiations over the Colorado River and any treaty terms all came to bear on his attitude toward the United States. Cárdenas wanted to strengthen Mexico, especially along the border, and hoped to develop more economic independence for the nation. Drawing on earlier revolutionary ideology, Cárdenas developed a nationalizing program both to minimize the number of foreign investors in Mexico and to redistribute property and wealth. He specifically focused on Baja California and eventually nationalized the holdings of the CRLC. Some of the land was sold to private landowners, and the rest was divided up and effectively leased to *ejidatarios* (peasants) for their use.[33]

As Cárdenas carried out his *mexicanización* plans for Baja California, he did try to obtain some water rights assurances from the United States.[34] Since no assurances were forthcoming, he developed plans to irrigate as much good farming acreage as possible. If and when any treaty negotiations with the United States did occur, he wanted to be in a strong position according to prior appropriation laws, with which the United States was very familiar. If the Mexicali Valley's irrigation needs were very clear, there was a much better chance to negotiate from a position of strength. Similarly, US states sought to appropriate as much of the Colorado River as possible to keep Mexico from appropriating it.

In *Border Oasis*, historian Evan Ward argues that in spite of his efforts, Cárdenas's revolution was incomplete, and agriculture and other enterprises continued to depend on American investment. Still, neither side could become completely independent of the other. While Mexican farms continued to rely on American banks and other businesses for success, US farmers to the north relied on cheap migrant labor from Mexico. In an attempt to appropriate as much water as possible, both sides overtaxed the Colorado River delta while at the same time finding it necessary to maintain a peaceful, if not always willing, relationship.[35]

During the ensuing years, immigration to the delta region increased, initially because of government efforts and later because of the need for labor in the United States. During the early 1940s, mobilization for wartime production caused labor shortages in the Imperial Valley, and in 1942, migrant workers were brought into the

United States on a temporary basis through what was called the Bracero Program. Ongoing stress over immigration enforcement issues combined with problems in predicting how much water would be available for farming in the Mexicali Valley. Dams upstream in the United States sometimes released large flows during floods that damaged crops across the border. During dryer times, more water was stored and less was released downstream, forcing Mexican farmers to beg for government intervention with the Americans to send more water downstream.[36]

In the early 1940s, talks about a US-Mexican water treaty began again, largely because of ongoing conflicts over the Rio Grande and President Franklin D. Roosevelt's interest in a united western alliance during wartime. As it had done before, Mexico tied Rio Grande negotiations directly to receiving a share of the Colorado River. In 1941 the United States offered 900,000 acre-feet of the Colorado each year in exchange for a 50/50 split of the Rio Grande. Mexico responded with an offer of 2 MAF, and talks once again halted.[37] Finally, in 1943 the United States agreed to give to Mexico 1.5 MAF of Colorado River water each year in exchange for rights to approximately one-third of the water in the Rio Grande. Mexican and US officials signed a treaty in 1944.[38]

Not everyone was happy, of course. Just prior to ratification of the treaty, Los Angeles's Metropolitan Water District published a pamphlet arguing against the agreement. Using maps and dramatic illustrations, the pamphlet depicted the treaty as a major security risk. During World War I, Californians had supported the All-American Canal to protect the United States from an Asian invasion. Now, in the final years of World War II, they argued that water itself was vital for the war effort and that giving it to Mexico would jeopardize wartime production.[39] Other basin states supported the treaty, seeing California's opposition as just another example of the state's greed.[40] In spite of the opposition, Congress ratified the treaty in 1945. Mexico would receive 1.5 MAF annually and a share of any possible surplus, up to a total of 1.7 MAF. Davis Dam, completed in 1951, was constructed to store water to meet the treaty obligation.[41] In 1950, Mexico completed Morelos Dam to divert the river west to Mexicali farmers. The completion of the dam enabled Mexican farmers to store water for times of drought instead of having to request it from the United States.

At least two potential problems remained unresolved, however. The treaty contained only a vague statement about sharing shortages, and this omission later stimulated demands to renegotiate the treaty. However, the problem of water quality emerged first. The treaty did not mention this issue, and neither the United States nor Mexico anticipated any of the problems that had arisen by the early 1960s. While no major issues developed for some time, scholars studied conditions in the delta and the impact of the treaty with Mexico. In 1949 the shifting water in the main delta channel again moved west to the Hardy riverbed, channeling the flow more tightly so that finally, for the first time since 1905, the Colorado River reached the sea.[42] While much of the river below the border was diverted into irrigation canals in the Mexicali Valley, annual flooding and surplus flows not used by US states helped keep the delta alive.

For almost a decade, the 1944 treaty kept US-Mexican relations in the borderlands along the Colorado River relatively peaceful, if not cooperative. This condition began to change in the early 1950s and reached a crisis point after 1963, when Glen Canyon Dam ended the flow of surplus water to the delta.[43] In addition, developments along the Gila River seriously affected water quality. The Wellton-Mohawk region of Arizona is a strip about fifty miles long and contains approximately 75,000 acres of arable land. Spanish Jesuits visited the area as early as 1700 and assiduously observed the farming habits of indigenous peoples there, who were growing vegetables, beans, and corn in the warm climate.[44] The area soon attracted other white settlers, who began farming the rich soils of the Mohawk Valley to the north of the river and the Antelope Valley on its south side. The small town of Wellton soon grew there, lending its name to the area. Originally "Well-Town," the place served as a watering station, first for the Butterfield Overland Mail stagecoaches in the 1860s and then for the Southern Pacific Railroad in the 1870s.

Although the Gila River was the main source of water for several thousand acres of farmland, development upstream began to cause shortages. By 1911, when the Bureau of Reclamation completed Roosevelt Dam on the Salt River (the major source for the Gila), shortages in the Wellton-Mohawk Valley had become severe. After 1915, farmers used electricity to pump well water for irrigation when the Gila ran low. The over-pumping of groundwater, combined with the geographic layout of the valley that prevented proper drainage, led to a dramatic increase in salt levels in the water, making it unfit for irrigation. By the 1930s many farmers had abandoned their fields; those who remained could grow only low-value crops such as alfalfa. Following the example of other agricultural regions in the arid West, farmers in the Wellton-Mohawk Valley asked the Bureau of Reclamation for help.[45]

In 1952 the bureau came to the valley's rescue by constructing the Gila Main Gravity Canal, which delivered Colorado River water the twenty-nine miles from Yuma. In 1953 the newly formed Wellton-Mohawk Irrigation and Drainage District assumed control of the water supply in the region.[46] Conditions rapidly improved and more land was returned to cultivation, but the lack of proper drainage meant salinity conditions once again increased to a critical level. If used return flows from irrigation were not drained away from farmland, salts would continue to build, rendering the land useless. In less than a decade, the fragile region needed assistance again. Pumping water out of drainage wells into the Gila River began to help, and in 1961 the Bureau of Reclamation finished construction of a drainage channel that would bring the salty agricultural runoff down the valley to the lower end, discharging it into the Gila channel, which then flowed into the Colorado River.[47]

This new fifty-mile, concrete-lined channel dramatically increased the salinity of the Colorado River just north of the Mexican border. Naturally very salty, the Colorado River starts out at its headwaters with a salinity of approximately 250 parts per million (ppm). By the time the water reaches the Hoover Dam, the salinity has increased to around 700 ppm because of agricultural runoff and naturally occurring salts from the landscape through which the river flows. By the end of 1961, salinity levels at the border had spiraled to 2,700 ppm, more than a 300 percent increase. At the time, the only water scheduled for release to Mexico during the winter months came from this saline runoff. While the quality had been adequate prior to 1961, farmers now found it unusable and had to let the water flow to the delta, losing the cultivation of about 100,000 acres.[48]

Farmers in Mexico began to protest and urged their government to take action, which it did in November 1961. US officials, for their part, denied that any violation of the 1944 treaty had occurred.[49] Fairly quickly, a debate in the media and among members of both governments swirled around the issue of water quality. Mexico claimed that the treaty guaranteed the country usable water, and some even argued that the treaty meant that the water Mexico received would contain no runoff or return flow. The United States argued that there was no guarantee in the treaty and that it was unreasonable to expect the water delivered to Mexico to be completely free of agricultural runoff.

As discussions stalled over the salinity issue, Mexican farmers and students in the Baja California Norte region began staging demonstrations against the United States, causing concern across the border.[50] The United States was still in the midst of the Cold War, and American officials observed the rise of the political activist Alfonso Garzón, who received the backing of the leftist party Movimiento de Liberación Nacional (MLN). Fidel Castro's revolution in Cuba two years earlier had had varying impacts on leftist groups in Latin America, and Mexico was not immune. While the MLN worked mostly to improve the economy of Mexico for Mexicans, Americans saw its left-leaning politics as a potential open door for communism on the US border.[51] The United States was also concerned about the leftist policies of Mexican president Adolfo López Mateos and the Partido Revolucionario Institucional (PRI). The MLN first supported Garzón for state representative in the Baja California Norte region and later Cárdenas's son Cuauhtémoc as president of the Mexicali Valley Irrigation District.[52] Having an MLN-affiliated, left-leaning politician near the border during the salinity talks seemed ominous. US ambassador Thomas C. Mann warned President Mateos that unless he could control the MLN and keep both the pro-Castro Cuauhtémoc and Garzón away from the US border, Mexico would not succeed in obtaining a $400 million loan for development projects.[53]

Meanwhile, discussions continued in the United States over the 1944 treaty and water quality. Some American officials argued that while the letter of the law did not require any particular quality of water, the spirit of the law did.[54] Secretary of the Interior Stewart Udall suggested that groundwater users on the Mexican side of the border needed to do more to drain their own salty water sources and that the United States could possibly build a bypass pipeline to send the Wellton-Mohawk drainage farther downstream, below the Mexican irrigation intakes. Indeed, drainage issues in Mexico along the border appeared similar to the Wellton-Mohawk region's problems, and perhaps as much as 50 percent of the salinity affecting farming had local causes. Arizona was particularly concerned that Mexico should solve its own problems without any further action by the United States. State officials feared that sending additional water to Mexico to dilute the salt could mean trouble in drought years. In a town hall meeting on water issues in 1964, attendees vowed to oppose any US agreement that accepted "any obligation to deliver to Mexico water of any particular quality."[55]

One of the dynamics that made an eventual agreement possible was western politicians' changing attitude toward Mexican water rights and environmental issues in general. In the past, western states had often opposed national and international issues because they viewed Mexican water users and Sierra Club members as enemies of western growth. By the late 1960s, however, this attitude had begun to change. Western politicians and water district representatives began to support reducing salinity levels and indicated support for addressing Mexican demands for better water quality. The change came primarily because western interests began to coincide with salinity reduction.[56]

In earlier years, members of the Imperial Irrigation District in California and residents of Yuma County in Arizona had protested any concessions to Mexico and blamed Mexican farmers for quality problems. It became obvious, however, that reducing the salinity of Mexican water delivery would also reap benefits north of the border. As more and more users tapped into the Colorado River, western leaders increasingly believed cooperation within the entire watershed was important. If there were to be enough water to go around, it was to their benefit to reduce salinity levels and maximize water resources. In addition, both western states and Mexico needed federal dollars to construct and maintain their water infrastructures, and both sides lost money to salinity damage on a regular basis.[57]

Finally, in 1965 the United States and Mexico agreed to a joint policy known as Minute 218. The United States agreed to build a bypass pipeline to send 50,000 acre-feet of salty water from Wellton-Mohawk to a spot below the Morelos Dam each year for a period of five years. This diversion reduced the salt levels in the Colorado River at the border from 2,700 ppm to 1,245 ppm.[58] The agreement was extended in 1972, and in 1973 the two countries signed what was to be a permanent agreement. The resulting Colorado River Basin Salinity Control Act (Minute 242 of the Mexican Water Treaty) included a thirty-seven-mile extension of bypass channels, canal-lining projects, the retiring of agricultural land in Wellton-Mohawk, and the construction of a desalination plant at Yuma.[59] Bypassing almost 100,000 acre-feet of water from Wellton-Mohawk to the delta brought salt levels down from a peak of 2,700 ppm to 1,000 ppm. In an effort to further lower salinity levels, the US government purchased and reduced acreage under irrigation in the valley to around 65,000 acres.[60]

These agreements addressed many of the conflicts surrounding the sharing of the Colorado River, but groundwater issues were not included. Competition for groundwater resources along the US-Mexico border had existed since 1928, when the Boulder Canyon Project Act authorized the All-American Canal. With a canal completely inside the border, seepage could be legally claimed and perhaps recovered.[61] When the 1944 treaty was negotiated, it contained a stipulation that states: "Mexico shall acquire no right beyond that provided by this subparagraph by the use of the waters of the Colorado River system, for any purpose whatsoever in excess of the 1,500,000 acre feet."[62] No real problems with groundwater arose until after completion of the Wellton-Mohawk drainage project, which increased the salinity of the Colorado River water Mexico received.

By the mid-1960s, new wells on both sides of the border had begun pumping well water as fast as possible. In 1965 Senator Hayden proposed a group of new wells in Yuma County to provide clean water for delivery across the border to Mexico. He hoped water upstream used to meet the terms of the Mexican Water Treaty could then be used for the Central Arizona Project (CAP) canal.[63] On the Mexican side of the border, water engineers had been pumping out of the transborder aquifer since the beginning of the decade. By the end of the decade, Mexico had lifted previous limits on groundwater pumping to compete with the new wells on the American side and grab as much water as possible. Both sides thus participated in a pumping war that further depleted the water available for both sides.[64]

These salinity issues led to the 1973 treaty, which contained the statement that "each country shall limit pumping of groundwater in its territory within five miles (eight kilometers) of the Arizona-Sonora boundary near San Luis to 160,000 acre-feet (197,358,000 cubic meters) annually."[65] The treaty also stipulated that both sides would consult with each other regarding future ground and surface water plans. However, in spite of language that sounded cooperative, the US government informed Mexico in 1983 that it was not entitled to any seepage from the All-American Canal. California was scrambling for more water resources and was worried about the consequences of the CAP canal when it began to operate. Then in 1988, as part of the San Luis Rey Indian Water Settlement, Title II reaffirmed ownership of any seepage water and authorized canal lining as an option for recovery of the water.[66]

Following completion of an environmental impact statement in 1994, the Bureau of Reclamation settled on a plan to construct parallel lined canals to recover 67,700 acre-feet of water per year. This plan was not implemented, however, until pressure on California to limit its share of Colorado River water to its allocated 4.4 MAF per year allowed the plan to move forward. The project began in 2006; by 2010 the Bureau of Reclamation had lined twenty-three miles of the All-American Canal with concrete, as well as thirty-five miles of California's Coachella Canal.[67] Research groups such as the Sonoran Institute fear these lining projects will severely impact the delta wetlands.[68] Seepage from the canals has recharged groundwater in the region, helping both farmers and delta wetlands farther south.

Mexican farmers protested the canal-lining projects, as expected, but other groups on both sides of the border opposed the plans as well. US environmental groups and Mexican business leaders organized and filed a lawsuit against the project in 2005. The groups involved were the Mexicali Economic Development Council, Citizens United for Resources and Environment, and Desert Citizens against Pollution. The lawsuit protested the damage the project would do to Mexican farmers along the border as well as to wetlands, threatening endangered species in the river delta.[69] At first, a federal judge dismissed the lawsuit, but the group appealed and halted work on the project for seven months. The group lost its appeal in 2007, and work on the lining projects continued until they were completed in 2010.[70] Current concerns are focused on safety issues, as more and more people try to cross the border and canal illegally. Some concerned citizens have asked that safety lines or ladders be installed, but nothing had been done at the time of a May 2, 2010, special on the CBS television series *60 Minutes,* which reported the deaths of more than 550 people.[71] In early 2011, groups such as Citizens for All-American Canal Safety had helped persuade the IID to put safety lines across the canal to help prevent drownings.[72]

Cooperation and Its Challenges

While rejuvenating the delta wetlands emerged as an unexpected outcome of efforts to reduce salinity in the Wellton-Mohawk region, one of the unintended consequences of the salinity control acts was the drying up of those wetlands. Efforts to line canals and to reduce irrigated acres in the district, combined with construction of the Yuma Desalting Plant (sometimes called the Yuma Desalination Plant; YDP), endangered important habitat that depended on the salty water. Designed to desalt agricultural runoff from the Wellton-Mohawk Canal, the plant was intended to bring the water in the Colorado River into compliance with the 1973 treaty before it crossed the border for Mexican use. Construction began in 1975, but budget cuts meant the project moved forward sporadically.[73]

In 1992, seventeen years after ground was broken and $250 million later, the reverse-osmosis plant was ready to run. It started up at about 35 percent power and ran for only eight months, when flooding damaged the plant. While repairs could have been made, they would have been expensive. Water flows in the Colorado River had increased because of a recent wetter-than-normal cycle, and meeting the requirements of the Mexico treaty did not present a problem. The plant sat idle until the results of drought in 2003 became severe. At that point, impending water shortages outweighed the costs of restarting the plant, which led to a successful test run in 2007 and a pilot run in 2010 that ended in April 2011. This one-year run processed around 30,000 acre-feet of water that was included in the water delivered to Mexico, conserving an equal amount in Lake Mead.[74] It would take approximately three years for the plant to reach full capacity, estimated at 78,000 acre-feet per year.[75]

The Colorado River delta has always been very fluid, susceptible to any changes in the river. Dams and diversion channels upstream affect where the water flows in the delta, moving it from one channel to another or completely out of channels. Periodic floods damage levees and earthen dams intended to keep farmland from flooding, thus allowing the river to shift or spread. In the delta, three major wetlands concerned environmentalists after the YDP was completed in 1992: the Rio Hardy wetlands, the Ciénega de Santa Clara (where most of the Wellton-Mohawk runoff is channeled), and the El Doctor wetlands. Although the delta region has been greatly reduced from its original 780,000 acres, these three areas provide important habitat for birds migrating along the Pacific flyway and for the only remaining populations of several endangered species.[76]

Another result of damming the river is that normal silt deposits that fanned out across the delta in earlier times now pile up behind Hoover and Glen Canyon Dams. Water that once flowed into the wetlands and to the sea no longer reaches beyond the agricultural diversion canals in the Mexicali Valley. Return flows through the Main Outlet Drain Extension (known as MODE) from the Wellton-Mohawk District to the delta wetlands have been the only steady source of water to the region. Still, life has continued in the wetlands, thanks to periodic flooding. Although this flooding was halted from 1963 to 1980 while Lake Powell was filling up behind Glen Canyon Dam, renewed high flows after 1981 returned parts of the delta wetlands that had virtually dried up. Some statistics show that between 1980 and 1993, the average flows of water across the Mexican border were three times the treaty allotment of 1.5 MAF.[77] The extra water, however, damaged some wetlands. The Rio Hardy wetlands existed because an earthen dam to control flooding allowed water to back up through the area. The 1983 flood destroyed the dam, and after 1986 the wetlands steadily began to drain.

Still, the changes in the delta during the 1980s mostly revitalized a previously declining ecosystem. When the YDP was completed in 1992, environmentalists took notice of the reviving delta and argued forcefully that the recovery should be maintained. If the YDP became fully operational, one of the most important flows of water to the delta would stop. Scientists knew the Colorado's high water levels likely represented a temporary cycle, and maintenance of the delta became an important aim for US environmentalists. Mexico was also interested in maintaining the recovery process in the delta because wetlands recovery could provide income from ecotourism.[78]

Consequently, in 1993 the Mexican government aided in the declaration of the delta as a biosphere reserve under the United Nations program for such reserves, established in 1970.[79] Called the Alto Golfo de California y Delta del Rio Colorado, this reserve encompasses around 2.3 million acres, including the entire Colorado River delta. Environmental groups in the United States applauded the creation of the reserve, and many began to provide advice and assistance in the delta recovery effort. One of these groups recommended that a new earthen dam and other levees be constructed throughout the delta to help preserve the wetlands areas. Another recommendation included redirecting agricultural drainage from the Imperial Valley away from the Salton Sea to the delta.[80]

In the late 1990s, the United States and Mexico began to talk and cooperate on plans to preserve the delta. In 1998 two bi-national study groups made proposals on how to maintain and even increase the revitalization of the wetlands. Both reports asserted that the most important part of the region's ecosystem was the 150,000-acre center of the delta and that it "could be sustained with relatively marginal annual instream flows."[81] All the delta wetlands needed were around 32,000 acre-feet of water annually and as much as 500,000 acre-feet every three to four years.

A few years earlier, officials with the US Fish and Wildlife Service and the Bureau of Reclamation had begun to talk about a larger program that would approach the delta and lower basin of the Colorado River as a complete ecosystem. The first restoration proposal for the delta appeared in a 1994 article in the *Colorado Journal of International Law and Policy*. The author argued that the 1973 Endangered Species

Act extended beyond the US border.[82] Other proposals followed, and in 1996 the US Fish and Wildlife Service and the Bureau of Reclamation formed the Lower Colorado River Multi-Species Conservation Program (MSCP), which included the three lower basin states, representatives from cities and environmental groups, and other interested parties. Because the program does not extend into Mexico to include the delta, some participants withdrew from the project.[83] In spite of these disagreements, the MSCP completed its studies in 2005 and is now implementing its program to preserve habitat along the lower Colorado River, stopping at the Mexican border.[84] The program has budgeted $70 million for the next five years alone. Encompassing more than fifty groups, agencies, and research programs, it intends to help mitigate ongoing river depletion and usage.[85]

Transborder Issues and Cooperation

Although sharing the Colorado River has caused tensions between the United States and Mexico over the decades, there are strong indications of new cooperation in the face of recent drought and other challenges. Some environmentalists argue that the best way to preserve and maintain the Colorado River watershed for all users would be to manage the watershed as a whole, disregarding both interstate and international boundaries. While this idea may not be realistic, a new era of cooperation between the United States and Mexico has characterized the late twentieth and early twenty-first centuries. In the face of population growth and climate change, such cooperation is necessary for the future health of the region.

Beginning in 1983, one of the first results of transborder cooperation was what became known as the La Paz Agreement. In August of that year, the Agreement between the United States of America and the United Mexican States on Cooperation for the Protection and Improvement of the Environment in the Border Area was signed by both countries.[86] In this agreement, the Environmental Protection Agency (EPA) and its Mexican counterpart, the Secretary for Natural Resources and the Environment (SEMARNAT), established a basis for future negotiation and coordination. Implemented as a result of this agreement, an Integrated Border Environmental Plan was formed that pinpointed several priorities: reducing air pollution, increasing transborder cooperation, and promoting more public awareness of border environmental issues. The plan also created the Border XXI Program, a five-year bilateral project to coordinate environmental programs.[87]

In 2003 the EPA and SEMARNAT created another program called Border 2012, a ten-year program that emphasized inclusion of citizen groups in the planning. This new emphasis on the participation of regional organizations, industry representatives, and local citizen groups helped create more awareness of environmental issues on the border. The inclusion of these stakeholders in the decision-making process was deemed a success by the participants, who worked together on air pollution and added other goals, including reduction of water contamination.[88] In 2007 the program reported ongoing challenges with inadequate waste processing and sewage contamination in Nogales, Sonora. However, it also reported the funding of twenty-two water projects through the cooperation of the EPA and the Comisión Nacional del Agua (National Water Commission). These efforts included work on infrastructure to provide clean water access for indigenous groups living on both sides of the border.[89]

In 2011 the EPA and SEMARNAT formulated Border 2020, a new plan that incorporates the best of the previous plans, continues to emphasize a "bottom-up" approach to decision-making, and includes indigenous groups in the region. Some of the work already completed includes chlorination units for drinking water on the Tohono O'odham Reservation, a sewer line for the Pechanga Tribe, and drinking water studies for Baja California Indians. Part of the funding for these projects came from Mexico's Comisión Nacional para el Desarrollo de los Pueblos Indígenas (National Commission for the Development of Indigenous Communities).[90] The ongoing objectives for Border 2020 include providing "at least 8,500 households with safe drinking water" and "at least 39,000 households with adequate wastewater sanitation."[91]

Another bi-national effort in March 2011 was the Border Sanitation and Wastewater Quality Summit in San Antonio, Texas. Convened by the International Boundary and Water Commission (IBWC), this meeting called together representatives from many border communities, the EPA, SEMARNAT, and other government officials from both the United States and Mexico. Both sides agreed to work toward better exchange of water-quality information and to promote the reuse of effluent.[92] Also in 2011, upgrades were made to the South Bay International Wastewater Treatment Plant in San Diego, which treats sewage from Tijuana. Both the United States and Mexico contributed money for the plant's construction and for ongoing operations and upgrades.[93] These upgrades came in the wake of two wastewater spills in Tijuana that crossed over the border into San Diego County. The two nations have also collaborated on wastewater infrastructure and treatment in Mexicali over the last several decades, although challenges with water quality in the New River continue.

Beginning in 2001 and extending through 2005 for some areas and into 2012 for others, a severe drought compelled the southwestern states to develop plans to cope with water shortages. By 2004, the levels of Lakes Mead and Powell were dropping precipitously. To address shortages, in 2003, California signed agreements with agricultural districts for large water transfers that would move water from farming to urban centers, and people began talking once again about the Yuma Desalting

Plant. Salvaging the salty runoff from Wellton-Mohawk might help meet shortages, although it would cut off the one steady supply of water the delta wetlands receive.[94] Furthermore, although the California Supreme Court ruled in early 2010 that the original agreement was unconstitutional, Los Angeles and San Diego continued to buy water from farmers in the Imperial Valley and to seek ways to adjust the original agreement.

In 2005 the Yuma Desalting Plant/Ciénega de Santa Clara Workgroup was formed to examine the environmental impacts on the delta wetlands if the YDP were restarted. Created by Sid Wilson, president of the Central Arizona Project, this group sought to find a balance between using the YDP for an additional water source and preserving the delta wetlands. The group recommended renegotiating the salinity agreement with Mexico to make it more flexible in times of shortage and creating a special fund that could be used to address water shortages. It further argued that Yuma-area groundwater should assist with preservation of the delta wetlands when operation of the YDP reduced flows to the region.[95] While some of its recommendations proved controversial, scholars found it remarkable that groups with conflicting aims (states and environmental groups) could reach any kind of agreement.[96]

In late 2006 the US Congress passed the United States–Mexico Transboundary Aquifer Assessment Act. This bill authorized a ten-year bi-national study of aquifers on the border to assist with future planning.[97] In 2007 the Sonoran Institute in Tucson published a study that outlined four possible scenarios for the Colorado River delta region. Examining what might happen if a drought extended through three decades, the group assessed the impact of legal conflict with no action, water marketing to meet shortages, long-term regional planning, and ecosystem-based management of water on the border. The report recommended community and water-use planning for the future to follow very closely what geologist and explorer John Wesley Powell had recommended at the turn of the twentieth century.[98] Instead of each southwestern state controlling its own water, a bi-national, basin-wide management system would work better to achieve long-term vitality through cooperative sharing of water. It is doubtful that two nation-states will surrender that much sovereignty for the sake of better resource management. However, bi-national cooperation on environmental pressures has paved the way for addressing water shortages.

What will happen in the future to the Colorado River and its delta is unclear. Although the situation eased in 2005, much of the Southwest remained in drought conditions throughout much of 2012.[99] A Drop 2 Reservoir called the Warren H. Brock Reservoir, authorized in 2005 to catch the water from canceled water orders before it crosses the border, was completed and filled with water in late 2010.[100] The reservoir holds approximately 8,000 acre-feet of water. Even though that is a small amount of water, the Bureau of Reclamation estimates that its use could save up to 70,000 acre-feet in annual withdrawals from Lake Mead.[101] Operation of the YDP is one solution supported by Arizona and other basin states north of the border. Some state officials say they will replace the flow to the Ciénega wetlands, but they have not said where the water will come from. As part of the ecological solution, several university and nonprofit groups formed the Colorado River Delta Water Trust in 2005. The trust is now purchasing water from agriculture to restore regular flows as well as periodic flooding necessary to mimic natural conditions.[102]

Currently, the United States and Mexico are engaged in ongoing discussions about how to deal with possible shortages. In June 2010 the two nations agreed to Minute 317, which established processes for joint management decisions.[103] In January 2011 the Colorado River Water Users Association held its annual meeting in Las Vegas. Doug Kenney, director of the Western Water Policy Program at the University of Colorado in Boulder, reported on his preliminary findings from a study called the "Colorado River Governance Initiative." He asserted that the only way for users of the river to manage future shortages is to find new ways to achieve inter-basin cooperation.[104] Any plans for river management changes need to include Mexico.

Missing in most of these conversations and restoration plans is much, if any, mention of the Cucapá.[105] Dependent on the delta's fragile environment, the Cucapá have traditionally fished the marshes and streams as their main source of food. While their relatives on the Cocopah Reservation near Somerton, Arizona, also struggled in the past with loss of land and water, they now have a casino and a golf course—both of which bring in much-needed revenue. Compared with their poor cousins in El Mayor, they are rich indeed. In the delta, the Cucapá get up early in the morning and drive an hour from their little shantytown to where there is water and some fish. There are ongoing struggles between the Cucapá and the Mexican government over the right to fish within the boundaries of the Delta Biosphere Reserve. Now that they are confined to fishing farther away from the mouth of the Colorado River, many Cucapá are seeking employment at tourist camps or finding it necessary to leave their native home.[106] Making matters even worse, El Mayor experienced a devastating earthquake on April 4, 2010. The 7.2-magnitude quake, referred to as the "Sierra El Mayor Earthquake," left the town without passable roads, electricity, or water.[107] The agreement—Minute 318—signed in December 2010 promises to help with water delivery and to store unused water while Mexico rebuilds its irrigation and water delivery infrastructure. Secretary of the Interior Ken Salazar hopes this agreement will help lay the groundwork for future cooperation with Mexico over Colorado River water. Part of the agreement is supposed to ensure delivery of water to the Ciénega de Santa Clara wetlands in the delta.[108]

Recently, anthropologist Shaylih Muehlmann argued that Mexican and environmentalist efforts to define indigeneity are complicating the Cucapás' struggle for fishing rights in the delta. Mexican government officials who seek to ban the Indians from fishing there argue that the Cucapá are not truly indigenous because they do not live or look like Indians should, holding on to a static condition of the ancient past. Since they do not dress or speak like Indians, their identity as a people is suspect.[109] Expedition Blue Planet North America, an investigatory and documentary team headed by Alexandra Cousteau, recently interviewed Cucapá elder Innocencia Gonzales. In the interview, Innocencia shared memories of a time when the Colorado ran right past her home and she used to swim in its waters when she was a young girl: "Sixty years ago we used to fish Toatoba in the Colorado River here . . . the mountains were green and now they are bare rock."[110] According to this elder, who was seventy-three years old when interviewed, there is a long way to go to restore the Colorado River delta.

Some scholars argue that current efforts to restore delta waters are not harming but instead are enhancing the livelihood of the Cucapá. Advocating water transfers through a water market would both maintain delta wetlands and help support fisheries on which the Cucapá depend. Although those interviewed by Cousteau and Muehlmann claimed the indigenous people are being ignored or sidelined, scholars working through the Sonoran Institute have included the Cucapá. They believe both nations and all people affected by conditions in the delta should participate in the solution.[111]

The good news is that both sides are talking. Water scholar Evan Ward wrote in *Border Oasis* that "ultimately, the ecological problems in the region are not the exclusive domain of either the United States or Mexico. Instead they are shared problems that demand complex solutions from both countries."[112] Mutually dependent on each other to some degree, the two nations have competed and sometimes cooperated, but now they must learn to share the Colorado River. In the current decade, the environment is receiving recognition as an important water user but also as a player that can affect the human story along the river's banks. Indigenous voices, often ignored, are gradually being heard as environmental groups such as the Sonoran Institute and Pronatura Mexico help share their stories and include them in delta restoration projects. Many issues remain, but continuing cooperation is a sine qua non on both sides of the border.

NOTES

1. Don Madaleno, interview, in Shaylih Muehlmann, "Where the River Ends: Environmental Conflict and Contested Identities in the Colorado Delta," PhD diss., University of Toronto, 2008, 98.

2. Ibid., 97.

3. For a good orientation to the Environmental Justice movement, see David Schlosberg, *Defining Environmental Justice: Theories, Movements, and Nature* (New York: Oxford University Press, 2007).

4. Ellen Wohl, *Virtual Rivers: Lessons from the Mountain Rivers of the Colorado Front Range* (New Haven, CT: Yale University Press, 2001).

5. For example, see research listed on University of Arizona's Department of Geosciences, Colorado River Delta Research Coordination Network website, updated through 2003, http://www.geo.arizona.edu/rcncrd/online_bibliography.html (accessed March 13, 2011). For more recent work, see University of Arizona's Water Resources Research Center, College of Agriculture and Life Sciences, WRRC Publications, http://ag.arizona.edu/azwater /publications.php?type=5 (accessed March 13, 2011). For Spanish resources, see Intercultural Center for the Study of Deserts and Oceans (El Centro de Estudios de Desiertos y Océanos; CEDO), http://www.cedointercultural.org/component/option,com_frontpage/Itemid,39 /lang,es/ (accessed March 13, 2011).

6. There are several spellings, including Cocopah, Cucapáh, Cocopa, and Cucupá. In most government references, the community on the US side of the border usually uses Cocopah. Most members of the Mexican community use Cucapá. See Muehlmann, "Where the River Ends," 9.

7. The Baja Peninsula was a single region of, first, Spanish-ruled and, later, independent Mexico until 1930, when it was administered as two separate territories: Baja California Norte (north) and Baja California Sur (south). In 1952 the northern division became an official state of Mexico; the southern division became a state in 1974.

8. Thomas R. McGuire, "The River, the Delta, and the Sea," *Journal of the Southwest* 45, no. 3 (Autumn 2003): 371–410.

9. Evan R. Ward, *Border Oasis: Water and the Political Ecology of the Colorado River Delta, 1940–1975* (Tucson: University of Arizona Press, 2003), 4.

10. For a recent work on the region's history, see William Vollmann, *Imperial* (New York: Penguin, 2010). See also George Kennan, *The Salton Sea: An Account of Harriman's Fight with the Colorado River* (Ithaca, NY: Cornell University Press, 2009 [1917]); Pat Laflin, *The Salton Sea: California's Overlooked Treasure* (Indio, CA: Coachella Valley Historical Society, 1995): 17–30; Donald Worster, *Rivers of Empire: Water, Aridity, and the Growth of the American West* (New York: Oxford University Press, 1985), 196; Marc Reisner, *Cadillac Desert: The American West and Its Disappearing Water* (New York: Penguin Books, 1993 [1986]), 122–23.

11. Norris Hundley Jr., *Dividing the Waters: A Century of Controversy between the United States and Mexico* (Berkeley: University of California Press, 1966), 32–34.

12. For more on Porfirio Díaz, see Paul H. Garner, *Porfirio Díaz: Profiles in Power* (Charlottesville: University of Virginia Press, 2001).

13. Evan Ward, "The Twentieth-Century Ghosts of William Walker: Conquest of Land and Water as Central Themes in the History of the Colorado River Delta," *Pacific Historical Review* 70, no. 3 (August 2001): 359–85; Frederick D. Kershner Jr., "George Chaffey and the Irrigation Frontier," *Agricultural History* 27, no. 4 (October 1953): 115–22.

14. C. P. Vetter, "The Colorado River Delta," October 1949, Box 5, Folder 3.1, Brody Collection, Water Resource Center Archives, University of California, Berkeley, 218.

15. Hundley, *Dividing the Waters,* 42–43.

16. Lowell L. Blaisdell, "Harry Chandler and Mexican Border Intrigue, 1914–1917," *Pacific Historical Review* 35, no. 4 (November 1966): 385–93.

17. Ward, "Twentieth-Century Ghosts of William Walker," 364–66. For a recent Marxist analysis, see Casey Walsh, " 'To Come of Age in a Dry Place': Infrastructures of Irrigated Agriculture in the Mexico-U.S. Borderlands," *Southern Rural Sociology* 24, no. 1 (2009): 21–43. See also Richard J. Orsi, *Sunset Limited: The Southern Pacific Railroad and the Development of the American West, 1850–1930* (Berkeley: University of California Press, 2005), 32, 40.

18. Michael Hiltzik, *Colossus: Hoover Dam and the Making of the American Century* (New York: Free Press, 2010), 59.

19. John M. Hart, *Empire and Revolution: The Americans in Mexico since the Civil War* (Berkeley: University of California Press, 2002), 269–304.

20. Ward, *Border Oasis,* 8–9.

21. Eric Boime, " 'Beating Plowshares into Swords': The Colorado River Delta, the Yellow Peril, and the Movement for Federal Reclamation, 1901–1928," *Pacific Historical Review* 78, no. 1 (2009): 32.

22. Ibid. See also Eugene Keith Chamberlin, "The Japanese Scare at Magdalena Bay," *Pacific Historical Review* 24, no. 4 (November 1955): 345–59.

23. George Hebard Maxwell, *Our National Defense: The Patriotism of Peace* (Washington, DC: Rural Settlements Association, 1915), 176–79.

24. Ward, *Border Oasis,* 9–10.

25. Colorado River Compact, Article III, Section C, 1922, http://www.usbr.gov/lc/region /pao/pdfiles/crcompct.pdf (accessed February 28, 2012).

26. *United States v. Texas,* US 162, US 1, 16 S. Ct. 725, 40 L. Ed. 867 (1896).

27. Stephen C. McCaffrey, "The Harmon Doctrine One Hundred Years Later: Buried, Not Praised," *Natural Resources Journal* 36 (Summer 1996): 549–90. See also Hundley, *Dividing the Waters,* 50–53; Ward, *Border Oasis,* 11–13.

28. Hundley, *Dividing the Waters,* 52–57.

29. Ward, *Border Oasis,* 12. See also Oscar Jáquez Martínez, *Troublesome Border* (Tucson: University of Arizona Press, 2006), 45–46. Historian Joseph Stout argues that Ashurst was also a sponsor of earlier filibustering expeditions to Mexico; see Joseph Allen Stout, *Schemers*

and Dreamers: Filibustering in Mexico, 1848–1921 (Fort Worth: Texas Christian University Press, 2002), 183.

30. Norris Hundley Jr., "The Politics of Water and Geography: California and the Mexican-American Treaty of 1944," *Pacific Historical Review* 36, no. 2 (May 1967): 209–26. See also Eric Boime, "Fluid Boundaries: Southern California, Baja California, and the Conflict over the Colorado River, 1848–1944," PhD diss., University of California, San Diego, 2002.

31. Hundley, *Dividing the Waters*, 68–69.

32. Charles J. Meyers and Richard L. Noble, "The Colorado River: The Treaty with Mexico," *Stanford Law Review* 19, no. 2 (January 1967): 368.

33. Ward, *Border Oasis*, 15–16. See also Michael J. Gonzales, *The Mexican Revolution, 1910–1940* (Albuquerque: University of New Mexico Press, 2002), 221–60.

34. Ward, *Border Oasis*, 16–17. See also José María Ramos García, *Relaciones México–Estados Unidos: Seguridad Nacional e Impactos en la Frontera Norte* (Mexicali: Universidad Autónoma de Baja California, 2005), 266–67.

35. Ward, *Border Oasis*, 20.

36. For information on the Bracero Program during World War II, see Ronald L. Mize and Alicia C.S. Swords, *Consuming Mexican Labor: From the Bracero Program to NAFTA* (Toronto: University of Toronto Press, 2011); Jorge Durand, *Braceros: Las Miradas Mexicana y Estadounidense: Antología (1945–1964)* (Senado de la República: LX Legislatura, 2007).

37. Meyers and Noble, "Colorado River," 376.

38. Treaty between the United States of America and Mexico, signed in Washington, DC, February 3, 1944, http://www.usbr.gov/lc/region/pao/pdfiles/mextrety.pdf (accessed February 28, 2012).

39. Metropolitan Water District, Mexican Treaty: An Exposition of the Proposed Mexican Water Treaty in Maps, Illustrations, and Text, 1944, Box 5, Folder 3.1, Brody Collection, Water Resource Center Archives, University of California, Berkeley.

40. Letter from Arizona Governor Sidney Osborn to President Franklin Roosevelt, March 16, 1944, MS CM MSS-1, Box 650, Folder 6, Colorado River/Central Arizona Project Collection, Hayden Library, Arizona State University, Tempe.

41. US Government Printing Office, Utilization of Waters of the Colorado and Tijuana Rivers and of the Rio Grande: Treaty between the United States of America and Mexico (Washington, DC: US Government Printing Office, 1946), 22–29.

42. Vetter, "Colorado River Delta," 219.

43. One of the best and most complete works on Glen Canyon Dam is Russell Martin, *A Story That Stands Like a Dam: Glen Canyon and the Struggle for the Soul of the West* (Salt Lake City: University of Utah Press, 1999 [1989]).

44. Byrd H. Granger, *Arizona's Names: X Marks the Place* (Tucson: Falconer, 1983).

45. H. M. Babcock, *Wellton-Mohawk Area, Yuma County, Arizona: Records of Wells, Well Logs, Water Analyses, and Maps Showing Location of Wells* (Tucson: US Geological Survey, 1948). In the 1940s, Arizona politicians representing the Yuma County region pushed for an irrigation project in the Wellton-Mohawk Valley. One possible purpose was to provide irrigated land on which to settle World War II veterans after the war. However, a very significant reason for political support for the project was increased water usage across the border. Prior to the treaty with Mexico, Arizona politicians sought to pump and thus claim priority rights to as much water as possible to prevent its use by Mexican farmers. See Ward, *Border Oasis*, 46–47.

46. Evan Ward, "'The Politics of Place': Domestic and Diplomatic Priorities of the Colorado River Salinity Control Act (1974)," *Journal of Political Ecology* 6 (1999): 31–56; Ward, *Border Oasis*, 52–61; Hundley, *Dividing the Waters*, 172–80. See also Shlomi Dinar, *Conflict and Cooperation along International Rivers: Scarcity, Bargaining Strategies, and Negotiation* (Boston: MIT Press, 2011), 176–77.

47. Hundley, *Dividing the Waters*, 172–74.

48. Ibid., 175.

49. Letter from Assistant Secretary of State Frederick G. Dutton to Senator Carl Hayden, March 8, 1962, CM MSS-1, Box 253, Folder 9, Colorado River/Central Arizona Project Collection, Arizona State University, Tempe; Letter from Science Adviser Roger Revelle to Interior Secretary Stewart Udall, July 14, 1962, AS 372, Box 164, Folder 2, Mexican Water Treaty, Morris K. Udall Papers, Special Collections, University of Arizona Library, University of Arizona, Tucson.

50. Ward, *Border Oasis*, 79.

51. Kate Doyle, "After the Revolution: Lázaro Cárdenas and the Movimiento de Liberación Nacional," National Security Archive Electronic Briefing Book no. 124, May 31, 2004, National Security Archive, Washington, DC, http://www.gwu.edu/~nsarchiv /NSAEBB/NSAEBB124/index.htm (accessed July 14, 2010).

52. Garzón began to lose influence when the two nations began working seriously on a bilateral solution. He was ultimately "co-opted" by federal politics. Ward, *Border Oasis*, 86.

53. Guillermo De La Peña, "Rural Mobilizations in Latin America since 1920," in Leslie Bethell, ed., *Latin America: Politics and Society since 1930* (New York: Cambridge University Press, 1998), 326–27. See also Ward, *Border Oasis*, 80–87.

54. Letter from Colorado Governor Steve McNichols to IBWC Commissioner J. F. Friedkin, May 17, 1962, AZ 372, Box 164, Folder 1, Mexican Water Treaty, Morris K. Udall Papers, Special Collections, University of Arizona, Tucson.

55. University of Arizona, "Fourth Arizona Town Hall on Arizona's Water Supply," April 6–8, 1964, Research Report (Phoenix: Arizona Academy, 1964), 151.

56. Ward, "Politics of Place," 38.

57. Ibid., 39.

58. Annual Project History, Colorado River Basin Salinity Control Project, Title I, Arizona-California, 1975–1995, Bureau of Reclamation Project Histories, Record Group 115, National Archives and Records Administration, Rocky Mountain Region, Denver, CO.

59. International Boundary and Water Commission, United States and Mexico, Text of Minute 242—Permanent and Definitive Solution to the International Problem of the Salinity of the Colorado River, August 30, 1973, http://www.usbr.gov/lc/region/pao/pdfiles /min242.pdf (accessed September 23, 2012).

60. Herbert Brownell and Samuel D. Eaton, "The Colorado River Salinity Problem with Mexico," *American Journal of International Law* 69, no. 2 (April 1975): 259, 265. For an excellent study of the origins and implications of the Salinity Control Act, see Ward, *Border Oasis*, esp. chapters 5 and 6.

61. Alfonso A. Cortez Lara, Megan K. Donovan, and Scott Whiteford, "The All-American Canal Lining Dispute: An American Resolution over Mexican Groundwater Rights?" *Frontera Norte* 21, no. 41 (January-June 2009): 130.

62. Text, the Mexican Water Treaty, 59 STAT. 1219, 1944, http://www.ibwc.state.gov/Files /1944treaty.htm (accessed September 23, 2012).

63. Ward, *Border Oasis*, 94–95.

64. Ibid., 93–97.

65. Text of Minute 242, 1973, http://www.usbr.gov/lc/region/pao/pdfiles/min242.pdf (accessed March 13, 2011).

66. Public Law 100-675, November 17, 1988, US Bureau of Reclamation Regional Programs, All-American Canal, http://www.usbr.gov/lc/region/programs/AAC/refdocs/PL100-675 .pdf (accessed March 13, 2011). See also Douglas L. Hayes, "The All-American Canal Lining Project: A Catalyst for Rational and Comprehensive Groundwater Management on the United States–Mexico Border," *Natural Resources Journal* 31 (Fall 1991): 803–27.

67. San Diego County Water Authority Fact Sheet, Canal Lining Project, July 19, 2010, http://www.sdcwa.org/all-american-canal-lining-project-named-apwa-public-works -project-year.pdf (accessed September 23, 2012); Imperial Irrigation District, All-American Canal Lining Project, 2009, http://www.iid.com/index.aspx?page=178 (accessed September 23, 2012).

68. Sharon Megdal, "Should Yuma Desalter Operate? Varied, Complex Issues Are Raised," Policy Statement, Appendix J, *Arizona Water Resource*, October 6, 2004, http://ag.arizona .edu/AZWATER/awr/mayjune04/policy.html (accessed July 18, 2010); Jeffrey P. Cohn, "Colorado River Delta," *Bioscience* 54, no. 5 (May 2004): 386–91.

69. Gig Conaughton, "Challenge to Canal Lining Could Undermine San Diego Water Plans," *North County Times* [San Diego], July 19, 2005, http://www.nctimes.com/news /local/article_ed405d17-75b0-5e4c-8538-51fea8059f6a.html (accessed March 13, 2011). See

also Imperial Group Info, Las Vegas, July 19, 2005, http://www.imperialgroup.info/PDF /PressRelease_CDEMvsUS.pdf (accessed March 13, 2011).

70. Cortez Lara, Donovan, and Whiteford, "All-American Canal Lining Dispute," 139.

71. CBS News, "The Deadly Passage of the All-American Canal," *60 Minutes*, May 2, 2010, http://www.cbsnews.com/stories/2010/04/30/60minutes/main6448631.shtml?tag =contentMain;contentBody (accessed March 13, 2011).

72. Eliot Spagat, "All-American Canal: Buoys Strung on Border Waterway to Prevent Drownings," *Huffington Post,* Los Angeles, January 4, 2011, http://www.huffingtonpost.com /2011/01/04/allamerican-canal-buoys-s_n_804405.html (accessed March 4, 2012).

73. Philip Fradkin, "US Plans for Immense Desalting Plant Assailed," *Los Angeles Times*, June 2, 1974; Terri Shaw, "US, Mexico to Move on Colorado River Issue," *Los Angeles Times*, June 17, 1972; "Mexico, US Reach River Peace," *Los Angeles Times*, September 3, 1973.

74. "Yuma Desalination Plant Pilot Run Concluded Successfully," *International Desalination and Water Reuse Quarterly,* April 1, 2011, http://www.desalination.biz/news /news_story.asp?id=5848&channel=0 (accessed March 4, 2012).

75. US Department of the Interior, Final Environmental Assessment: Yuma Desalting Plant Pilot Run (Yuma, AZ: US Bureau of Reclamation, Yuma Office, June 2010), http:// www.usbr.gov/lc/yuma/environmental_docs/cityofyuma/finalea.pdf (accessed September 4, 2010).

76. Edward P. Glenn et al., "Effects of Water Management on the Wetlands of the Colorado River Delta, Mexico," *Conservation Biology* 10, no. 4 (August 1996): 1176.

77. Ibid., 1177–78.

78. David H. Getches, "Impacts in Mexico of Colorado River Management in the United States," in Henry F. Diaz and Barbara J. Morehouse, eds., *Climate and Water: Transboundary Challenges in the Americas* (Boston: Kluwer Academic, 2003), 173.

79. United Nations Educational, Scientific and Cultural Organization, Biospheres, 1993, http://www.unesco.org/mab/doc/faq/brs.pdf (accessed July 18, 2010).

80. Glenn et al., "Effects of Water Management," 1184.

81. S. Joshua Newcom, "Deciding about the Colorado River Delta: Rejuvenated Wetlands Raise New Issues about Where Flood Flow Should Go," *River Report: Water Education Foundation* (Spring 1999): 1–11.

82. Frank Wilson, "A Fish out of Water: A Proposal for International Instream Flow Rights in the Lower Colorado River," *Colorado Journal of International Law and Policy* 5 (1994): 249–72.

83. McGuire, "The River, the Delta, and the Sea," 401–2. See also http://www.defenders .org/ and http://www.lcrmscp.gov/ (accessed July 18, 2010).

84. Black Schmidt, "Helping Species Survive," *Knight-Ridder/Tribune Business News,* June 8, 2006.

85. Ben Preston, "Something for Everyone," *Miller-McCune*, March 8, 2010, http://www .miller-mccune.com/science-environment/something-for-everyone-10470/ (accessed March 4, 2012).

86. Environmental Protection Agency, La Paz Agreement, August 1983, http://www.epa .gov/Border2012/docs/LaPazAgreement.pdf (accessed March 6, 2012).

87. Ibid.

88. Environmental Protection Agency, Results Report, Border 2012, September 2002, http://www.epa.gov/usmexicoborder/ (accessed September 23, 2012).

89. Interim Report, 2007, Border 2012, http://www.epa.gov/border2012/docs /implementation_2007_eng.pdf (accessed March 6, 2012).

90. Ibid.

91. Environmental Protection Agency, Border 2020: U.S.-Mexico Environmental Program, June 2012, http://www.epa.gov/border2020/index.html (accessed September 23, 2012).

92. International Boundary and Water Commission, "Summit Convened to Improve Sanitation and Water Quality," *Boundary Marker* (Spring 2011), http://www.ibwc.gov/Files /BM_Spring_2011.pdf (accessed March 6, 2012).

93. IBWC, "San Diego Treatment Plant Upgrade Begins Operations," *Boundary Marker* (Spring 2011), http://www.ibwc.gov/Files/BM_Spring_2011.pdf (accessed March 6, 2012).

94. Steve Cornelius et al., "Recent Developments on the Colorado River: Implications for Delta Conservation," Sonoran Institute Memorandum, August 1, 2003, http://cleartheair .edf.org/documents/2931_RecentDevelopmentsColoradoRiver.pdf (accessed September 23, 2012).

95. Yuma Desalting Plant/Ciénega de Santa Clara Workgroup, "Balancing Water Needs on the Lower Colorado River: Recommendations," April 22, 2005, http://ag.arizona.edu /azwater/publications/YDP%20report%20042205.pdf (accessed July 7, 2010).

96. Joe Gelt, "Opposing Sides Find Common Ground in Yuma Desalter Controversy," *Arizona Water Resource* 13, no. 6 (2005): 1–3.

97. United States–Mexico Transboundary Aquifer Assessment Act, 2006, http://www .govtrack.us/congress/billtext.xpd?bill=s109-214 (accessed August 22, 2010). See also Sharon Megdal, "Front-Row View of Federal Water Lawmaking Shows Process Works: US-Mexico Transboundary Aquifer Assessment Act Pondered, Passed, and Signed," *Arizona Water Resource* 15, no. 3 (January-February 2007): 1–2.

98. Mark Lellouch, Karen Hyun, and Sylvia Tognetti, *Ecosystem Changes and Water Policy Choices: Four Scenarios for the Lower Colorado River Basin* (Washington, DC: Island, 2007).

99. Bureau of Reclamation, http://www.usbr.gov/uc/water/crsp/cs/gcd.html (accessed July 18, 2010).

100. Bureau of Reclamation, Boulder Canyon Project—All-American Canal System, February 1, 2012, http://www.usbr.gov/projects/Project.jsp?proj_Name=Boulder+Canyon

+Project+++All-American+Canal+System (accessed September 23, 2012); Joe Gelt, "Drop 2—End-of-the-Line Reservoir Salvages Colorado River Water," *Arizona Water Resource* 16, no. 5 (2008): 1–2.

101. "Arizona's Gamble—Conserve Water Now, Prevent Deeper Cuts Tomorrow," Circle of Blue: Reporting the Global Water Crisis, December 29, 2010, http://www.circleofblue .org/waternews/2010/world/arizonas-gamble-conserve-water-now-prevent-deeper-cuts -tomorrow/ (accessed March 14, 2011).

102. "How to Restore the Colorado River," *Grist Magazine*, September 2, 2010, http:// grist.org/living/2010-07-30-how-to-restore-the-colorado-river/ (accessed March 4, 2012).

103. "Commission Signs Agreement for U.S.-Mexico Cooperation on Colorado River Management," International Boundary and Water Commission, United States and Mexico, July 1, 2010, http://www.ibwc.gov/Files/Press_Release_070110.pdf (accessed March 14, 2011).

104. Cited in Bret Walton, "Prompted by Scarcity, Colorado River Basin States Examine Their Lifeline," Circle of Blue: Reporting the Global Water Crisis, January 24, 2011, http:// www.circleofblue.org/waternews/2011/world/prompted-by-scarcity-colorado-river-basin -states-examine-their-lifeline/ (accessed March 14, 2011).

105. Some research groups have discussed the role the Cucapá can play in plans for the delta and how they can benefit from those plans. The Sonoran Institute proposed a Borderlands Initiative that created a cooperative group to address delta issues and included the Cucapá in that group. See Robert J. Glennon and Peter W. Culp, "The Last Green Lagoon: How and Why the Bush Administration Should Save the Colorado River Delta," *Ecology Law Quarterly* 28 (2002): 963–66.

106. "A Tribe without a River," *Tucson Citizen*, June 20, 2006.

107. Miriam Raftery, "Cucapa Indians in Baja Devastated by Quake in Mexico," *Censored News: Indigenous Peoples and Human Rights News Publisher*, April 10, 2010.

108. US Department of the Interior, News Release, December 20, 2010, http://www.doi .gov/news/pressreleases/Salazar-Elvira-Announce-Water-Agreement-to-Support-Response -to-Mexicali-Valley-Earthquake.cfm (accessed March 14, 2011).

109. Muehlmann, "Where the River Ends," 80–84.

110. Anne Casselman, "Remembering the River That Was," *US Policy*, Embassy of the United States, Brussels, Belgium, November 9, 2010, http://www.uspolicy.be/headline /remembering-river-was (accessed September 23, 2012). The Toatoba fish is indigenous to the Gulf of California and is currently listed as an endangered species.

111. Glennon and Culp, "Last Green Lagoon," 903–92.

112. Ward, *Border Oasis,* 152.

8

The Water Market

Banking and Selling the Colorado River

THROUGHOUT THE HISTORY OF THE Colorado River, the humans who depend on it have often competed with each other over how to divide these contested waters. Metropolitan areas have grown beyond what the river can reasonably support, and unsettled indigenous claims create additional demands on a meager resource. Ongoing drought and climate change add even more pressure to this over-tapped lifeline, and solutions are anything but clear. Conservation techniques, recycling, and transferring water between users have become some of the important answers to problems in the Colorado River basin.[1]

As the world's population is rapidly approaching 7 billion, the scarcity of water now occupies more attention than most other environmental issues. In 2008 the award-winning documentary film *Flow (For the Love of Water)*, directed by Irena Salina, demonstrated this worldwide interest. Proclaiming that the wars of the future would be fought over water, Salina highlighted the pressures of water scarcity and the effects of privatization. According to water experts she interviewed, the commoditization of water is already widespread, and its continued growth is inevitable. Although there are significant problems in marketing water, this system could work to address growing water shortages in a fair and equitable way, albeit only with government regulations.

In many regions, the failure of publicly administered utility companies to adequately supply their constituents with clean and safe water has made it possible for private companies to grow and take over market share within a wide variety of conditions. For a decade and a half, Australia has used a kind of water market to address its shortages between supply and demand. Since 1997 it has operated a water exchange through several entities (such as the National Water Exchange) that transact both temporary trades and permanent sales of water rights.[2] Assessments of these exchanges seem generally positive, with analysts citing increased income for farmers who sold or leased water rather than growing crops.

Similar water exchanges have also been successful in Chile, India, Pakistan, and China, to name a few. Currently, China might face the worst water crisis worldwide, with an ongoing drought that has steadily worsened since it began in 2009. As early as the mid-1990s, Chinese farmers informally marketed their water; after 2002 the government made significant efforts to treat and recycle wastewater as part of its growing water market.[3] The questions in these countries are not about whether water should be marketed but about where the water comes from and who reaps the benefits. As large water companies such as Suez, Vivendi, and Thames vie for market share in far-flung places such as India and South Africa, the discussion about the commoditization of water seems almost moot. Globally, water underwent privatization during much of the twentieth century.

Scholarship on water marketing appeared as early as 1968.[4] Many scholars voice the concern that the growing southwestern water market will deprive some people of a basic right to water and have unintended impacts on third parties.[5] In this chapter I seek answers to several questions. What are the long-term impacts of water marketing on the environment, agriculture, and urban populations? Who reaps the economic benefits and who loses? How can negative impacts of water marketing be addressed to preserve everyone's right to water? An examination of these issues in the American southwestern context can serve as a microcosm in which to seek answers to the global questions dominating water marketing discussions: can and should water be considered a commodity that can be owned by an individual? Or should water be preserved as unique, a social good not owned by any one person? Is there a middle ground between the two views? These are not simple questions, and no single study will likely be able to answer them adequately for all circumstances. Yet the example of water marketing in the American Southwest seems to support its benefits, although only with important restrictions through government regulation. While a regulated market has not yet been perfected, this study indicates that a middle ground might actually be successful in both conserving water and allocating it fairly.

How the Market Functions

It is not particularly easy to trade or market a river's flow. Such an effort requires agreed-upon measurements and innovative exchanges. So far, the majority of southwestern water transactions have been conducted through water banking, or a so-called spot market. These trades are generally short-term or onetime sales. One type of water banking that is becoming more common is the dry-year option. In this type of water sale, the buyer who anticipates future water shortages can purchase the "option" to draw a certain amount of water from the seller at a future time. Other water transactions through "water trusts" occur when an action group or board purchases water for environmental uses to restore stream flow or water levels in wetlands. Recently, some long-term sales have stretched over decades. These are usually controversial because of their more "permanent" redistribution of water resources. By far the most common kind of water trading that occurs in the American West, and certainly trading that involves the Colorado River, is done through a water bank.

To operate a water market or bank, excess water is created in a variety of ways. One of the most common methods today is through land fallowing. Farmers agree to leave some fields fallow and sell the water they would normally use to irrigate those fields to other users, usually urban water districts. Such water trades involve many challenges. One is the difficulty of measuring the amount of water a farmer actually uses and thus has available to sell, excluding any return flow that reenters the river or aquifer. The water exclusive of return flow is called *consumptive use* and remains very difficult to quantify. Another way to create excess water is by recycling wastewater, which is normally easy to transfer, or "wheel," to irrigators or industries but less easy to sell to urban users. Often, pumping groundwater substitutes for surface water, or a certain user with rights to surface water agrees to pump groundwater and sell the surface water. This transfer does have the negative impact of possibly overdrawing groundwater resources. In fact, many water users engage in conjunctive use, the combination of surface and groundwater.[6]

Another common transfer method in the American West during the past decade or more comes through water storage. A user will store unused water in underground aquifers during wet years and pump the water out during shortfalls. Sometimes this kind of water banking or transfer takes place when an entity such as the Los Angeles Metropolitan Water District (MWD) pays someone else to store water in its underground aquifer, with the city paying wheeling and storage costs as well as pumping and delivery costs later when the stored water is needed. Other ways to find excess water for transfers consist of water conservation initiatives, sometimes subsidized by the prospective water buyer. A metropolitan area might pay for water-saving irrigation and pumping infrastructure with an agreement that any water saved will be sold back to the city. In this scenario, water otherwise wasted is used in a city without the need to reduce the amount of agriculture and with no unintended income loss for the region.

Development of a US Water Market

The case of water markets in the American West is an interesting study of reluctance to commoditize water and concern about how to prevent such marketing from damaging third parties to the transactions. Yet a market has functioned in the region since the early 1990s in some form or another. To understand the growth of and challenges with marketing the Colorado River's water, the story must begin with California.[7] During the first half of the twentieth century, water users in that state and the rest of the arid West relied on the Bureau of Reclamation—that is, the federal government—to find and utilize new sources of water. When water shortages seemed imminent, a state or municipality requested federal help to build another dam or canal. Los Angeles's efforts to procure water from a distance first occurred at the turn of the twentieth century, when William Mulholland and his team bought up land and its water rights in the Owens Valley, 250 miles from Los Angeles. Piping all that water away from Mono Lake and the valley turned the area into a dust-filled wasteland (see chapter 1). This water grab became consistently linked with any later efforts the city made to obtain more water. In many ways, the Owens Valley experience is still a formidable obstacle for Los Angeles and other Southern California urban areas.[8]

Even though Los Angeles was successfully procuring water from that valley by 1913 and added the all-important pipeline from the Colorado River between 1933 and 1941, Los Angeles's water supply was anything but secure. The 1922 Colorado River Compact had allocated only 4.4 million acre-feet (MAF) per year to the state, but it was consistently using more than 5.2 MAF while other states were not using all of their allocations.[9] This condition was not a problem until the 1960s, when California lost the Supreme Court case *Arizona v. California* in 1963 and Arizona managed to obtain long-sought funding for the Central Arizona Project pipeline (CAP). Two effects of this case are relevant to the present discussion: the impending loss of excess Colorado River water California needed, and the legalization of federal water transfers written into the judgment. For the first time, Bureau of Reclamation water contracts could be traded to other thirsty buyers on a temporary basis.

As the Los Angeles area's population continued to boom in the late 1960s and early 1970s, the city struggled to find more water. The last major water project in California was the State Water Project, built in 1960. Consisting of a series of dams, reservoirs, pumping stations, and canals, this is the largest state-built water project in the United States. The California Aqueduct brings water from the northern to the southern part of the state, watering the San Joaquin Valley with most of the water going to support agriculture. Los Angeles still needed water, but times had changed in the 1970s with the arrival of the environmental movement. Any proposals for new dams or canals received much scrutiny and serious opposition; with the creation of the Environmental Protection Agency in 1970, the dam- and large project–building days were mostly over.

In addition to the challenges posed by the environmental movement in the 1970s, the West experienced a severe drought in 1976–77. During this drought crisis the Bureau of Reclamation stepped in to provide emergency water in California through a temporary water bank. This experience paved the way for future water markets, especially in times of crisis. Because the water bank worked reasonably well, some argued it should be the wave of the future for western water. Several reports advocated the development of water transfers to address shortages in California, but many remained skeptical that an operating water market could be successful.[10] Still, in 1980, California passed a law that made trades easier by declaring that water transfers of any kind were not an abandonment of a right. In other words, just because someone with water rights was not personally using that water, they did not lose the right under appropriative water law.[11]

Growing out of ad hoc rules miners had used to keep the peace in the gold rush days of the American West, the doctrine of prior appropriation meant that whoever arrived first and mined the gold or land or water had priority rights to it. These miners' codes eventually became legal statutes. As this law evolved, whoever had physical control of water had the right to divert it to any "beneficial use." This stipulation was meant to ensure that individuals or companies did not hoard water, a scarce resource in the arid West. One could not simply "own" the water. Beneficial use dictated that whoever diverted or claimed the water had to actually use it for some tangible and beneficial purpose.

By 1982, it was clear in California that some kind of water transfer through a type of market would be one of the few options for future water resources in cities such as Los Angeles and San Diego. While its population continued to expand exponentially, Arizona was completing major portions of the Central Arizona Canal and would soon be taking its full share of water from the Colorado River. Plans to build a peripheral canal to bring Northern California water around the delta region to the south were defeated in 1982, a significant loss in Los Angeles's hunt for more water. Clearly, the only way to obtain more water would be through conservation and buying someone else's water.[12]

As the debate over water markets heated up in the mid-1980s, another severe drought hit the arid West—particularly California—and lasted through 1993. In the midst of water shortages, discussions about water markets again moved past the theoretical. By 1988 the drought, which was partly a result of an El Niño event in the Pacific during 1986 and 1987, had become one of the worst in US history, spreading over 36 percent of the country that summer. As the rest of the US states began to recover, California's drought dragged on, leading to shortages.

The most interesting water marketing during this drought took place in 1988. In Southern California, just north of the border with Mexico, vast agricultural enterprises and fields of crops cover the Imperial Valley. The Imperial Irrigation District (IID) is the primary utility company and has some of the most senior rights to Colorado River water, wheeled to farmers through the All-American Canal. The 1988 agreement set up a thirty-five-year water transfer from the IID to Los Angeles through a water conservation program that included canal lining and reservoir regulation. Los Angeles's MWD paid for the program in exchange for the water saved— approximately 100,000 acre-feet per year. Ten years later the IID signed a similar agreement with the San Diego County Water Authority (SDCWA) for almost double that amount over a ten-year period.[13]

In both 1991 and 1992, California's Department of Water Resources set up emergency drought water banks. In both years the majority of the water sold to the water banks came from fallowing agricultural land. The first such agreement was the 1992 transfer from the Palo Verde Irrigation District to the MWD, which fallowed approximately 20,000 acres for a period of two years. The prices were set to account for the cost of wheeling the water, and cities happily paid the price for a secure water supply.[14] Most analysts agreed that water banking seemed to work well, especially during the 1992 experiment.[15] Yet many predicted that future water shortages could not be met through temporary and intrastate trades alone.[16] During the 1990s, analysts saw that the completion of the Central Arizona Project could potentially take Arizona's full 2.8 MAF out of the main stream of the Colorado River.

Other developments in water marketing began to appear in the late 1990s. In 1996, Arizona created the Arizona Water Banking Authority and was soon working on water trade and storage agreements with Nevada. In the late 1990s, Las Vegas experienced exponential population growth and was desperately seeking more water resources. Arizona's CAP was running at only about 73 percent capacity in 1996, partly because of the high costs of the water. Farmers preferred to pump cheaper groundwater instead of paying for CAP water, so Arizona began storing some of its 2.8 MAF in underground aquifers through recharge. Arizona agreed to sell some of its water allocation to Nevada, which would take the purchased water out of Lake Mead. Arizona would take an equal amount of water out of the underground aquifer, instead of from main-stream flow, for use. In this way, an interstate market functioned, although some preferred to call it a water bank.[17]

The largest and still the most important water transfer agreement of Colorado River water had its origins sometime around 1996. In that year both Arizona and Nevada started using their full allocations, which meant California was drawing water considered "surplus." Five years later the Department of the Interior initiated what were called Interim Surplus Guidelines that allowed California to draw a certain amount of extra water from Lake Mead, but with the understanding that it would begin reducing that amount. California needed to demonstrate that it was making reasonable progress in reducing its demands for Colorado River water; as long as it did so, the Department of the Interior was willing to gradually wean the state off the extra water.[18]

Without major water transfers, however, California's use of Colorado River water continued to exceed the 4.4 MAF allocation, and in early 2003 the secretary of the interior declared California in violation of the Interim Surplus Guidelines.[19] This meant that California would not be allowed to draw surplus water from Lake Mead but would have to make do with the 4.4 MAF immediately. This declaration sent shockwaves through state government and the leaders of the many water district organizations in the southern part of the state. Much of the overdraft of Colorado River water went to Los Angeles and San Diego, whose populations had rapidly expanded. Leaders of the urban water districts had begun talks in 2001 with some of the agricultural water districts that had priority rights to Colorado River water. These talks had ended in stalemate but were resumed with new urgency in 2003.

In October of that year the Quantification Settlement Agreement was struck, which, if implemented, would constitute the largest agriculture-to-urban water transfer in US history. The agreement actually consisted of thirty-five separate agreements, and the major result would be a transfer of approximately 30 MAF of Colorado River water from the Imperial Valley and Coachella Valley Water Districts to Los Angeles and San Diego over a period of seventy-five years. In effect, this meant a transfer of around 2.5 MAF per year from agriculture to urban water users. The major signers of the agreement were two government bodies (the US Department of the Interior and the state of California), two agricultural water districts (Imperial Irrigation District and Coachella Valley Water District), and two urban water districts (San Diego County Water Authority and Metropolitan Water District). These two water-hungry cities agreed to pay for and receive water transfers and to pay millions of dollars to address resulting environmental impacts, such as the declining water levels and quality of the Salton Sea.[20]

The major problem with the agreement was that it committed the state of California to an unknown amount of debt that would come from the Salton Sea restoration project and other water-saving projects such as canal lining. While water districts such as the MWD would also pay large portions of these bills, state authorities questioned the legality of assuming such a large and unspecified amount of debt. Litigation ensued, although water transfers from fields to cities began right away. Almost seven years passed before the California Superior Court ruled the Quantification Settlement Agreement invalid in February 2010. It is uncertain what will happen with this agreement in the future, but appeals have been filed and some expect the agreement to eventually become binding, with some modifications.[21]

Meanwhile, the MWD and the SDCWA continue to receive and pay for water transfers from the Imperial and Coachella Valleys. These water transfers have been made particularly vital by the ongoing drought that reduced the Colorado River's flow. In 2005, California traded 1.2 MAF of water, about 3 percent of the state's annual water use.[22] The IID sells approximately 80,000 acre-feet of water from farms each year to San Diego. The controversial All-American Canal lining project, begun in 2007 and completed in early 2010, is supposed to save approximately 67,700 acre-feet of water each year; all the lining projects combined should save nearly 100,000 acre-feet.[23] A mitigation fund can be tapped to address other negative impacts, including the environmental impact on wetlands.[24] None of the parties, however, is paying attention to the impact canal lining will have on groundwater, which is so important to farmers across the border in Mexico.

Benefits and Costs

During its development over the past fifty years, many analysts have debated the costs and benefits of marketing water. By the mid-1980s a significant body of scholarship existed supporting reallocation of water through a market. The first and most beneficial result usually cited was a better division of water to the most efficient user. Economist B. Delworth Gardner argued in his 1986 essay "The Untried Market Approach to Water Allocation" that the Colorado River in particular should be allocated through a water market that gives water to the "highest valued users."[25]

The river's limited and over-allocated resource, Gardner asserted, was often used to grow low-profit crops on over-watered farmland instead of being used more efficiently.[26]

Some scholars argued that transfers of river water from the upper basin to the lower basin, where it produces higher-value crops, would be the most beneficial distribution of the water. In Colorado, for example, the average elevation is 6,800 feet, making the average growing season around 150 days a year. The largest crops grown in Colorado and Wyoming are alfalfa and feed grains. In contrast, large areas of California and Arizona have nearly year-round growing seasons, yielding much higher-value crops for a similar amount of water. An interstate water market would also include more water in the lower basin for hydroelectric generation and mitigation of high salinity levels. Although the upper basin states would be forced to take on some of the costs of such transfers, some economists have argued that the upper basin should share more of the negative effects of heavy Colorado River water use. An interstate market redistributes both benefits and costs in a more equitable way.[27]

From many perspectives, the most important benefit of water marketing is a more realistic price that would better reflect the actual value of water and encourage conservation. Some municipalities, such as Tucson, Arizona, have experimented with raising the price of water to help motivate people to conserve. Such plans face considerable resistance from urban water users, but they might work better with agriculture by distributing costs and thus encouraging conservation. Redistributing water to higher-value uses would also give more to industry. For example, a common comparison is made between cotton and microchips. For one acre-foot of water, the value of irrigated cotton or alfalfa is approximately $60 and 3 full-time jobs, while the same amount of water can produce $1 million worth of microchips and 9,000 jobs.[28] More generally, one acre-foot of water can yield approximately $350 in profits from farm produce. Conversely, the same amount of water could produce $300,000 in income from industrial products.[29] From these perspectives, a water market would reallocate resources to the highest-value uses.

Another way to think about market reallocations is to consider the impact on ownership. Some economists suggest that individuals, rather than water districts or states, should hold all water rights. If an individual obtained his or her water through a market purchase or sold excess water to the highest bidder, he or she would be more likely to conserve and use water efficiently.[30] Sellers of water rights would benefit from the increased value, and buyers would be motivated to both conserve and apply water to the highest possible economic benefit. In this way, a water market impacts water law by placing ownership in users' hands. Such a system would require the removal of all barriers to interstate water trades.[31]

Today, the primary beneficiaries of water marketing are urban areas. As the populations of metropolitan areas such as Los Angeles, Phoenix, and Las Vegas increase exponentially, cities look for new resources. Since cyclical drought and other pressures have caused the over-allocation of the Colorado River and groundwater aquifers, cities cannot simply build a new canal or reservoir. The only place to find new water is through purchases, usually from agriculture. In this way, the water market redistributes resources away from agriculture to urban and industrial uses and from regions with lower populations to more densely populated areas. Consequently, states move water from one area to another to correspond to changing market demand.

Many analysts argue that water marketing can also provide a substantial benefit to the environment. One example is pollution. While government bureaucracies seek pollution mitigation with uniform laws and regulations, a marketplace allows for a more nuanced approach to a problem that is anything but uniform. It is easier to control some points of pollution than it is others, so some economists advocate buying and selling pollution credits. If a watershed is treated as a single unit, some kind of uniform total standard for pollution can be established. Within the watershed, some users who find it very expensive to mitigate pollution such as salinity can buy credits from other users for whom it is much cheaper to control contamination. Those who can better afford the costs are thus encouraged to mitigate even more pollution to build up tradable credits.[32]

Other potential environmental benefits include the ability to purchase water for in-stream flows. Some theorists assert that individuals should be able to purchase water to preserve stream flow if they so desire. Charles W. Howe has suggested that the old "beneficial use" principle attached to appropriative water rights in the West be discarded. People should be able to purchase a long-term water right in a market they might not use for a very long time. Hotels and restaurants that might be affected by a reduced stream flow should be allowed to purchase water rights to maintain that flow.[33] Others argue that since many regions benefit specifically from recreation and tourism, an open water market would actually help preserve in-stream flows. Economic development will then be automatically regulated by the economic interest in environmental preservation.[34]

Another region experiencing some benefits from water marketing is the Colorado River delta in Mexico. Bureau of Reclamation dams and reservoirs that make urban areas such as Las Vegas and Los Angeles possible deprive the delta wetlands of water, further stressing already endangered species. US efforts in the early 1970s to reduce salinity levels from agricultural runoff led to the construction of a bypass canal that took salty water south of the Mexican border into the Ciénega de Santa Clara wetlands. When drought and lowering reservoirs led to further water conservation and the restarting of a long-dormant desalting plant in Yuma, flow to the wetlands decreased and may continue to decrease following completion of a new reservoir on the California-Mexico border. As discussed earlier, as part of a solution to reduced water flow, several university and nonprofit groups formed the Colorado River Delta Water Trust in 2005. This trust is currently purchasing water from agriculture to restore regular flows and the periodic flooding necessary to mimic natural conditions in the delta.[35]

While water markets have a number of clear benefits, there are also many problems that dampen the enthusiasm of some experts. In 2001 scholar Kenneth Frederick eloquently assessed the ongoing barriers to water markets. "The fugitive nature of the resource," he argued, "the variety of services it provides, and interdependence among users limit the potential for efficient water marketing."[36] He also acknowledged that third-party impacts from water transfers would have to be addressed but that doing so would be difficult since such impacts would be very hard to measure.

Perhaps the most significant barrier is the impact a water market often has on farming communities. If water is leased or sold from one state to another, farmworkers lose jobs. Third-party impacts in farming communities go far beyond the simple loss of agricultural labor employment. These small towns develop around the farmer, and many kinds of businesses would feel the impact, including farm equipment dealers, seed and fertilizer sellers, and bankers. Community restaurants, retailers, and schools suffer from declining property values and taxes. Area lakes could lose important water needed for recreational activities and tourist dollars. Rural water users also continue to see the specter of Mono Lake in the dry Owens Valley and fear that agricultural transfers to urban users will eventually empower entities such as the Metropolitan Water District to take more water, effectively killing local economies. While urban users are willing to pay higher prices for water and such cost increases might encourage water efficiency, rural politics sees agriculture-to-urban water transfers as a threat.

Perhaps not often considered by US economists is the fact that Mexican farmers also experience third-party impacts. Spurred in part by serious drought in 2004, state and federal governments looked for ways to conserve water through canal-lining projects and desalination. However, seepage from the unlined canals recharged groundwater in Mexico, helping both farmers and the delta wetlands farther south.[37] In 2009 the Bureau of Reclamation lined twenty-three miles of the All-American Canal with concrete to prevent seepage, and by early 2010, thirty-five miles of California's Coachella Canal were also complete. A significant source of groundwater recharge along the border was thus lost, causing international strain.

A similar conservation project, the so-called Drop 2 Reservoir, was authorized in 2005 and completed in 2010 as the Warren H. Brock Reservoir. It catches excess water at the border, shares of which are sold to Arizona, Nevada, and California. In the past, farmers might have ordered a water delivery from Lake Mead upstream, which took two or three days to reach its destination. If rain came before delivery, a farmer might have canceled the order to save money. Previously, this water would have flowed across the border to Mexican farmers. The state of Nevada funded construction of the new reservoir to buy water for Las Vegas. In the end, Las Vegas paid $115 million for 400,000 acre-feet of water. Nevada is now entitled to 40,000 acre-feet of water annually through 2036. Arizona and California each paid $28.6 million for a share of 100,000 acre-feet, to be withdrawn between 2016 and 2036 at a maximum amount of 36,000 acre-feet per year.[38] This plan will work well for conservation and water marketing inside the United States but will have significant negative impacts on Mexican farmers and delta wetlands that relied on the approximately 70,000 acre-feet of water each year.[39]

Another important third-party impact or externality is the effect of marketing on groundwater. Economist Mason Gaffney identified what he saw as "unconstrained pumping." As long as farmers could pump unlimited amounts of groundwater, prices for surface water would remain low, interfering with water marketing.[40] If water that is not used by a person or group of people is sold and transferred away from its current stream or aquifer, local water users may resort to over-pumping groundwater, leading to subsidence, increased salinity, and the lowering of the water table. Often, a water user who trades surface water on the market replaces that water by pumping groundwater, creating other third-party impacts in the region.[41] Known as *conjunctive use*, many water sales today combine surface trades with groundwater to replace traded water.

Some economists argue that until the costs for transfers are addressed and shared between buyers and sellers, those costs will discourage participants in a water market. One study illustrated that while transaction costs that came from administering a market were usually shared equally (establishing prices, negotiating delivery schedules and wheeling methods), policy-induced costs such as compliance with state laws and mitigating third-party impacts were almost exclusively borne by the seller.[42] Some of these problems might be addressed by including the costs of such impacts in the price of the transactions.[43] The only way to actually monitor and control third-party impacts or externalities, some argue, is through government regulation of the water market. Thus a completely "free" market governed solely by the laws of supply and demand is not practical. Creating a working balance between open marketing and government regulation, however, is still a considerable challenge.[44]

While some economists have argued persuasively in favor of water markets, others argue simultaneously that water should not be commoditized. In the same volume as

Gardner's advocacy of water markets in 1986, scholars Helen M. Ingram, Lawrence A. Scaff, and Leslie Silko countered with a discussion of the dangers of water marketing. Because of the third-party impacts of a water market, the authors argued that water is a "social good," not owned by any individual. According to this public trust doctrine, the government should act as a trustee for water supplies and allocate water according to an "equity doctrine."[45] The major principles of this doctrine would be reciprocity and good-faith obligations for all water users.

More scholars are paying attention to issues of equity and water. *Sustainability* is perhaps the most recognized term in environmental conversations today, but a close second might be *environmental justice*. If people have a basic human right to water, then some see growing commoditization as very troubling. Most water sales are made by small farmers to either municipal entities or larger agribusiness, and this condition does transfer control of an important commodity to a smaller elite. Yet some scholars argue that such a market actually improves equity issues by giving small farmers access to revenue they would not have otherwise.[46]

Current Challenges and Possible Solutions

The major barrier to water marketing today was clearly identified in a 1999 California Legislative Analyst's Office report.[47] The authors argued that the most difficult part of marketing water in the state was the existence of unclear and overlapping laws regulating water transfers. Some transfers needed federal review if they involved water from a federal system, such as the State Water Project. Others required only state oversight, and still others just needed agreement between local water districts. To address this problem, the Legislative Analyst's Office urged the passage of a single water transfer act that would cover all important circumstances in a coordinated way. It also suggested that water rights in the state be better quantified, that information on all water transfers be published in a central location so everyone can see and better participate in the process, and that the law address third-party impacts fairly.[48]

One reason the authors of the report considered new legislation important was the forecast for severe water shortages in the future. According to statistics cited in the report, California was using approximately 79.5 MAF of water per year. Environmental uses accounted for the largest portion, at 46 percent; agriculture used 43 percent, and urban areas consumed only 11 percent. This 79.5 MAF was met as follows: 20 percent from various federal projects, 8 percent from state projects, 13 percent from the Colorado River, only 1 percent from recycling and desalination, 27 percent from local surface projects, and 31 percent from groundwater. Unless the state found ways to increase recycled water and redistribute surface waters, by 2020 shortages of 2.4 MAF would occur in good years, with a staggering shortfall of as much as 6.2 MAF in times of drought. If all states in the Colorado River basin did a similar analysis, the predicted shortfalls would likely be equally worrying.[49]

While many see the need to redraw the laws to allow for a free water market, not everyone agrees on how that market should operate and who can participate in it. Privatization of water is one of the proposed ways to encourage more long-term water transfers through a fair market. A proposed fifty-year agreement between the Cadiz Water Project and the MWD was one such transfer. Founded in 1983, Cadiz, Inc., is a land and water resource company that discovered a large aquifer under its landholdings in the Cadiz Valley of eastern San Bernardino County. Excess Colorado River water from the aqueduct would be stored in an underground aquifer owned by Cadiz. For storage and wheeling costs, Los Angeles could tap these stored waters in times of drought.[50] Delays and a lengthy five-year environmental impact analysis cooled MWD's interest in the project, however, and it withdrew from the agreement in 2003. The company continued to pursue its plans without the MWD and in June 2009 signed water delivery agreements with five other Southern California water distributors. The large aquifer and plans to replenish pumped water through stored recharge are expected to provide enough water for around 400,000 Southern California residents per year.

Several other private water companies are participating in California's growing water market. Azurix Corporation purchased land in Madera County, California, in 1999 to create a large underground storage unit for water. Azurix, however, was a subsidiary of Enron Corporation and found its stocks falling apart in connection with Enron's crisis. In 2001, Azurix–North America was sold to American Water Works.[51]

Still, privatization has its own challenges and drawbacks. If a company owns water for sale, there is minimal chance for any kind of public or governmental oversight. A company can sell to the highest payer or offer reduced rates to someone with power or influence. Thus privatization strengthens the chances for an unequal distribution of water in favor of an elite. Likewise, there is little incentive for private water companies to mitigate environmental externalities or respond to public interests. Arizona law professor Robert Glennon argues that no US government unit should hand over water ownership to a private company. While such actions might make sense in countries where the government has failed to deliver water, the US market needs continued governmental oversight and regulation to keep powerful private corporations from creating an unbreakable monopoly over water. Although Glennon concedes that most water in the American West is already privatized (mostly owned by private agricultural interests), he believes ongoing governmental oversight can help a water market operate efficiently.[52]

In view of these challenges, most water marketing and transfers in the near future will continue to be conducted between public water districts. Since agricultural water districts have the largest water rights to the Colorado River, most marketing today consists of agriculture-to-urban transfers. Even though the largest such transfer—the 2003 Quantification Settlement Agreement—is still under litigation and modification, the Imperial Irrigation District did receive confirmation that it could legally make such a large and long-term water transfer, and it has done so since 2005. As recently as April 2010, IID president Brian J. Brady testified to the US Congress that the IID would continue the transfers. However, he admitted that there were problems. "There can be no enduring settlement of longstanding disputes among Colorado River water users within California," he asserted, "without a credible resolution of the transfer mitigation question and impacts to the Salton Sea." Agreeing with many others that third-party impacts had to be addressed, he argued that the effects of transfers on the Imperial Valley also needed attention. "And the water transfer paradigm of the future is only viable if it works for the Imperial Valley today," he concluded.[53] Many farmers and farmworkers protested the transfers, fearing their land values would fall and jobs would disappear while a few made money on cheap water. Others argued that there was no other practical way the IID could have modernized its water delivery systems and no other place Los Angeles and San Diego could find more water sources.[54]

Although challenges to a water market in the West remain, the recent drought has encouraged intrastate transfers, most of which have been legally formalized.[55] Ongoing discussions about third-party costs, however, have continued to create doubt about the long-term effectiveness of a market for western water, especially in California. Some have argued that to mitigate these impacts, especially in agricultural regions that might lose jobs and other sources of income, costs could be exacted through transfer fees. Another way to address such impacts would be to schedule gradual transfers of water out of an area, allowing rural areas to find other economic revenue streams before all of it was transferred.[56]

While most experiments in water marketing have been intrastate transactions, recent years have seen a growing interest in and need for interstate transfers.[57] In 2006, faced with ongoing drought, the seven states of the Colorado River basin created what was the first cooperative document since 1922 (and Arizona did not agree to that one until 1944). The 2006 Seven Basin States' Preliminary Proposal Regarding Colorado River Interim Operations was a response to Secretary of the Interior Gale A. Norton's plans to deal with drought when allocating Colorado River water among the states. Demands had long exceeded allocations in Southern California and Nevada. Along with many other recommendations, the proposal

outlined an innovative method to encourage water conservation. To allow reallocations through water transfers, the states recommended a "credit system" they called
an "Intentionally Created Surplus" (ICS). This "surplus" would be created through
conservation measures that included land fallowing, canal lining, desalination, and
other innovations to conserve water.[58]

To use these credits, the states agreed to strict guidelines on measuring conserved
water, and each state was given a limit to the number of credits it could use in a year.
According to Robert Glennon and water lawyer Michael J. Pearce, this credit system
was a radical departure:

> The ICS credit system is an extraordinary change in how Colorado River waters are used.
> Although it does not formally change the allocation of water from one state to another,
> it effectively operates that way. Nothing in the program prevents one state from buying
> water from a contractor in another state who has engaged in an activity that qualifies for
> ICS credits.[59]

Nevada can and does draw more than its 300,000-acre-foot annual allowance
from Lake Mead through this credit system, as do other states. Arizona currently
receives credits for storing Colorado River water in its underground aquifers under
the operation of the Arizona Water Banking Authority. In 2009 the Arizona Water
Bank stored around 182,670 acre-feet of water, 40,000 of which was for the state of
Nevada. In 2012 it expected to store approximately 119,000 acre-feet of water.[60]

Although most of the discussion of water markets in the West centers on California,
the state of Colorado has long been trading water. The state has experienced heated
water wars—especially between the western slope, where most of the water is located,
and the Front Range, where most of the population lives. Plans for more pipeline
projects are being debated that would satisfy the water cravings of the Front Range
by taking all of the state's Colorado River allotment except for about 160,000 acre-
feet each year. One proposal is to build a 540-mile pipeline from the Flaming Gorge
reservoir to the Denver area. Another is for a 250-mile pipeline from the Yampa River,
one of the last free-flowing rivers in the region. Local and environmental groups are
exerting tremendous pressure against both projects, essentially stalling them for the
moment. The problem is that even if the controversial projects were constructed,
they would provide only a mere fraction of the water Denver needs for the next
twenty years.[61]

While Colorado and other states such as Nevada contemplate building aqueducts
and reservoirs, still others are looking to technology for solutions. One of the innovations that might free up more Colorado River water for the market is source metering, a new kind of meter to measure the amount of water a user diverts from a river,

stream, canal, or reservoir. First developed in 1993, source metering was established in Kansas, Texas, and Washington. These meters help users withdraw only the amount of water they need through precise measuring. According to some calculations, approximately thirty-six states will suffer water shortages by 2013, so source metering might be "an effective mechanism for responding to climate change."[62] Another recent innovation is the measuring of evapotranspiration, water loss through a combination of evaporation and plant transpiration. Seeking to help farmers make better decisions about where and when to plant certain crops, the Idaho State Water Department developed metering that uses two orbiting satellites to send data back to analysts. The hope is that one day individual farmers will be able to access these data with handheld devices.[63]

These new technologies will not solve all western water problems, however. Although third-party costs in water transfers need solutions, one of the most pressing issues is climate change. While Colorado River stream flows have varied a great deal over the centuries, recent drought has created a significant conversation about the possible length of dry conditions in the West and why they might extend much longer than a few years. A 2004 report indicated that snowpack, the primary source of Colorado River water (and indeed of the water in all western rivers), was declining. Since the 1950s the global temperature has risen approximately 0.8°C. This seemingly small increase has created a significant decrease in snowpack in the American West. If even the most moderate predictions for global warming come true over the next several decades, snowpack in the West might be reduced by 60 percent. If this happens, stream flows in the region could decline by as much as 50 percent. With snowpack melting earlier in the spring as well as decreasing in volume, water shortages in the summer will impact everything from agriculture to forests to fish runs.[64]

Some scientists argue that not only will global warming turn more snowpack into rain but that much more evaporation of rainfall will occur.[65] More rain instead of snowpack means more runoff and water loss through flooding. As both flooding and evaporation reduce the amount of water flowing into rivers, already overdrawn reservoirs may sink to new lows. In 2008 the Scripps Institute of Oceanography predicted that by 2021, Lake Mead could be virtually unusable.[66] If such dire predictions come true, then no amount of water marketing can fix the problem, but some argue that a free market for water today will conserve enough to stave off possible disaster. In 2009 another group of scientists published a study that suggested more hope. While climate change and overuse of water resources continued to be very real threats, they argued that "flexibility in current management practices" could likely stave off some of the dire predictions.[67]

Basin-Wide Planning and the Future

Because of growing water demands in the Southwest and the potential impact of climate change, river users in the upper basin are also talking about a larger project: an interstate water bank for the upper basin, which some are calling a Compact Water Bank. The fear is that one day the lower basin will issue a "compact call," or demand, that the upper basin meet its obligation under the 1922 Colorado River Compact to deliver a minimum of 7.5 MAF below Lees Ferry. The amounts have been met in recent years in part because of water storage in Lakes Mead and Powell. When the most recent severe drought hit in 2002, all basin states watched in fear as Lake Powell water levels declined by two-thirds by the end of 2005. In a four-year period, the basin that had taken seventeen years to fill had lost around 16 MAF.[68]

In the face of this major decline in supply, people in Colorado began to worry, especially those in Denver and other Front Range cities. People in the state are basically talking about two kinds of water rights: pre-1922 and post-1922 rights. Those who had water rights allocated prior to the signing of the 1922 compact are considered senior under prior appropriation water doctrine. Denver and many other cities in Colorado have water rights more junior than those on much of the agricultural western slope. Water managers fear that a "call" on the river from the lower basin will mean a compact curtailment. Under a curtailment, those with senior water rights would receive their water first, leaving Denver and many other urban areas dry. One possible solution is a water bank for emergencies such as a compact curtailment, and this idea has been under discussion in various forms since 2005. Post-1922 water rights holders in Colorado might purchase the more senior water rights and bank them against a future compact curtailment.[69]

The conversations are ongoing, however, and it is not clear if a water bank will be established. Some argue that the water bank would only postpone a water crisis while still resulting in costs for post-1922 rights holders for a useless plan. Others argue that the lower basin would never issue a compact call because of the threat of litigation. Some say that if water levels in the Colorado River drop drastically, all basin states will have to join together and redo the compact, redistributing a much smaller amount of water and spreading the cutbacks fairly. So far, no such discussions have occurred, in spite of the encouraging 2007 seven–basin state agreement on interim operations. Still, that agreement might have set a necessary precedent for future negotiations.

Regardless of the challenges and controversies, the western water market is here to stay, and the Colorado River will remain a vital part of that market. It seems clear that a water market must be regulated by the government in some way to ensure that everyone has access to it, providing some kind of middle ground between free markets and public trusts. Perhaps the best designs for this regulation will come from basin-wide planning in which governors and water districts cooperate with the Department of the Interior. In 1986 water analyst B. Delworth Gardner stated that "water markets can be our salvation."[70] Yet as many regions globally begin using water markets to meet demand, many important questions remain unanswered. What are the long-term impacts of water marketing? How can negative impacts of water marketing be addressed to preserve everyone's right to water? Do we assert the validity of the public trust doctrine, or do we find no other realistic answer than to commoditize water in a free market?

As economists, hydrologists, politicians, and historians ponder these questions, the developing water market in the American West can serve as a microcosm for larger global questions. Do we seek a marketplace for water in which there will be winners and losers? Do we want equality, with all users treated the same but managed through government mandates? These questions beg answers as the entire power structure of water continues to change. Robert Glennon argued recently that a state-regulated water market may be the best solution to water shortages:

> A regulated market makes eminent sense in the case of water, a resource with cultural, spiritual, religious, environmental, and economic value. Water is a shared resource, widely but unevenly distributed, used and reused, in constant movement through the hydrologic cycle. As a shared resource owned by the state and used by its citizens, water requires stewardship by the state. A state-regulated market makes sense for a public good such as water in order to prevent externalities, an example of what economists call market failure.[71]

Glennon sees state-regulated marketing as a way to avoid the pitfalls of political and bureaucratic control of water. He further asserts that "water allocated through the political process inevitably goes to the most powerful economic and political interests in the state." A regulated market, he argues, can be more equitable.[72]

In the case of the Colorado River, power has shifted toward new players in the politics of water. Some formerly powerful cities and water districts are paying more and controlling less. Even state governments find themselves challenged by the growing power of Native American communities with the most senior water rights in the West. Private corporations are taking larger pieces of the water market while agriculture slowly sells off some of its jealously protected rights. Water in the West still flows toward money, and having water still equals power. The Colorado River is still serving humans who control it and who divide its waters like a giant plumbing system. Yet the river still shapes its users by reminding us of our dependency on nature.

NOTES

1. A version of this chapter was previously published and is reprinted here with kind permission from Springer Science+Business Media; April R. Summitt, "Marketing the Colorado River: Water Allocations in the American Southwest," *Water History* 3, no. 1 (February 2011): 45–62.

2. Ereney Hadjigeorgalis, "A Place for Water Markets: Performance and Challenges," *Review of Agricultural Economics* 31, no. 1 (2008): 55.

3. US Department of Commerce, *Water Supply and Wastewater Treatment in China* (Washington, DC: International Trade Commission, January 2005), 1–95, National Technical Information Service, 5285 Port Royal Road, Springfield, VA 22161. See also Fereidoun Ghassemi and Ian White, *Inter-Basin Water Transfer: Case Studies from Australia, United States, Canada, China, and India* (Cambridge, UK: Cambridge University Press, 2007).

4. B. Delworth Gardner and H. H. Fullerton, "Transfer Restrictions and Misallocations of Irrigation Water," *American Journal of Agricultural Economics* 50, no. 3 (August 1968): 556–71. Other works that focus on US experiments with marketing include L. M. Hartman and Don Seastone, *Water Transfers: Economic Efficiency and Alternative Institutions* (Baltimore: Johns Hopkins University Press, 1970); Charles J. Meyers and Richard A. Posner, *Market Transfers of Water Rights* (Washington, DC: National Water Commission, 1971); H. Stuart Burness and James P. Quirk, "Water Law, Water Transfers, and Economic Efficiency: The Colorado River," *Journal of Law and Economics* 23, no. 1 (April 1980): 111–34; Terry L. Anderson, ed., *Water Rights: Scarce Resource Allocation, Bureaucracy, and the Environment* (San Francisco: Pacific Institute for Public Policy Research, 1983); J. J. Vaux Jr. and R. E. Howitt, "Managing Water Scarcity: An Evaluation of Interregional Transfers," *Water Resources Research* 20, no. 7 (July 1984): 785–92. Recent works that promote water marketing include Terry L. Anderson and Pamela Snyder, *Water Markets: Priming the Invisible Pump* (Washington, DC: CATO Institute, 1997); Brent M. Haddad, *Rivers of Gold: Designing Markets to Allocate Water in California* (Washington, DC: Island, 2000); Robert Glennon, "Water Scarcity, Marketing, and Privatization," *Texas Law Review* 83, no. 1873 (2005): 1873-1902; Glennon, *Unquenchable: America's Water Crisis and What to Do about It* (Washington, DC: Island, 2009).

5. Helen M. Ingram, Lawrence A. Scaff, and Leslie Silko, "Replacing Confusion with Equity: Alternatives for Water Policy in the Colorado River Basin," in Gary D. Weatherford and F. Lee Brown, eds., *New Courses for the Colorado River: Major Issues for the Next Century* (Albuquerque: University of New Mexico Press, 1986), 186–95.

6. See B. Delworth Gardner, "The Untried Market Approach to Water Allocation," in Gary D. Weatherford and F. Lee Brown, eds., *New Courses for the Colorado River: Major Issues for the Next Century* (Albuquerque: University of New Mexico Press, 1986), 155–76.

7. One of the best studies of California water issues is Norris Hundley Jr., *The Great Thirst: Californians and Water, a History* (Berkeley: University of California Press, 2001). Two classics on western water in general are Marc Reisner, *Cadillac Desert: The American West and Its Disappearing Water* (New York: Penguin Books, 1993 [1986]), and Donald Worster's groundbreaking *Rivers of Empire: Water, Aridity, and the Growth of the American West* (New York: Oxford University Press, 1985). One of the most complete examinations dedicated exclusively to the Colorado River is Philip L. Fradkin, *A River No More: The Colorado River and the West* (Berkeley: University of California Press, 1993 [1968]).

8. For an excellent overview of the Owens Valley water transfer and lessons to be learned for current water transfers, see Gary D. Libecap, *Owens Valley Revisited: A Reassessment of the West's First Great Water Transfer* (Stanford: Stanford Economics and Finance, 2007).

9. US Bureau of Reclamation, "Final Environmental Impact Statement," Volume 1, October 2002, 6, www.usbr.gov/lc/region/g4000/FEIS/Vol1/2-chps1and2.pdf, (accessed January 2, 2013).

10. For example, see Charles E. Phelps, Nancy Y. Moore, and Morlie Hammard Graubard, *Efficient Water Use in California: Water Rights, Water Districts, and Water Transfers* (Santa Monica, CA: R-2386-CSA/RF RAND Corporation, 1978).

11. Josh Newcom and Elizabeth McCarthy, *The Layperson's Guide to Water Marketing* (Sacramento: Water Education Foundation, 2000), 6.

12. For more information on the peripheral canal and California water issues during this era, see Hundley, *Great Thirst*, chapter 5; Stephanie S. Pinceti, *Transforming California: A Political History of Land Use and Development* (Baltimore: Johns Hopkins University Press, 2003), 206.

13. Kevin M. O'Brien and Robert R. Gunning, "Water Marketing in California Revisited: The Legacy of the 1987–92 Drought," *Pacific Law Journal* [Sacramento] 25, no. 3 (1993–94): 1053.

14. J. F. Booker and R. A. Young, "Economic Impacts of Alternative Water Allocation Institutions in the Colorado River Basin," Research Project Technical Completion Report, US Geological Survey, August 1991, Bureau of Reclamation Library and Archives, Denver, CO.

15. Richard E. Howitt, "Empirical Analysis of Water Market Institutions: The 1991 California Water Market," *Resource and Energy Economics* 16, no. 4 (1994): 357–71; Morris Israel and Jay R. Lund, "Recent California Water Transfers: Implications for Water Management," *Natural Resources Journal* 35, no. 1 (1995): 1–32.

16. J. F. Booker and R. A. Young, "Modeling Intrastate and Interstate Markets for Colorado River Water Resources," *Journal of Environmental Economics and Management* 26, no. 1 (1994): 66–87.

17. "Water in the West: Buying a Gulp of the Colorado," *The Economist* 346, no. 8052 (January 24, 1998): 28.

18. Bureau of Reclamation, Department of the Interior, "Interim Surplus Guidelines," *Federal Register* 66, no. 17 (January 25, 2001): 7772–82.

19. In fairness, the Department of the Interior should have also prodded the Bureau of Reclamation to meet agreements, made as early as 1992, that the bureau would develop a reasonable process for transferring water from agricultural to municipal uses. Such processes might have helped California reach its reduction requirement within the deadline. See Michael P. Colombo, "Municipal and Industrial Water Transfers: Bureau of Reclamation," Report no. W-FL-BOR-0121-2002, November 2003, Bureau of Reclamation Library and Archives, Denver, CO.

20. San Diego County Water Authority, Quantification Settlement Agreement for the Colorado River, March 2010, http://www.sdcwa.org/sites/default/files/files/publications /qsa-sf.pdf (accessed October 15, 2012).

21. Elizabeth Varin, "The Cost of the Quantification Settlement Agreement: Imperial County Governments Spend Millions in Legal Battles," *Imperial Valley Press Online*, August 22, 2010, http://www.ivpressonline.com/news/ivp-news-cost-qsa-county-legal-battles -20100822,0,7076864.story (accessed October 15, 2012). Litigation was ongoing as of early March 2012; Superior Court of California, Quantification Settlement Agreement Cases, http://www.saccourt.ca.gov/coordinated-cases/qsa/qsa.aspx (accessed March 1, 2012).

22. San Diego County Water Authority, "All-American Canal Lining Project Named APWA Public Works Project of the Year," July 19, 2010, http://www.sdcwa.org/all-american -canal-lining-project-named-apwa-public-works-project-year (accessed February 26, 2012); Ellen Hanak, "Stopping the Drain: Third-Party Responses to California's Water Market," *Contemporary Economic Policy* 23, no. 1 (January 2005): 60.

23. "All-American Canal Lining Project Reaches Critical Milestone," San Diego County Water Authority Newsletter, April 30, 2009, http://www.sdcwa.org/all-american-canal -lining-project-reaches-critical-milestone (accessed October 15, 2012).

24. Ellen Hanak, "Finding Water for Growth: New Sources, New Tools, New Challenges," *Journal of the American Water Resources Association* 43, no. 4 (August 2007): 1031.

25. Gardner, "Untried Market Approach," 156.

26. For earlier works advocating water markets, see B. Delworth Gardner and H. H. Fullerton, "Transfer Restrictions and Misallocations of Irrigation Water," *American Journal of Agricultural Economics* 50, no. 3 (August 1968): 556–71; L. M. Hartman and Don Seastone, *Water Transfers: Economic Efficiency and Alternative Institutions* (Baltimore: Johns Hopkins University Press, 1970); Charles J. Meyers and Richard A. Posner, *Market Transfers of Water Rights* (Washington, DC: National Water Commission, 1971); H. Stuart Burness and James P. Quirk, "Water Law, Water Transfers, and Economic Efficiency: The Colorado River," *Journal of Law and Economics* 23, no. 1 (April 1980): 111–34; Terry L. Anderson, ed., *Water Rights: Scarce Resource Allocation, Bureaucracy, and the Environment* (San Francisco: Pacific

Institute for Public Policy Research, 1983); J. J. Vaux Jr. and R. E. Howitt, "Managing Water Scarcity: An Evaluation of Interregional Transfers," *Water Resources Research* 20, no. 7 (July 1984): 785–972.

27. Booker and Young, "Modeling Intrastate and Interstate Markets," 84–85.

28. Robert Glennon, "Water Scarcity, Marketing, and Privatization." *Texas Law Review* 83, no. 1873 (2005): 1887.

29. Frank Ackerman and Elizabeth A. Stanton, "The Last Drop: Climate Change and the Southwest Water Crisis" (Stockholm Environment Institute–US Center, 2011), http://sei-us .org/Publications_PDF/SEI-WesternWater-0211.pdf (accessed October 15, 2012), 7. See also Carl Baronskey and Warrren J. Abbot, "Water Conflicts in the Western United States: The Metropolitan Water District of Southern California," *Studies in Conflict and Terrorism* 20 (1997): 143.

30. Glennon, "Water Scarcity," 1888. See also Gardner, "Untried Market Approach," 163–65; Terry L. Anderson and Pamela Snyder, *Water Markets: Priming the Invisible Pump* (Washington, DC: Cato Institute, 1997), 96–97.

31. Glennon, "Water Scarcity," 1887.

32. Anderson and Snyder, *Water Markets,* 96–97.

33. Charles W. Howe, "Increasing Efficiency in Water Markets: Examples from the Western United States," in Terry L. Anderson and Peter J. Hill, eds., *Water Marketing: The Next Generation* (New York: Rowman and Littlefield, 1997), 79–99.

34. Jordan A. Clayton, "Market-Driven Solutions to Economic, Environmental, and Social Issues Related to Water Management in the Western USA," *Water* 1, no. 1 (2009): 19–31.

35. Colorado River Delta Water Trust, http://sonoraninstitute.org/images/stories/delta _water_trust_10-22-10_130pm_lores.pdf (accessed February 26, 2012).

36. Kenneth D. Frederick, "Water Marketing: Obstacles and Opportunities," *Forum for Applied Research and Public Policy* 16, no. 1 (Spring 2001): 60.

37. Sharon Megdal, "Should Yuma Desalter Operate? Varied, Complex Issues Are Raised," *Arizona Water Resource* (May-June 2004), http://wrrc.arizona.edu/sites/wrrc.arizona.edu /files/AWRColumn/Operating.the.Yuma.Desalter_MayJun04.pdf (accessed October 15, 2012).

38. Warren H. Brock Storage Reservoir, Bureau of Reclamation, Lower Colorado Region, October 2012, http://www.usbr.gov/lc/yuma/facilities/Brock/yao_brock.html (accessed October 15, 2012); Shaun McKinnon, "Yuma Reservoir Is a Water Saver," *Arizona Republic* [Phoenix], November 26, 2010, http://www.azcentral.com/arizonarepublic/news/articles /2010/11/26/20101126yuma-reservoir-water.html (accessed February 26, 2012).

39. Ibid. See also Joe Gelt, "Drop 2—End-of-the-Line Reservoir Salvages Colorado River Water," *Arizona Water Resource* 16, no. 5 (2008): 1–2.

40. Mason Gaffney, "What Price Water Marketing? California's New Frontier," *American Journal of Economics and Sociology* 56, no. 4 (October 1997): 486, 501–2.

41. Hanak, "Stopping the Drain," 73. See also Glennon, "Water Scarcity," 1873–1902; Steven P. Erie, *Beyond Chinatown: The Metropolitan Water District, Growth, and the Environment in Southern California* (Stanford: Stanford University Press, 2006), 169–204.

42. Sandra O. Archibald and Mary E. Renwick, "Expected Transaction Costs and Incentives for Water Market Development," in K. William Easter, Mark W. Rosegrant, and Ariel Dinar, eds., *Markets for Water: Potential and Performance* (Boston: Kluwer Academic, 1998), 101.

43. Some scholars argue that transaction costs are best borne by water rights holders, who should thus be given transfer rights. See Megan Hennessy, "Colorado River Water Rights: Property Rights in Transition," *University of Chicago Law Review* 71, no. 4 (Autumn 2004): 1664. For a discussion of third-party impacts of water marketing, see Brent M. Haddad, *Rivers of Gold: Designing Markets to Allocate Water in California* (Washington, DC: Island, 2000).

44. Hadjigeorgalis, "A Place for Water Markets," 50–67; Bonnie G. Colby, "Transaction Costs and Efficiency in Western Water Allocation," *American Journal of Agricultural Economics* 72, no. 5 (December 1990): 1184–92; and K. William Easter, Mark W. Rosegrant, and Ariel Dinar, eds., *Markets for Water: Potential and Performance* (Boston: Kluwer Academic, 1998).

45. Ingram, Scaff, and Silko, "Replacing Confusion with Equity," 186–95.

46. Hadjigeorgalis, "A Place for Water Markets," 50–67.

47. The Role of Water Transfers in Meeting California's Water Needs, September 8, 1999, Legislative Analyst's Office, Sacramento, CA, http://www.lao.ca.gov/1999/090899_water _transfers/090899_water_transfers.html (accessed October 15, 2012), 9. More than three decades earlier, the federal government commissioned a national water survey. The commission's report, completed in 1973, clearly outlined the overlapping layers of policy and control extending from federal policy and legislation to state regulations to community politics and governance. See United States, National Water Commission, "Water Policies for the Future: Final Report to the President and to the Congress of the United States" (Port Washington, NY: Water Information Center, 1973).

48. The Role of Water Transfers in Meeting California's Water Needs, 9.

49. Ibid., 2.

50. Ibid., 4.

51. Western Water Company, Cherry Creek Project, http://www.azwaterbank.gov /documents/2009FinalPlanofOperation_000.pdf (accessed October 15, 2012).

52. Glennon, "Water Scarcity," 1900–1902.

53. Brian J. Brady, "Collaboration on the Colorado River: Lessons Learned to Meet Future Challenges," Testimony of Brian J. Brady, GM of Imperial Irrigation District, to US House of Representatives Subcommittee on Water and Power, Committee on Natural Resources,

April 9, 2010, 3–4, http://naturalresources.house.gov/uploadedfiles/bradytestimony04.09.10 .pdf (accessed October 15, 2012).

54. Robert Glennon, *Unquenchable: America's Water Crisis and What to Do about It* (Washington, DC: Island, 2009), 263.

55. Jedidiah Brewer et al., *Water Markets in the West: Prices, Trading, and Contractual Forms* (Cambridge, MA: National Bureau of Economic Research, 2007).

56. Charles W. Howe and Christopher Goemans, "Water Transfers and Their Impacts: Lessons from Three Colorado Water Markets," *Journal of the American Water Resources Association* 39, no. 5 (October 2003): 1055–65.

57. For recent overviews of water transfers, see the California Water Plan Report 2005 and previous years at http://www.waterplan.water.ca.gov/ (accessed September 15, 2010).

58. Bureau of Reclamation, Department of the Interior, "Seven Basin States' Preliminary Proposal Regarding Colorado River Interim Operations," http://www.usbr.gov/lc/region /programs/strategies/consultation/Feb06SevenBasinStatesPreliminaryProposal.pdf (accessed October 15, 2012).

59. Robert Glennon and Michael J. Pearce, "Transferring Mainstem Colorado River Water Rights: The Arizona Experience," Arizona Legal Studies Discussion Paper no. 07-05 (Tucson: University of Arizona College of Law, 2007), 18.

60. Arizona Annual Plan of Operation, Arizona Water Banking Authority, 2009, http:// www.azwaterbank.gov/documents/2009FinalPlanofOperation_000.pdf (accessed October 15, 2012); Sandra A. Fabritz-Whitney, "Arizona Water Banking Authority Annual Plan of Operation 2012," Arizona Water Banking Authority, December 2011, http://www.azwaterbank .gov/Plans_and_Reports_Documents/documents/Final2012PlanofOperation.pdf (accessed October 15, 2012).

61. Eric Hecox, "Water 2012: Can Flaming Gorge Meet Future Water Needs?" *Valley Courier* [Alamosa, CO], September 12, 2012; Colorado Bar Association, *Water Transactions* (Denver: Continuing Legal Education, 2009), 5.

62. Stephanie Lindsay, "Counting Every Drop: Measuring Surface and Ground Water in Washington and the West," *Environmental Law* [Northwestern School of Law] 39, no. 193 (2009): 211.

63. John Grimond, "A Special Report on Water," *The Economist* (May 22, 2010): 9.

64. Robert F. Service, "Water Resources: As the West Goes Dry," *Science* 303, no. 5661 (February 20, 2004): 1124–27.

65. Peter Friederici, "The Next Market Crunch: Water," *Pacific Standard* [Santa Barbara, CA], July 14, 2008, http://www.psmag.com/magazines/2008-08-01/the-next-market -crunch-water-4424/ (accessed October 15, 2012).

66. Scripps Institute of Oceanography, "Scripps News: Lake Mead Could Be Dry by 2021," February 12, 2008, http://scrippsnews.ucsd.edu/Releases/?releaseID=876 (accessed May 11, 2010).

67. B. Rajagopalan, K. Nowak, J. Prairie, M. Hoerling, B. Harding, J. Barsugli, A. Ray, and B. Udall, "Water Supply Risk on the Colorado River: Can Management Mitigate?" *Water Resources Research* 45 (2009), W08201, doi:10.1029/2008WR007652, http://www.agu.org/journals/ABS/2009/2008WR007652.shtml (accessed September 5, 2011).

68. Allen Best, "Tapped Out," *Colorado Biz Magazine,* October 1, 2009, http://www.cobizmag.com/articles/tapped-out-front-range-cities-get-up-to-half-their-water-from-the-Western-S (accessed October 15, 2012).

69. Tom Iseman, "Banking on Colorado Water," *PERC Reports: The Property and Environment Research Center* 28, no. 1 (Spring 2010), http://www.perc.org/articles/article1227.php (accessed October 15, 2012).

70. Gardner, "Untried Market Approach," 163–65. See also Anderson and Snyder, *Water Markets,* 96–97.

71. Glennon, *Unquenchable,* 310.

72. Ibid., 314.

Conclusion

THE STORY OF THE COLORADO RIVER is as convoluted as it is long, and defining its many rivulets is a complicated process. To fully understand this river and its past, one must examine many separate pieces of history scattered throughout two nations, in seven US and two Mexican states, and sort through large amounts of scientific data. One needs to be part hydrologist, geologist, economist, sociologist, and anthropologist, as well as a historian, to fully understand the entire story. Considering its narrow size and meager flow, this river's tale is very large indeed. One of the world's most important rivers, the Colorado ranks first in litigation and regulation. To the American Southwest, it means everything.

What will happen to the Colorado River in the future is hard to predict and foreboding, but we have a clue from past human consumption and research that predicts long-term climate change. New players have entered the western water struggle and shifted the entire environment from one of power contests between elites to efforts to achieve equity in the midst of increasing demands and decreasing resources. There are no easy formulas for success in efforts to both use the river and preserve its ecosystems. Likewise, there are no easy solutions to the disconnect between increasing demands and declining flows, between prior appropriation law and region-wide planning needs. Managing the future will require *real* sacrifices and serious conservation efforts from everyone.

Present and Future Challenges

Two overriding concerns dominated the river's past and persist in the present: quality and quantity. Both exigencies still challenge users and planners looking toward the future of the Colorado River. Each issue impacts the other, creating a maze of possible answers to any single problem. While the challenge of quantity in shares of the river is perhaps the oldest issue, water quality has become perhaps the most poignant. In the longue durée of the earliest recorded years of human interaction with the river, its water went first to those who used it first, and smaller shares went to everyone who came later. Since the introduction of water law, the western water doctrine of prior appropriation might seem outdated today, but it remains very much alive in spite of agreements that set it aside. The 1922 Colorado River Compact, the first attempt to establish a fair distribution of the river's water among the basin states, has only partially readjusted that distribution or clarified its meaning. In 1944 the United States finally recognized Mexico's right to a fair share of the river's flow.

The problem of water quality first received serious consideration when Mexico protested increasingly high levels of salinity from agricultural runoff in the early 1960s. As a solution, the 1973 Salinity Control Act set levels of acceptable salinity at three points along the stream: below Hoover Dam, below Parker Dam, and at Imperial Dam on the border. Today, these target levels remain the same, but efforts to keep salinity at or below those standards require more and more direct intervention. Wastewater treatment plants up and down the river remove tons of salt every year, and regulating effluent is an ongoing project throughout the basin.

Water quantity remains a difficult challenge, and the 1,450-mile river dictates its own terms here. Earlier experiments with weather modification failed to bring more rain to the basin, and augmentation plans collapsed because other river basin consumers were unwilling to risk their own resources. To address increasing population pressures, Colorado River basin states have over-tapped and diverted so much of the flow that there is no more surplus to bank against future shortfalls. Cyclical drought has always been a reality in the Southwest, but short human memory led users to assume that temporary abundance was the norm. Seeing shortages on the horizon, states claimed and diverted as much water as possible before others took it all. Today, climate change threatens to decrease the river's flow, perhaps by as much as 20 percent, making any previous divisions of its water obsolete. To answer this and all other challenges, approaches vary widely. Wrapped up within these two issues are ongoing contests between agricultural and urban users, state governments and Indian communities, the United States and Mexico, environmentalists and water districts. Clearly, there is no simple binary conflict between two sides in this insolvable contest over water.

Basin States and Individual Solutions

Just as the Colorado River passes through and affects life in seven US states and two in Mexico, these nine separate entities affect the river and impact its future. Each region formulates its own plan for dealing with quality, quantity, and equity—often creating similar approaches to those of its neighbors that are sometimes divergent and conflicting. California was the first state to divert the river and still claims the largest share of its waters. Of the 4.4 million acre-feet (MAF) to which it is entitled each year, California uses all of that amount and has often used much more. Growth in Los Angeles, San Diego, and other metropolitan areas has increased demands on the Colorado River, but the only way to stretch the over-allocated resource is through redistribution and conservation. Both Los Angeles and San Diego have instituted vigorous water recycling and use reduction plans through rebates and price incentives.

Other urban areas are also finding innovative solutions. Orange County, California, provides one of the best examples of the power of public education in conservation efforts. In 2007, Orange County began recycling sewer water by injecting treated water into the ground to recharge the aquifer, filtering it through soil and mixing it with other water.[1] Following a vigorous public education campaign, Orange County officials found widespread support for the plan. Now processing 70 million gallons per day, the processing plant purifies enough water for 500,000 residents and visitors to Disneyland.[2] This water, which costs approximately $600 per acre-foot, will look very good in the future if prices of imported water in California reach $800 per acre-foot in two years, as expected.

Other cities are following Orange County's example. While sewage recycling was first proposed in San Diego back in 1989, the first serious attempt occurred ten years later during a drought. Opponents labeled the project "Toilet to Tap," which went a long way toward defeating the idea. San Diego did not manage to educate its population sufficiently to accept the plan until recently. Facing ongoing water shortages and increased demand, the city approved a sewer recycling pilot project in July

2010.[3] At present, similar programs in Tucson and elsewhere recharge aquifers with recycled water. One day cities will forgo efforts to bypass the "yuck factor" and pump the water directly into their systems. Not widely addressed, however, is the issue of various contaminants that are not easily removed in recycling processes. Some endocrine-disrupting chemicals such as estrogen are extremely difficult to remove. Other problematic contaminants come from pesticides and industrial chemicals. Recycling processes will need to keep improving to make recycled water safe for drinking.[4]

Other recent conservation efforts in California include voluntary water conservation through better target delivery of water and seepage recovery projects. Continuing agriculture-to-urban water transfers help conserve water usage while maintaining adequate supplies in cities. Recent California studies predict that even with climate change conditions, water supplies to cities will meet demand through ongoing transfers.[5] In Arizona, water-use plans also include agriculture-to-urban transfers, increasingly seen as the best and only real solution to shortages. Agriculture contributes just 1 percent of the state's economic output but uses 70 percent of its water resources. Farmers, however, jealously guard that 5 MAF of water. In 2009 the *Arizona Republic* quoted Tom Davis, general manager of the Yuma County Water Users Association: "We don't want to get into a situation of saying 'my use is better than yours,' but there needs to be a better way than just whoever has the most money gets the most water."[6] Clearly, water transfers are of concern to farmers, and even urban dwellers worry about the environmental impact of transporting food to the state. In 1965, 80 percent of Salt River Project water went to agriculture, in contrast to only 15 percent in 2009. The unintended consequences of reducing farmland throughout the Colorado River basin states will remain controversial but will not likely stop anytime soon.

Another, less controversial conservation effort in Arizona is industry reductions. Intel Corporation uses large amounts of water to fabricate and package its silicon chips. According to its reports, one Intel plant in Chandler used more than 600 million gallons in 2007. The plant, however, is working hard to recycle almost 75 percent of its salty wastewater, pumping it into the aquifer after processing in its desalination plant.[7] Efforts to recharge groundwater throughout the Phoenix metropolitan area are sincere, but there are other realities, such as energy requirements. To pump water to consumers or even to pump water through recycling plants, a great deal of electricity is required. Using a range of 5.5 to 9.8 kilowatt-hours per 1,000 gallons, the Central Arizona Project (CAP) used 2.8 million megawatt-hours of electricity in 2009 to deliver 1.5 MAF of water.[8] To further complicate the numbers, the Palo Verde Nuclear Plant uses 20 billion gallons of water for cooling each year. Both Intel and the CAP pipeline use the plant's electricity, which is not usually figured into the cost of water.[9]

To meet growing municipal needs, the city of Phoenix is working in several directions—including recycling wastewater, leasing water from Indian communities, and purchasing additional water from the main stem of the Colorado River. In 2011, city officials listed six major approaches to meet projected water needs, including reclaiming water use and leasing water from the White Mountain Apache community.[10] A recent point of pride is that water use per person has dropped by 20 percent between 1999 and 2012, leaving water consumption at 1999–2000 levels.[11] In Tucson, ongoing conservation efforts to maintain renewable groundwater levels have been exemplary. There are still pressures for more water, however, and Tucson accesses and stores additional Colorado River supplies through the CAP pipeline.[12]

In Nevada, one finds another mixed approach to water. The city of Las Vegas continues to use all of the state's Colorado River allocation and has long needed more. Water transfer agreements with Arizona have helped conditions, and various conservation efforts continue. The city gives rebates for grass removal and other conservation techniques and reported in 2009 that per capita water use was declining. The city hopes to reach its goal of 199 gallons per person per day by 2035, but in the meantime it must construct a lower, third intake line into Lake Mead because of dropping water levels.[13] Water-banking agreements with Arizona and California are designed to keep saved water allocations for future shortfalls.

Similar to the lower basin states, Colorado has also developed complex plans to ensure adequate water supplies and to bank against future shortfalls. Not all reports are optimistic, however. Although most metropolitan areas do not want to address growth limitations, some Denver analysts fully recognize the problem. In a recent report, projections of population growth for the state as a whole were estimated at an increase of 3 million people by 2020. If that growth actually occurs, the city of Denver expects to be short at least 630,000 acre-feet of water per year, almost twice the amount of Nevada's entire Colorado River allocation.[14] While some states, including Nevada, are calling for a renegotiation of the Colorado River Compact, Colorado sees that notion as a threat. In 2008, Arizona senator John McCain was quoted as saying, "The water compact that Colorado and other upper basin states have with California and Arizona should be renegotiated." In response, Colorado senator Ken Salazar said a renegotiation would happen "over my dead body."[15] Today, Salazar is President Barack Obama's interior secretary and helps negotiate drought management regulations. Perhaps not surprising, no renegotiation of the Colorado River Compact has taken place.

Yet in spite of such fears, similar to other basin cities, Denver is concentrating mostly on conservation—seeking an increased supply through water saving, rebate programs, and public education efforts. Although the city is hoping for higher participation numbers in the future, many Denver residents are trying to save water. Approximately 15 percent of the city's water customers are taking advantage of rebates for conservation, saving an estimated 960 acre-feet of water per year—enough to provide a year's supply of water to 2,400 households.[16]

While New Mexico, Utah, and Wyoming have less obvious immediate pressure on their water supplies from the Colorado River, all three states have done more conservation planning by educating public water users and offering tax incentives or rebates to consumers who are willing to make changes.[17] In fact, all of the basin states seem to realize that public education is key to addressing future water shortages. No one method has been most successful, however, and many state officials fear changes will be made only when they become mandatory. Such forced change will only be accepted when no other alternatives exist.

Basin-Wide Planning and Negotiation

Although most efforts to address water shortages have been local or statewide, some inter-basin planning is beginning to take place. The Bureau of Reclamation advised that basin-wide planning begin as early as the 1960s, but competing state interests were too strong to overcome. As climate change and drought make watershed planning more important than ever, however, states are finally recognizing their interdependence and talking to each other about a larger plan. Some current predictions about the impact of climate change over time are dire. A projected 10 percent to 20 percent drop in future water levels of the Colorado River compounds the existing concern about water levels in Lakes Mead and Powell, at less than 50 percent capacity. Some hydrologists argue that "droughts of the past will become the norm of the future."[18]

In a recent study by the Western Resource Advocates group and the Environmental Defense Fund, the authors call for specific actions in the arid western states but also for a nationwide climate policy that would focus on clean energy sources that use little or no water. One statistic cited is that thermoelectric plants in five western states (Arizona, Colorado, Nevada, New Mexico, and Utah) consumed approximately 292 billion gallons of water every day during 2005. Colorado estimates that such water use for power generation in the Rocky Mountain region will increase by 200 million gallons per day by 2030. The encouraging news is that many cities are trying to conserve the water used by electric plants. For example, Albuquerque's conservation plans saved more than 19 billion gallons of water in 2008.[19]

To survive future shortfalls, the report's authors recommended seven specific measures: passing comprehensive national climate and clean energy legislation, implementing energy-efficient policies for building codes and appliances, expanding water-efficient sources of energy, offering more incentives for conservation, increasing the use of recycled water, sponsoring new technologies, and working collaboratively on all water-related issues. These recommendations are predicated on the assumption that population growth will continue in spite of the economic recession that began in December 2007. Population in the seven Colorado River basin states totaled around 18 million in 2005. That number is expected to increase by 9 million, to a total of 27 million people, by 2030.[20]

In June 2009 the Bureau of Reclamation proposed a basin-wide study to the Department of the Interior following extensive planning with all seven basin states. Costs of the study—an estimated $2 million—would be shared, with 50 percent paid from federal resources and 50 percent by the basin states and their corresponding water districts and organizations. The report lists the stakeholders, which include 30 million people, 4 million acres of irrigated farmland, 14 American Indian communities, 7 wildlife refuges, 4 national recreation areas, and 5 national parks. Mexico is mentioned briefly: "The Colorado River is also vital to Mexico to meet both agricultural and municipal water needs."[21] This small mention shows little change in attitude over time, in spite of transboundary cooperation between the Environmental Protection Agency and its Mexican counterpart, the Secretary for Natural Resources and the Environment (see chapter 7).

Among the various stages for the basin study, scheduled through late 2012, stakeholder involvement was an important one. The bureau held regular meetings throughout the basin to consult with water users, districts, and city municipality systems. The final report was published December 12, 2012, and predicted a water shortage of at least 3.2 MAF by 2060. The bureau plans to work with many groups to find water conservation strategies.[22] A basin-wide salinity control program has operated since 1974 in one form or another. In 2005, Secretary of the Interior Gale Norton issued interim guidelines for operation of Colorado River basin reservoirs under drought conditions. The following year, the seven basin states created one of the few existing cooperative agreements regarding use of the river. In their proposal, the state representatives announced plans for cooperative activities to address drought shortages, augment water supplies, and implement weather modification efforts.[23]

In response to the Bureau of Reclamation's 2007 environmental impact statement on the interim guidelines, the basin states recommended a sharing agreement between Nevada and Arizona in times of shortage and the development of non–Colorado River water sources by the lower basin states, among other things. By far the most strongly emphasized recommendation, however, was reduced deliveries to Mexico. Reminding the secretary of the interior of agreements already carried out, the states urged that deliveries across the border be reduced by 400,000 acre-feet at first, then up to 600,000 acre-feet when levels in Lake Mead dropped below certain levels.[24] While players in the politics of the river have changed and shifted in strength, some of the attitudes from the twentieth century still influence interactions between the United States and Mexico regarding water.

According to the decision about drought operations, published in December 2007, the secretary of the interior will declare a shortage if the amount of water available to the lower basin is less than 7.5 MAF. When Lake Mead levels are projected to be below 1,050 feet in elevation, the lower basin will receive 7.167 MAF, a shortfall of 0.333 acre-feet. Arizona's share will then drop from 2.8 to 2.48 MAF and Nevada's share from 300,000 to 287,000 acre-feet. If levels drop even lower, Arizona's and Nevada's shares drop correspondingly, while California's allocation remains steady, at 4.4 MAF. The rules also created specific regulations for each state to receive Intentionally Created Surplus water credits. Each state has specific limits on the number of credits it can use in a year, but the system is meant to reward water-use efficiency.[25]

On August 12, 2010, Lake Mead levels were measured at 1,087 feet, the lowest since 1956. When hydrologists predicted that the level would drop another 3 feet by the end of the year, efforts were doubled to complete the Warren H. Brock Reservoir near the border to catch unused releases. Another effort centered on negotiations with Mexico to leave some of its allocation in Lake Mead, since a 7.2-magnitude earthquake on April 4, 2010, meant some Mexicali farmers would not use irrigation water that summer.[26] Both sides recognized the potential benefits of such an agreement. Mexico would be able to store water for future use, and Lake Mead levels would remain high enough—it was hoped—to avoid a declaration of a shortage that would trigger allocation reductions.[27] However, while this example illustrates transboundary cooperation, an equitable relationship between the United States and Mexico over water is highly unlikely.

One ongoing issue that could impact future river shortages is a water quantification settlement with the Navajo and Hopi Nations. While all other Indian communities in the basin have negotiated water rights settlements, this potentially large claim is the only one left undone. In March 2009, New Mexico settled water rights with the Navajo through congressional legislation that also provided $870 million for a water delivery system. In September 2010 the Navajo Nation voted on a proposed agreement with Arizona called the Little Colorado River Water Rights Settlement Agreement. Under that agreement, the reservation would be entitled to 10,000 acres of irrigated land along the Little Colorado, 40,000 acre-feet of its annual flow, and 32,000 acre-feet of fourth-priority Colorado River flow.[28] Much of the current settlement was based on a 2006 Bureau of Reclamation study assessing water needs in north-central Arizona.[29] Although some members of the Navajo Nation protested the settlement, which gave them a very small share of Colorado River water, most members agreed to it, realizing that it was a necessary exchange for federal money to build water delivery infrastructure. In late 2012, however, the proposed settlement collapsed because of ongoing opposition in the Navajo and Hopi Nations (see chapter 6).

A Cycle Complete

There is no question that the Colorado River is in trouble and suffers from over-allocation, careless use, and ongoing climate change. It is possible that the river could disappear or become seasonal, appearing aboveground only during rare flooding cycles. Many a desert river has become a "lessening stream" or even a "river no more."[30] The Colorado River could stop flowing, but the fact that so many people depend on its water brings some optimism that something will change. Nothing remains static; as a river flows, so do human interactions with the natural environment. People have been careless and greedy, but people can and do change.

So does nature, of course. There is no static, perfect state to which it must be returned. In some ways, all rivers altered by humans become "organic machines," hybrids of natural and built environments.[31] Likewise, many rivers around the world are at times subdued by power elites who make them "rivers of empire."[32] Certainly, modern society can make them seem like "virtual" or "disconnected" rivers.[33] The Colorado River has, at one time or another, been each of these. It is a plumbing system, a garden hose, and a boundary. It is above all a lifeline, endangered but with power remaining. Its long story illustrates the larger human experience with the natural world.

In the early days of human interaction with the river, it held all the power. Humans labored to harness its power but it broke free, resisted channeling, and flowed down the path of least resistance. Its periodic floods cleared away anything humans left in its path, and its cyclical droughts sometimes removed people as well. After much effort, American settlers in the West began to tame the river and harness its strength. Western water law encouraged at first private companies and then government entities to channel, divert, dam, and store the river's waters. During the early history of the Bureau of Reclamation, the motivation for dams and aqueducts to be built was to help individuals carve out small farms and make them flourish with irrigation water. Americans believed it was their Manifest Destiny to conquer the frontier, tame its rivers, and make its deserts bloom. Man would triumph over nature through strength and growing technology.

By the mid-twentieth century, however, it had become clear that big business would win over the individual and that agribusiness and urbanization would swallow up the original aims for the Colorado River basin. As the Bureau of Reclamation very slowly gave up its version of the agrarian myth, southwestern settlers eagerly accepted a new vision of large, irrigated agribusiness and unlimited urban growth. As the states' populations grew, each struggled for larger shares of the river's water. The law of prior appropriation contained a fatal flaw, as it encouraged both individuals and states to "use it or lose it." Even as the river showed signs of stress and over-allocation, states continued to plan water projects to utilize the river before someone else took all the water that was left.

Thus, during most of the twentieth century and up to the present, human interaction with the river was and continues to be governed by political power relationships. Humans had apparent control over the river, and those with the most money had more of its water. Although the United States and Mexico competed with each other for influence over the river before Hoover Dam was built, the relationship shifted, and US power over the river turned into almost complete control. By the late twentieth century, however, new players had arrived on the scene, wielding influence over the river that challenged the traditional power of state governments and big water districts. American Indian communities, long dissatisfied with "paper water," sought "wet water" through negotiated settlements. By the turn of the twenty-first century, these Indian water allocations had become very important and powerful tools. While water rights did not end poverty and all of its side effects, they did empower a previously ignored people. Today, communities sell and lease their water to thirsty cities and irrigate new farms that grow traditional crops that might help curb rampant health problems. Another player joined the game in the late twentieth century as the environmental movement spread and impacted policy. Now, the ecosystem itself holds power as cities and states are mandated to restore habitat to fish, fowl, and other wildlife.

Eventually, over-allocation took its toll, along with climate change and cyclical drought. As water levels in the reservoirs rise and fall in response to these factors, human interaction with the Colorado River takes on a greater sense of urgency. Now, we seek to coax the river back to health, from the headwaters to its delta wetlands. We still gulp huge amounts of its water and show no signs of stopping. Yet we find new ways to negotiate and share both abundance and shortages with humans and wildlife. We have to do this so we can stay where we are, along the Colorado's banks. We cooperate because the river has the real power and is reminding us of that fact as the cycle of power has returned to where it started. Although we change it, drain it, divert it, and deplete it, the river is actually controlling us.

Notes

1. Randal C. Archibold, "From Sewage, Added Water for Drinking," *New York Times,* November 27, 2007.

2. Jennifer Lance, "Toilet to Tap: Orange County Turning Sewage Water into Drinking Water," *Simple Earth Media,* March 14, 2009, http://bluelivingideas.com/2009/03/14 /toilet-to-tap-orange-county-turning-sewage-water-into-drinking-water/ (accessed September 17, 2010).

3. Maureen Cavanaugh and Gloria Penner, "Political Analysis: The Legacy of Toilet to Tap," *KPBS Online,* August 4, 2010, http://www.kpbs.org/news/2010/aug/04/political -analysis-legacy-toilet-tap/ (accessed September 17, 2010).

4. World Health Organization, Health Risk in Aquifer Recharge with Recycled Water, August 2003, http://www.who.int/water_sanitation_health/wastewater/wsh0308chap3.pdf (accessed March 14, 2011).

5. Department of Water Resources, *California Water Plan, Update 2009,* vol. 3: *Colorado River Integrated Water Management,* http://www.waterplan.water.ca.gov/docs /cwpu2009/0310final/v3_coloradoriver_cwp2009.pdf (accessed September 17, 2010).

6. Shaun McKinnon, "Farms Looked at as Water Resources Vanish," *Arizona Republic* [Phoenix], October 25, 2009.

7. Matthew Power, "Peak Water: Aquifers and Rivers Are Running Dry. How Three Regions Are Coping," *Wired Magazine,* April 4, 2008, http://www.wired.com/science /planetearth/magazine/16-05/ff_peakwater?currentPage=all (accessed October 17, 2012).

8. "The Water-Energy Nexus," *Arroyo* (Tucson: University of Arizona Water Resources Research Center, 2010), 2, http://ag.arizona.edu/azwater/files/Arroyo_2010.pdf (accessed September 17, 2010).

9. Power, "Peak Water."

10. City of Phoenix, 2011 Water Resources Plan, http://phoenix.gov/waterservices/wrc /yourwater/newsupplies.html (assessed October 19, 2012).

11. Ibid.

12. City of Tucson, Water Plan, 2000–2050, http://cms3.tucsonaz.gov/water/waterplan (accessed September 18, 2010). See also Drought Plan Update, 2012, http://cms3.tucsonaz .gov/sites/default/files/water/drought_plan_update_spring_2012.pdf (accessed March 16, 2012).

13. Las Vegas Valley Water District, Water Resources, 2012, http://www.lvvwd.com/about /wr.html (accessed October 19, 2012).

14. Denver magazine *5280.com,* April 2010, http://www.5280.com (accessed September 18, 2010).

15. Bob Ewegen, "McCain Suggests Raiding Colorado's Water," *Denver Post,* August 16, 2008.

16. Denver Water, Solutions: Saving Water for the Future, 2010, http://www.denverwater
.org/docs/assets/DCC8BD7A-E2B9-A215-2D2FDDC3D6C736E7/Solutions2010.pdf
(accessed September 18, 2010).

17. New Mexico State Water Plan, December 3, 2003, http://www.ose.state.nm.us/water
-info/NMWaterPlanning/2003StateWaterPlan.pdf; Utah Division of Water Resources, July
2003, http://www.conservewater.utah.gov/WhyConserve/; Wyoming State Water Plan,
October 2007, http://waterplan.state.wy.us/plan/statewide/execsummary.pdf (all accessed
September 18, 2010).

18. Allen Best, "The Climate-Water Connection in the West," *Planet-Profit Report,* July 13,
2010, http://www.planetprofitreport.com/index.php/articles/the-climate-water-connection
-in-the-west/ (accessed September 18, 2010).

19. Western Resource Advocates and Environmental Defense Fund, Protecting the
Lifeline of the West: How Climate and Clean Energy Policies Can Safeguard Water, 2010,
http://www.westernresourceadvocates.org/water/lifeline/lifeline.pdf (accessed September
18, 2010).

20. Ibid.

21. Bureau of Reclamation, Colorado River Basin Water Supply and Demand Study, June
2009, 7, http://www.usbr.gov/lc/region/programs/crbstudy/CRBasinStudy.pdf (accessed
September 18, 2010).

22. *Bureau of Reclamation News*, December 21, 2012, "Secretary Salazar Releases
Colorado River Basin Study," http://www.usbr.gov/newsroom/newsrelease/detail.
cfm?RecordID=41645 (accessed January 2, 2013).

23. Seven Basin States Preliminary Proposal Regarding Colorado River Interim
Operations, February 3, 2006, http://www.usbr.gov/lc/region/programs/strategies
/consultation/Feb06SevenBasinStatesPreliminaryProposal.pdf (accessed October 19,
2012).

24. Governor's Representatives on Colorado River Operations, Letter to Secretary Dirk
Kempthorne, April 30, 2007, http://www.ose.state.nm.us/PDF/ISC/BasinsPrograms
/Colorado/1-TransmittalLetter.pdf (accessed September 18, 2010).

25. Interim Guidelines for the Operation of Lake Powell and Lake Mead, December
2007, http://www.usbr.gov/lc/region/programs/strategies/RecordofDecision.pdf (accessed
September 18, 2010).

26. Shaun McKinnon, "Lake Mead at 54-Year Low, Stirring Rationing Fear," *Arizona
Republic*, August 12, 2010.

27. "Mexico, U.S. Discuss Colorado River Water," *Durango* [CO] *Herald,* August 15, 2010.

28. "Secret Negotiations Released on Navajo Water Rights Settlement," *Censored News*,
September 18, 2010, http://bsnorrell.blogspot.com/2010/09/secret-negotiations-released
-on-navajo.html (accessed September 18, 2010).

29. US Department of the Interior, Bureau of Reclamation, North Central Arizona Water Supply Study, October 2006, http://www.usbr.gov/lc/phoenix/reports/ncawss /NCAWSSP1NOAPP.pdf (accessed September 18, 2010).

30. See Michael F. Logan, *The Lessening Stream: An Environmental History of the Santa Cruz River* (Tucson: University of Arizona Press, 2002); Philip L. Fradkin, *A River No More: The Colorado River and the West* (Berkeley: University of California Press, 1995 [1968]).

31. Richard White, *The Organic Machine: The Remaking of the Columbia River* (New York: Hill and Wang, 1995).

32. Donald Worster, *Rivers of Empire: Water, Aridity, and the Growth of the American West* (New York: Oxford University Press, 1985).

33. Ellen Wohl, *Virtual Rivers: Lessons from the Mountain Rivers of the Colorado Front Range* (New Haven, CT: Yale University Press, 2001); Wohl, *Disconnected Rivers: Linking Rivers to Landscapes* (New Haven, CT: Yale University Press, 2004).

ARCHIVAL COLLECTIONS

Archives and Public Records, Arizona State Library, Phoenix.

Bureau of Reclamation Library and Archives, Denver, Colorado.

Carl T. Hayden Papers, Special Collections, Hayden Library, Arizona State University, Tempe.

Colorado River/Central Arizona Project Collection, Hayden Library, Arizona State University, Tempe.

Department of the Interior Papers, William J. Clinton Presidential Library, Little Rock, AK.

Elmer K. Nelson Papers, Water Resource Center Archives, University of California, Berkeley. Special Collections, Stanford University Library, Palo Alto, CA.

J. Willard Marriot Library, University of Utah, Salt Lake City.

Morris K. Udall Papers, University of Arizona Library, Special Collections, University of Arizona, Tucson.

National Archives and Records Administration, Rocky Mountain Region, Denver, Colorado.

Papers of Charles C. Fisk, Water Resources Archive, Colorado State University, Fort Collins.

Papers of Delph E. Carpenter and Family, Water Resources Archive, Colorado State University, Fort Collins.

Water Resource Center Archives, University of California, Berkeley.
Wayne Aspinall Papers, Penrose Library, University of Denver.

SELECTED GOVERNMENT DOCUMENTS

Bureau of Reclamation, Department of the Interior. *Central Arizona Project.* Washington, DC: Department of the Interior, 1995.

Bureau of Reclamation, Department of the Interior. Colorado River System Consumptive Uses and Losses Report, 1976–1980, 1981–1985, and 1991–1995. Bureau of Reclamation Library and Archives, Denver, CO.

Bureau of Reclamation, Department of the Interior. "Interim Surplus Guidelines." Federal Register 66, no. 17 (January 25, 2001): 7772–7782.

Castro, Raul H., Governor. "Arizona State Water Plan, Phase II." State of Arizona, February 1977. Bureau of Reclamation Library, Denver, Colorado.

Colorado River Compact. 1922. Article VII. Colorado River/Central Arizona Project Collection, Hayden Library, Arizona State University, Tempe. Western Waters Digital Library, www.westernwaters.org.

Colorado River Compact Commission. Minutes of Colorado River Compact Commission, 1922. Arizona Historical Foundation Collection, Hayden Library, Arizona State University, Tempe.

Department of Justice. Appropriation Act of July 1952. 66 Stat. 560, http://www.uspto.gov /web/trademarks/PL107_273.pdf (accessed January 28, 2012).

Grand Canyon Protection Act. 1992. http://www.gcmrc.gov/library/reports/LawoftheRiver /GCPA1992.pdf (accessed July 29, 2010).

"Hualapai Contract." August 30, 1960. Folder 4, Box 8, Carl Hayden Papers, Special Collections, Hayden Library, Arizona State University, Tempe, Arizona.

International Boundary and Water Commission, United States and Mexico. Text of Minute 241. July 14, 1972, El Paso, Texas. http://www.ibwc.state.gov/Files/Minutes/Min241.pdf (accessed July 15, 2010).

International Boundary and Water Commission, United States and Mexico. Text of Minute 242—Permanent and Definitive Solution to the International Problem of the Salinity of the Colorado River. August 30, 1973. http://www.usbr.gov/lc/region/pao/pdfiles /min242.pdf (accessed September 23, 2012).

Public Law 90-537, Colorado River Project Basin Act. September 30, 1968. http://usbr.gov /lc/region/pao/pdfiles/crbproj.pdf (accessed August 16, 2010).

Public Law 97-293, Reclamation Reform Act. 1982. http://www.usbr.gov/rra/Law_Rules /public%20law%2097-293.pdf (accessed September 9, 2012).

San Carlos Irrigation Project. Bill S. 966. 68th Cong., 1st sess., House of Representatives, Report No. 618. Washington, DC: Government Printing Office, 1924.

Senate Report No. 755. 82nd Cong., 1st sess. Washington DC: Government Printing Office, 1951.

Report in the Matter of the Investigation of the Salt and Gila Rivers—Reservations and Reclamation Service. 62nd Cong., 3rd sess., House of Representatives, 1912–13.

Washington, DC: Government Printing Office, 1913. http://water.library.arizona.edu
/body.1_div.39.html (accessed January 28, 2012).

United States Congress. "An Act to Provide for the Sale of Desert Lands in Certain States
and Territories." *United States Statutes at Large* 19, ch. 107. Washington DC: Government
Printing Office, 1877.

United States Congress. *Water Treaty with Mexico: Hearings before the Committee on Foreign
Relations, United States Senate, Seventy-Ninth Congress, First Session, on Treaty with
Mexico Relating to the Utilization of the Waters of Certain Rivers, January 22–26, 1945.*
Washington, DC: US Government Printing Office, 1945.

"United States—Mexico Agreement on Colorado River Salinity: Exchange of Identic Notes
of August 30, 1973 between the United States Ambassador to Mexico and the Mexican
Secretary of Foreign Relations." *American Journal of International Law* 68, no. 2 (April
1974): 376–79. http://dx.doi.org/10.2307/2199693.

United States, National Water Commission. "Water Policies for the Future: Final Report
to the President and to the Congress of the United States." Port Washington, NY: Water
Information Center, 1973.

US Department of Commerce. *Water Supply and Wastewater Treatment in China.*
Washington, DC: International Trade Commission, January 2005, 1–95. National
Technical Information Service, 5285 Port Royal Road, Springfield, VA 22161.

US Department of the Interior. Final Environmental Assessment: Yuma Desalting Plant
Pilot Run. Yuma: US Bureau of Reclamation, Yuma Office, June 2010.

US Government Accounting Office. Colorado River Basin Water Problems: How to Reduce
Their Impact. May 4, 1979. Bureau of Reclamation Library and Archives, Denver, CO
(GAO Report 1979).

US Government Printing Office. *Utilization of Waters of the Colorado and Tijuana
Rivers and of the Rio Grande: Treaty between the United States of America and Mexico.*
Washington, DC: Government Printing Office, 1946.

NEWSPAPERS

Albuquerque Journal
Arizona Farmer [Phoenix]
Arizona Republic [Phoenix]
Arizona Star [Tucson]
Californian [Bakersfield]
Chandler Arizonan
Colorado Springs Gazette
Cortez [CO] *Journal*
Daily Courier [Prescott, AZ]
Daily Times [Farmington, NM]
Denver Post
Deseret News [Salt Lake City]
Durango [CO] *Herald*

Green Valley [AZ] *News and Sun*
Hartford [CT] *Courant*
Herald-Examiner [Los Angeles]
High Country News [Paonia, CO]
Imperial Valley Press [El Centro, CA]
Imperial Valley Press Online
Indian Country Today [New York City]
Inglewood [CA] *News*
Lima News [Lima, OH]
Limelight News [Palm Springs, CA]
Los Angeles Daily News
Los Angeles Times
The Mirror [Los Angeles]
Native American Times [Tahlequah, OK]
Navajo Times [Window Rock, AZ]
Nevada State Journal [Reno]
New York Times
North County Times [San Diego]
Pasadena Star News
Reno Gazette
Sacramento Chronicle
Salt Lake Tribune
Sunday Herald [Provo, UT]
Tucson Citizen
US News & World Report
US Water News Online
Valley Courier [Alamosa, CO]
Washington Post
Yuma [AZ] *Sun and Sentinel*

OTHER PRIMARY SOURCES

Arizona Annual Plan of Operation, Arizona Water Banking Authority. http://www
.azwaterbank.gov/Plans_and_Reports_Documents/Annual_Plan_of_Operation.htm,
2000 through 2010.
Arizona v. California, 376 US 340 (1964).
Babcock, H. M. *Wellton-Mohawk Area, Yuma County, Arizona: Records of Wells, Well Logs,
Water Analyses, and Maps Showing Location of Wells*. Tucson: US Geological Survey, 1948.
Blake, William Phipps, and Harry Thomas Cory. *The Imperial Valley and the Salton Sink*.
Charleston, SC: Nabu Press, 2010.
Blanchard, C. J. "Home-making by the Government: An Account of the Eleven Immense
Irrigating Projects to Be Opened in 1908." *National Geographic Magazine* 19 (April 1908):
250–87.

Brady, Brian J., General Manager. "Collaboration on the Colorado River: Lessons Learned to Meet Future Challenges." Testimony of Brian J. Brady GM of Imperial Irrigation District to U.S. House of Representatives Subcommittee on Water and Power, Committee on Natural Resources. April 9, 2010, 1–4.

California Water Plan Reports. http://www.waterplan.water.ca.gov/. Most recent update is 2005.

California's Water: An LAO Primer. Legislative Analyst's Office, October 22, 2008. http://WWW.Lao.ca.gov/laoapp/main.aspx. LAO Office, Sacramento, California.

Clark, David. *Uncompahgre Project*. Denver: Bureau of Reclamation History Program, 1994. http://www.usbr.gov/projects/Project.jsp?proj_Name=Uncompahgre%20Project&pageType=ProjectHistoryPage (accessed August 8, 2012).

Cody, Betsy A., and Nicole T. Carter. 35 Years of Water Policy: The 1973 National Water Commission Report and Present Challenges. CRS Report for Congress, May 11, 2009. Congressional Research Service, 7-5700. http://aquadoc.typepad.com/files/r40573-final-crs-nwc-report-1.pdf.

Colombo, Michael P. "Municipal and Industrial Water Transfers: Bureau of Reclamation." Report No. W-FL-BOR-0121-2002 (November 2003). Bureau of Reclamation Library, Denver, Colorado.

Cornelius, Steve, Peter Culp, Jennifer Pitt, and Francisco Zamora. "Recent Developments on the Colorado River: Implications for Delta Conservation," Sonoran Institute, Memorandum. August 1, 2003. http://www.swhydro.arizona.edu/archive/V3_N1/feature2.pdf (accessed September 23, 2012).

Dominy, Floyd. Oral History by Brit Storey. April 6, 1994. Boyce, Virginia, 75. Bureau of Reclamation Library, Denver, Colorado.

Duvall, C. Dale. Oral Interview by Brit Allan Storey. 1993. Department of Veterans Affairs, Washington, DC. Printed and filed at the Bureau of Reclamation Library, Denver, Colorado.

Fikes, Bradley. "Water Agencies Pondering How to Share 'the Lifeline of the West'." *San Diego Business Journal* 16, no. 12 (March 20, 1995): 1–2.

Fisk, Charles C. "The Metro Denver Water Story: A Memoir." Unpublished manuscript, Papers of Charles C. Fisk, Water Resources Archive, Colorado State University. Digitized by Water Resources Archive, http://digitool.library.coloradostate.edu///exlibris/dtl/d3_1/apache_media/L2V4bGlcmlzL2RobC9kM18xL2FwYWNoZV9tZWRpYS81NjQ5.pdf.

Grand Canyon Trust. "Colorado River Basin Management Study." April 1997. Dividing the Waters Collection, Box 21, Folder E-3-003, Water Resource Center Archives, University of California, Berkeley.

Guenther, Herbert R., Chairman. Arizona Water Banking Authority Annual Plan of Operation 2010. December 2009.

Higginson, R. Keith. Oral History Interview with Brit Allan Storey. March 22, 1995. Bureau of Reclamation Library and Archives, Denver, CO.

Hoover, Herbert. *The Memoirs of Herbert Hoover*. New York: Macmillan, 1965.

Karamouz, Mohammad, ed. "Water Resources Planning and Management: Saving a Threatened Resource—In Search of Solutions." *Proceedings of the Water Resources Sessions at Water Forum '92*. Bureau of Reclamation Library and Archives, Denver, CO.

Kennan, George. *The Salton Sea: An Account of Harriman's Fight with the Colorado River*. Ithaca, NY: Cornell University Press, 2009 (original work published 1917).

Keys, John. "The Future of Western Water Developments." Natural Resource Law Center Symposium, Dams: Water and Power in the New West, Boulder, CO, June 2–4, 1997. Dividing the Waters Collection, Box 5, Folder A-2-067, Water Resource Center Archives, University of California, Berkeley.

Laflin, Pat. "The Salton Sea: California's Overlooked Treasure." *Periscope*. Indio, CA: Coachella Valley Historical Society, 1995. http://www.sci.sdsu.edu/salton/PeriscopeSaltonSeaCh5-6.html (accessed September 10, 2010).

Lellouch, Mark, Karen Hyun, and Sylvia Tognetti. *Ecosystem Changes and Water Policy Choices: Four Scenarios for the Lower Colorado River Basin*. Washington, DC: Island Press, 2007.

Luccke, Daniel F. "Dams, Their Costs and Benefits." Natural Resources Law Center Symposium, Dams: Water and Power in the New West, University of Colorado, Boulder, June 2–4, 1997. Dividing the Waters Collection, Box 5, Folder A-2-067, Water Resource Center Archives, University of California, Berkeley.

Maxwell, George Hebard. *Our National Defense: The Patriotism of Peace*. Washington, DC: Rural Settlements Association, 1915.

Metropolitan Water District. "Mexican Treaty: An Exposition of the Proposed Mexican Water Treaty in Maps, Illustrations and Text." Brody Collection, Box 5, Folder 3.1, Water Resource Center Archives, University of California, Berkeley.

Morton, Larry D. Oral Interviews, Vols. I and II, by Brit Allan Storey. 1996. Bureau of Reclamation, Denver, CO.

Nash, Linda L., and Peter H. Gleick. "The Colorado River Basin and Climate Change." A Report Prepared for the United States Environmental Protection Agency, December 1993. Bureau of Reclamation Library, Denver, CO.

Newcom, Josh, and Elizabeth McCarthy. *The Layperson's Guide to Water Marketing*. Sacramento: Water Education Foundation, 2000.

Newell, Frederick H. "What May Be Accomplished by Reclamation." *Annals of the Academy of Political and Social Science* 33, no. 3 (May 1909): 174–79.

Pontius, Dale. *Colorado River Basin Study, for the Western States Water Policy Review Advisory Commission*. Tucson: SWCA, Inc. Environmental Consultants, 1997.

Porter, Eliot. *The Place No One Knew: Glen Canyon on the Colorado*, ed. David Brower. San Francisco: Sierra Club, 1966.

Powell, John Wesley. *Report on the Lands of the Arid Region of the United States*. Boston: Harvard Common Press, 1983 [1878].

Powell, John Wesley, Eliot Porter, and Don D. Fowler. *Down the Colorado: Diary of the First Trip through the Grand Canyon, 1869*. New York: E. P. Dutton, 1969.

Powell, John Wesley, Eliot Porter, and Don D. Fowler. Report on the Lands of the Arid Region of the United States: With a More Detailed Account of the Lands of Utah: With Maps. Charleston: Nabu Press, [1978] 2010 (1878).

Robison, Roland. Oral Interview by Brit Allan Storey. 1993. Bureau of Reclamation Library, Denver, CO.

The Role of Water Transfers in Meeting California's Water Needs. Legislative Analyst's
    Office, September 8, 1999. http://www.lao.ca.gov/1999/090899_water_transfers/090899
    _water_transfers.html (accessed October 15, 2012).
San Diego County Water Authority. "Quantification Settlement Agreement for the
    Colorado River." March 2010. www.sdcwa.org/sites/default/files/files/publications/qsa-fs
    .pdf (accessed October 15, 2012).
Smythe, William E. *The Conquest of Arid America*. Charleston, SC: Nabu, [1905] 2010.
Southern Nevada Water Authority Resource Plan. 2009. http://www.snwa.com/assets/pdf
    /wr_plan_exec_summary.pdf (accessed January 29, 2012).
Sprouse, Terry W. "Water Issues on the Arizona-Mexico Border: The Santa Cruz, San Pedro
    and Colorado Rivers, An Issue Paper." Water Resources Research Center College of
    Agriculture and Life Sciences, University of Arizona. http://ag.arizona.edu/azwater/files
    /terrypaper1.pdf (accessed July 7, 2010).
Tarlock, A. Dan. "River Management in the Twenty-First Century: The Vision Thing."
    Natural Resources Law Center Symposium, Dams: Water and Power in the New West,
    University of Colorado, Boulder, June 2–4, 1997. Dividing the Waters Collection, Box 5,
    Folder A-2-067, Water Resource Center Archives, University of California, Berkeley.
Udall, Stewart. Oral History Interview by Joe B. Frantz, May 19, 1969. Bureau of
    Reclamation Library, Denver, CO.
United Nations Environment Programme. Agenda 21. 1992. http://www.unep.org/Documents
    .Multilingual/Default.asp?Documentid=52 (accessed July 21, 2010).
United Nations Resolution 38/161. "Process of Preparation of the Environmental Perspective
    to the Year 2000 and Beyond." http://www.un-documents.net/a38r161 (accessed July 21,
    2010).
United States v. Texas, US 162, US 1, 16 S. Ct. 725, 40 L. Ed. 867 (1896).
University of Arizona. "Arizona's Water Future: Challenges and Opportunities." 85th Arizona
    Town Hall, October 31–November 3, 2004, Tucson. http://ag.arizona.edu/azwater
    /publications/townhall/th_report.pdf (accessed July 7, 2010).
University of Arizona. "Fourth Arizona Town Hall on Arizona's Water Supply," April 6–8,
    1964. Research Report. Phoenix: Arizona Academy, 1964.
US Geological Survey. Climatic Fluctuations, Drought, and Flow in the Colorado River
    Basin. August 2004. http://pubs.usgs.gov/fs/2004/3062 (accessed March 17, 2011).
US Geological Survey. Evaporation from Lake Mead, Arizona and Nevada. 1997–1999.
    Scientific Investigations Report, 2006-5252. http://pubs.usgs.gov/sir/2006/5252/pdf
    /sir20065252.pdf (accessed January 26, 2012).
Vetter, C. P. "The Colorado River Delta." October 1949. Box 5, Folder 3.1, Brody Collection,
    Water Resource Center Archives, University of California, Berkeley.
Wagner, Henry Raup. *California Voyages, 1539–1541: Translation of Original Documents.* San
    Francisco: J. Howell, 1925.
Western Water Assessment. Colorado River Streamflow. 2009. http://treeflow.info/lees/
    index.html (accessed January 2, 2013).
Western Water Policy Review Advisory Commission. *Water in the West: The Challenge for
    the Next Century*. Boulder: Western Water Assessment, June 1998.

Winters v. United States, 207 US 546 (1908). http://caselaw.lp.findlaw.com/cgi-bin/getcase
.pl?court=us&vol=207&invol=564 (accessed January 28, 2012).

Yuma Desalting Plant/Cienega de Santa Clara Workgroup. "Balancing Water Needs on the
Lower Colorado River: Recommendations." April 22, 2005. http://ag.arizona.edu
/azwater/publications/YDP%20report%20042205.pdf (accessed July 7, 2010).

BOOKS, ARTICLES, AND DISSERTATIONS

Abbot, Carl, Stephen J. Leonard, and Thomas J. Noel. *Colorado: A History of the Centennial
State.* 4th ed. Boulder: University Press of Colorado, 2005.

Abrams, Robert H. "Reserved Water Rights, Indian Rights and the Narrowing Scope of
Federal Jurisdiction: The Colorado River Decision." *Stanford Law Review* 30, no. 6 (June
1978): 1111–48.

Adler, Robert W. *Restoring Colorado River Ecosystems: A Troubled Sense of Immensity.*
Washington, DC: Island Press, 2007.

Adler, Robert W., Jessica C. Landman, and Diane M. Cameron. *The Clean Water Act 20
Years Later.* Washington, DC: Island Press, 1993.

Agthe, Donald E., R. Bruce Billings, and Nathan Buras. *Managing Urban Water Supply.*
London: Springer Press, 2003.

Ainlay, Thomas, and Judy Dixon Gabaldon. *Las Vegas: The Fabulous First Century.* Mt.
Pleasant, SC: Acadia Publishing, 2003.

Anderson, Terry L., ed. *Water Rights: Scarce Resource Allocation, Bureaucracy, and the
Environment.* San Francisco: Pacific Institute for Public Policy Research, 1983.

Anderson, Terry L., and Peter J. Hill, eds. *Water Marketing: The Next Generation.* New York:
Rowman & Littlefield Publishers, 1997.

Anderson, Terry L., and Donald R. Leal. "Priming the Invisible Pump." In *Water Ethics:
Foundational Readings for Students and Professionals*, ed. Peter G. Brown and Jeremy J.
Schmidt, 91–104. Washington, DC: Island Press, 2010.

Anderson, Terry L., and Pamela Snyder. *Water Markets: Priming the Invisible Pump.*
Washington, DC: CATO Institute, 1997.

Anker, Debby, John De Graaf, and Antonio Castro. *David Brower: Friend of the Earth.* New
York: Twenty-first Century Books, 1993.

Ansari, Abid A., Sarvajeet Singh Gill, Guy R. Lanza, and Walter Rast, eds. *Eutrophication:
Causes, Consequences and Control.* New York: Springer Press, 2011. http://dx.doi.org
/10.1007/978-90-481-9625-8.

Archibald, Sandra O., and Mary E. Renwick. "Expected Transaction Costs and Incentives
for Water Market Development." In *Markets for Water: Potential and Performance*, ed. K.
William Easter, Mark W. Rosegrant, and Ariel Dinar, 95–118. Boston: Kluwer Academic,
1998.

Arreola, Daniel D., and James R. Curtis. *The Mexican Border Cities: Landscape Anatomy and
Place Personality.* Tucson: University of Arizona Press, 1994.

Ashley, Perry J. *American Newspaper Journalists, 1926–1950.* Detroit: Gale Research,
1984.

August, Jack L., Jr. *Dividing Western Waters: Mark Wilmer and* Arizona v. California. Fort Worth: Texas Christian University Press, 2007.

August, Jack L., Jr. *Vision in the Desert: Carl Hayden and Hydropolitics in the American Southwest.* Fort Worth: Texas Christian University Press, 1999.

Bailey, James M. "Reconsideration and Reconciliation: Arizona's 'Brothers Udall' and the Grand Canyon Dams Controversy, 1961–1968." *New Mexico Historical Review* 80, no. 2 (Spring 2005): 133–62.

Bailey, Lynn R. *Bosque Redondo: An American Concentration Camp.* Pasadena: Socio-Technical Books, 1970.

Bain, David Haward. *The Old Iron Road: An Epic of Rails, Roads, and the Urge to Go West.* New York: Viking, 2004.

Baley, Charles W. *Disaster at the Colorado: Beale's Wagon Road and the First Emigrant Party.* Logan: Utah State University Press, 2002.

Baronskey, Carl, and Warren J. Abbot. "Water Conflicts in the Western United States: The Metropolitan Water District of Southern California." *Studies in Conflict and Terrorism* 20 (1997): 137–66.

Bartis, James T., Tom LaTourrette, Lloyd Dixon, D. J. Peterson, and Gary Cecchine. *Oil Shale Development in the United States.* Santa Monica: RAND Corporation, 2005.

Bennett, Lynne Lewis. "The Integration of Water Quality into Transboundary Allocation Agreements: Lessons from the Southwestern United States." *Agricultural Economics* 24, no. 1 (December 2000): 113–25. http://dx.doi.org/10.1111/j.1574-0862.2000.tb00097.x.

Berman, David. *Interstate Compact Water Commissions: Selected Case Studies: A Report.* Washington, DC: Washington Center for Metropolitan Studies, 1962.

Best, Allen. "Tapped Out." *Colorado Biz Magazine,* cobizmag.com (October 2009).

Bial, Raymond. *Great Journeys: The Long Walk—The Story of Navajo Captivity.* New York: Benchmark Books, 2003.

Billington, David P., Donald C. Jackson, and Martin V. Melosi. *The History of Large Federal Dams: Planning, Design, and Construction.* Denver: US Department of the Interior, Bureau of Reclamation, 2005.

Biven, W. Carl. *Jimmy Carter's Economy: Policy in the Age of Limits.* Chapel Hill: University of North Carolina Press, 2002.

Black, Peter E. *Conservation of Water and Related Land Resources.* Boca Raton: Lewis Publishers, 2001.

Blaisdell, Lowell L. "Harry Chandler and Mexican Border Intrigue, 1914–1917." *Pacific Historical Review* 35, no. 4 (November 1966): 385–93.

Boime, Eric. "'Beating Plowshares into Swords': The Colorado River Delta, the Yellow Peril, and the Movement for Federal Reclamation, 1901–1928." *Pacific Historical Review* 78, no. 1 (2009): 27–53. http://dx.doi.org/10.1525/phr.2009.78.1.27.

Boime, Eric. Fluid Boundaries: Southern California, Baja California, and the Conflict over the Colorado River, 1848–1944. PhD diss., University of California, San Diego, 2002.

Booker, J. F., and R. A. Young. "Economic Impacts of Alternative Water Allocation Institutions in the Colorado River Basin." Research Project Technical Completion Report, U.S. Geological Survey, August 1991, Bureau of Reclamation Library, Denver, CO.

Booker, J. F., and R. A. Young. "Modeling Intrastate and Interstate Markets for Colorado River Water Resources." *Journal of Environmental Economics and Management* 26, no. 1 (1994): 66–87. http://dx.doi.org/10.1006/jeem.1994.1005.

Borne, Peter G. *Jimmy Carter: A Comprehensive Biography from Plains to Post-Presidency.* New York: Scribner, 1997.

Boronkay, Carl, and Warren Abbott. "Water Conflicts in the Western United States: The Metropolitan Water District of Southern California." *Studies in Conflict and Terrorism* 20, no. 2 (1997): 137–66. http://dx.doi.org/10.1080/10576109708436030.

Brewer, Jedidiah, Robert Glennon, Alan Ker, and Gary D. Libecap. *Water Markets in the West: Prices, Trading, and Contractual Forms.* Cambridge, MA: National Bureau of Economic Research, 2007.

Brinkley, Douglas. *The Wilderness Warrior: Theodore Roosevelt and the Crusade for America.* New York: HarperCollins, 2009.

Brooks, Nathan. "Indian Reserved Water Rights: An Overview." CRS Report for Congress. Washington, DC: Congressional Research Service, January 24, 2005. http://www.policyarchive.org/handle/10207/bitstreams/1917.pdf (accessed January 28, 2012).

Brothers, Kay, and Terry Katzer. "Water Banking through Artificial Recharge, Las Vegas Valley." *Journal of Hydrology* (Amsterdam) 115, nos. 1–4 (1990): 77–103. http://dx.doi.org/10.1016/0022-1694(90)90199-8.

Brower, David Ross Brower. *For the Sake of the Earth: The Life and Times of David Brower.* New York: Peregrine Smith Books, 1990.

Brown, Robert Leaman. *The Great Pikes Peak Gold Rush.* Caldwell, ID: Caxton Printers, 1985.

Brown, Stephen R., ed. *Water Rights Sales and Transfers.* Eau Clair, WI: Lorman Education Services, 2009.

Brownell, Herbert, and Samuel D. Eaton. "The Colorado River Salinity Problem with Mexico." *American Journal of International Law* 69, no. 2 (April 1975): 255–71. http://dx.doi.org/10.2307/2200267.

Brundtland, Gro Harlem. *Madam Prime Minister: A Life in Power and Politics.* New York: Farrar, Straus and Giroux, 2002.

Brundtland, Gro Harlem. *Report of the World Commission on Environment and Development: Our Common Future.* New York: United Nations, 1987.

Bruvold, William H. "Residential Water Conservation: Policy Lessons from the California Drought." *Public Affairs Report* 19, no. 6 (December 1978): 1–7.

Burness, H. Stuart, and James P. Quirk. "Water Law, Water Transfers, and Economic Efficiency: The Colorado River." *Journal of Law & Economics* 23, no. 1 (April 1980): 111–34. http://dx.doi.org/10.1086/466954.

Butts, Kent Hughes. "The Strategic Importance of Water." *Parameters* (Spring 1997): 65–83.

Camín, Héctor Aguilar, and Lorenzo Meyer. *In the Shadow of the Mexican Revolution: Contemporary Mexican History, 1910–1989.* Trans. Luis Alberto Fierro. Austin: University of Texas Press, 1993.

Cannon, Brian Q. *Reopening the Frontier: Homesteading in the Modern West.* Lawrence: University of Press Kansas, 2009.

Car, Kathleen Marion, and James D. Crammond, eds. *Water Law: Trends, Policies, and Practice*. Chicago: American Bar Association, 1995.

Carlson, Allen. *Nature, Aesthetics, and Environmentalism: From Beauty to Duty*. New York: Columbia University Press, 2008.

Caylor, Ann. "'A Promise Long Deferred': Federal Reclamation on the Colorado River Indian Reservation." *Pacific Historical Review* 69, no. 2 (May 2000): 193–215.

Chamberlin, Eugene Keith. "Baja California after Walker: The Zerman Enterprise." *Hispanic American Historical Review* 34, no. 2 (May 1954): 175–89. http://dx.doi.org/10.2307/2509323.

Chamberlin, Eugene Keith. "The Japanese Scare at Magdalena Bay." *Pacific Historical Review* 24, no. 4 (November 1955): 345–59.

Chamberlin, Eugene Keith. "Mexican Colonization versus American Interests in Lower California." *Pacific Historical Review* 20, no. 1 (February 1951): 43–55.

Chase, Richard Allen. *The Pioneers of '47: Migration along the Mormon Trail*. Logan, UT: Watkins Print Co., 1997.

Childs, Craig. *House of Rain: Tracking a Vanished Civilization across the American Southwest*. New York: Little, Brown, 2007.

Clarkson, Robert W., and Michael R. Childs. "Temperature Effects of Hypolimnial-Release Dams on Early Life States of Colorado River Basin Big-River Fishes." *Copeia* 2000, no. 2 (May 8, 2000): 402–12. http://dx.doi.org/10.1643/0045-8511(2000)000[0402:TEOHR D]2.0.CO;2.

Clayton, Jordan A. "Market-Driven Solutions to Economic, Environmental, and Social Issues Related to Water Management in the Western USA." *Water* 1, no. 1 (2009): 19–31. http://dx.doi.org/10.3390/w1010019.

Clow, Richmond L., and Imre Sutton, eds. *Trusteeship in Change: Toward Tribal Autonomy in Resource Management*. Boulder: University Press of Colorado, 2001.

Cohen, Michael P. *The History of the Sierra Club, 1892–1970*. San Francisco: Sierra Club Books, 1988.

Cohn, Jeffrey P. "Colorado River Delta." *Bioscience* 54, no. 5 (May 2004): 386–91. http://dx.doi.org/10.1641/0006-3568(2004)054[0386:CRD]2.0.CO;2.

Colby, Bonnie G. "Transaction Costs and Efficiency in Western Water Allocation." *American Journal of Agricultural Economics* 72, no. 5 (December 1990): 1184–92. http://dx.doi.org/10.2307/1242530.

Colby, Bonnie G., and Katharine L. Jacobs, eds. *Arizona Water Policy: Management Innovations in an Urbanizing Arid Region*. Washington, DC: Resources for the Future Publishing, 2007.

Collins, Kay. The Transmountain Diversion of Water from the Colorado River: A Legal-Historical Study. MA thesis, University of New Mexico, Albuquerque, 1965.

Colorado Bar Association. *Water Transactions*. Denver: Continuing Legal Education, 2009.

Copeland, Claudia. *The Clean Water Act: Current Issues and Guide to Books*. Hauppauge, NY: Nova Science Publishers, 2003.

Copulos, Milton R. "Enforcement of an Anachronism: The 160 Acre Limitation." Policy Archive for Non-Partisan Public Policy Research. December 15, 1977. http://www .policyarchive.org/handle/10207/bitstreams/8434.pdf (accessed August 12, 2010).

Cortez Lara, Alfonso A. "Irrigation and Transboundary Water Management in the Lower Colorado River: The Changing Role of Agriculturalists in the Mexicali Valley, Mexico." PhD diss., Michigan State University, Lansing.

Cortez Lara, Alfonso A., Megan K. Donovan, and Scott Whiteford. "The All-American Canal Lining Dispute: An American Resolution over Mexican Groundwater Rights?" *Frontera Norte* 21, no. 41 (January–June 2009): 127–50.

Cross, Coy F. *Go West, Young Man!: Horace Greeley's Vision for America*. Albuquerque: University of New Mexico Press, 1995.

Dean, Robert. "Dam Building Still Had Some Magic Then: Stewart Udall, the Central Arizona Project, and the Evolution of the Pacific Southwest Water Plan, 1963–1968." *Pacific Historical Review* 66, no. 1 (February 1997): 81–98.

DeCoteau, Jerilyn. "The Effects of Development on Indian Water Rights: Obstacles and Disincentives to Development of Indian Water Rights." In *Water and Growth in the West: Natural Resources Law Center Symposium*. Boulder: University of Colorado School of Law, 2000, 1–17.

Dejong, David H. "'Abandoned Little by Little': The 1914 Pima Adjudication Survey, Water Deprivation and Farming on the Pima Reservation." *Agricultural History* 81, no. 1 (Winter 2007): 36–69. http://dx.doi.org/10.3098/ah.2007.81.1.36.

Dejong, David H. "Forced to Abandon Their Farms: Water Deprivation and Starvation among the Gila River Pima, 1892–1904." *American Indian Culture and Research Journal* 28, no. 3 (2004): 29–56.

Dellenbaugh, Frederick Samuel. *The Romance of the Colorado River*. New York: G. Putnam and Sons, 1902.

Dernbach, John C. *Stumbling toward Sustainability*. Washington, DC: Environmental Law Institute, 2002.

Diaz, Henry F., and Barbara J. Morehouse, eds. *Climate and Water: Transboundary Challenges in the Americas*. Boston: Kluwer Academic Publishers, 2003.

Dinar, Shlomi. *Conflict and Cooperation along International Rivers: Scarcity, Bargaining Strategies, and Negotiation*. Boston: MIT Press, 2011.

Dolnick, Edward. *Down the Great Unknown: John Wesley Powell's 1869 Journey of Discovery and Tragedy through the Grand Canyon*. New York: HarperCollins, 2001.

Duchemin, Michael. "Water, Power, and Tourism: Hoover Dam and the Making of the New West." *California History* 86, no. 4 (2009): 60–78.

Dunaway, Finis. *Natural Visions: The Power of Images in American Environmental Reform*. Chicago: University of Chicago Press, 2005.

Durand, Jorge. *Braceros: Las Miradas Mexicana y Estadounidense: Antología (1945–1964)*. Bogotá: Senado de la República, LX Legislatura, 2007.

Dwyer, John Joseph. *The Agrarian Dispute: The Expropriation of American-Owned Rural Land in Postrevolutionary Mexico*. Durham: Duke University Press, 2008.

Dziegielewski, Benedykt, Hari P. Garbharran, and John F. Langowski, Jr. *Lessons Learned from the California Drought (1987–1992): National Study of Water Management during Drought*. Darby, PA: Diane Publishing, 1997.

Easter, K. William, Mark W. Rosegrant, and Ariel Dinar, eds. *Markets for Water: Potential and Performance*. Boston: Kluwer Academic Publishers, 1998.

Engstrand, Iris Wilson. *San Diego: California's Cornerstone*. San Diego: Sunbelt Publications, 2005.

Erie, Steven P. *Beyond Chinatown: The Metropolitan Water District, Growth, and the Environment in Southern California*. Stanford: Stanford University Press, 2006.

Espeland, Wendy Nelson. *The Struggle for Water: Politics, Rationality, and Identity in the American Southwest*. Chicago: University of Chicago Press, 1998.

Etulain, Richard W. *Western Lives: A Biographical History of the American West*. Albuquerque: University of New Mexico Press, 2004.

Felice, Jack San. *When Silver Was King: Arizona's Famous 1880's Silver King Mine*. Mesa, AZ: Millsite Canyon, 2006.

Finkhouse, Joseph, and Mark Crawford. *A River Too Far: The Past and Future of the Arid West*. Reno: University of Nevada Press, 1991.

Fishman, Charles. *The Big Thirst: The Secret Life and Turbulent Future of Water*. New York: Simon & Schuster, 2011.

Fixico, Donald L. *Termination and Relocation: Federal Indian Policy, 1945–1960*. Albuquerque: University of New Mexico Press, 1986.

Forbes, Jack D. "Melchior Díaz and the Discovery of Alta California." *Pacific Historical Review* 27, no. 4 (November 1958): 351–57.

Fogelson, Robert M. *The Fragmented Metropolis: Los Angeles, 1850–1930*. Berkeley: University of California Press, 1993.

Fradkin, Philip L. *A River No More: The Colorado River and the West*. Berkeley: University of California Press, 1995 (original work published 1968).

Frederick, Kenneth D. "Water Marketing: Obstacles and Opportunities." *Forum for Applied Research and Public Policy* 16, no. 1 (Spring 2001): 54–61.

Friederici, Peter. "The Next Market Crunch: Water." *Business & Economics Magazine*, July 14, 2008.

Frisch, Scott A., and Sean Q. Kelly. *Jimmy Carter and the Water Wars: Presidential Influence and the Politics of Pork*. Amherst, NY: Cambria Press, 2008.

Gaffney, Mason. "What Price Water Marketing? California's New Frontier." *American Journal of Economics and Sociology* 56, no. 4 (October 1997): 475–520. http://dx.doi.org/10.1111/j.1536-7150.1997.tb02656.x.

García, José María Ramos. *Relaciones México–Estados Unidos: Seguridad Nacional e Impactos en la Frontera Norte*. Mexicali: Universidad Autónoma de Baja California, 2005.

Gardner, B. Delworth. "The Untried Market Approach to Water Allocation." In *New Courses for the Colorado River: Major Issues for the Next Century*, ed. Gary D. Weatherford and F. Lee Brown, 155–176. Albuquerque: University of New Mexico Press, 1986.

Gardner, B. Delworth. "Water as an Equity Problem." *Center Magazine* (November-December 1980): 32–41.

Gardner, B. Delworth, and H. H. Fullerton. "Transfer Restrictions and Misallocations of Irrigation Water." *American Journal of Agricultural Economics* 50, no. 3 (August 1968): 556–71. http://dx.doi.org/10.2307/1238259.

Gardner, Richard L., and Robert A. Young. "Assessing Strategies for Control of Irrigation-Induced Salinity in the Upper Colorado River Basin." *American Journal of Agricultural Economics* 70, no. 1 (February 1988): 37–49. http://dx.doi.org/10.2307/1241974.

Garner, Paul H. *Porfirio Díaz: Profiles in Power*. Charlottesville: University of Virginia Press, 2001.

Garrick, Dustin, Katharine Jacobs, and Gregg Garfin. "Models, Assumptions, and Stakeholders: Planning for Water Supply Variability in the Colorado River Basin." *Journal of the American Water Resources Association* 44, no. 2 (April 2008): 381–98. http://dx.doi .org/10.1111/j.1752-1688.2007.00154.x.

Garstang, Michael, Roelof Bruintjes, Robert Serafin, Harold Orville, Bruce Boe, William Cotton, and Joseph Warburton. "Weather Modification: Finding Common Ground." *American Meteorological Society* (May 2004): 647–55.

Gelt, Joe. "Drop 2—End-of-the-line Reservoir Salvages Colorado River Water." *Arizona Water Resource* 16, no. 5 (2008): 1–2.

Gelt, Joe. "Opposing Sides Find Common Ground in Yuma Desalter Controversy." *Arizona Water Resource* 13, no. 6 (2005): 1–3.

Gelt, Joe. "Study Says Northern Arizona's Water Supplies Unsustainable." *Arizona Water Resource* 15, no. 3 (2007): 1–2.

Getches, David. "Defending Indigenous Water Rights with the Laws of a Dominant Culture." In *Liquid Relations: Contested Water Rights and Legal Complexity*, ed. Dik Roth, Rutgerd Boelens, and Margreet Zwarteveen. Piscataway, NJ: Rutgers University Press, 2005, 44–65.

Getches, David H. "Impacts in Mexico of Colorado River Management in the United States." In *Climate and Water: Transboundary Challenges in the Americas*, ed. Henry F. Diaz and Barbara J. Morehouse. Boston: Kluwer Academic Publishers, 2003. http://dx.doi.org /10.1007/978-94-015-1250-3_8.

Ghassemi, Fereidoun, and Ian White. *Inter-Basin Water Transfer: Case Studies from Australia, United States, Canada, China, and India*. Cambridge, UK: Cambridge University Press, 2007. http://dx.doi.org/10.1017/CBO9780511535697.

Gleick, Peter H. "The Effects of Future Climatic Changes on International Water Resources: The Colorado River, the United States, and Mexico." *Policy Sciences* 21, no. 1 (1988): 23–39. http://dx.doi.org/10.1007/BF00145120.

Glenn, Edward P., Christopher Lee, Richard Felger, and Scott Zengel. "Effects of Water Management on the Wetlands of the Colorado River Delta, Mexico." *Conservation Biology* 10, no. 4 (August 1996): 1175–86. http://dx.doi.org/10.1046/j.1523-1739.1996.10041175.x.

Glennon, Robert. *Water Follies: Groundwater Pumping and the Fate of America's Fresh Waters*. Washington, DC: Island Press, 2002.

Glennon, Robert. "Water Scarcity, Marketing, and Privatization." *Texas Law Review* 83, no. 1873 (2005): 1873–1902.

Glennon, Robert. *Unquenchable: America's Water Crisis and What to Do about It*. Washington, DC: Island Press, 2009.

Glennon, Robert J., and Peter W. Culp. "The Last Green Lagoon: How and Why the Bush Administration Should Save the Colorado River Delta." *Ecology Law Quarterly* 28 (2002): 903–92.

Glennon, Robert, and Michael J. Pearce. "Transferring Mainstem Colorado River Water Rights: The Arizona Experience." Arizona Legal Studies Discussion Paper No. 07-05. Tucson: University of Arizona College of Law, 2007.

Gober, Patricia. *Metropolitan Phoenix: Place Making and Community Building in the Desert.* Philadelphia: University of Pennsylvania Press, 2006.

Gonzales, Michael J. *The Mexican Revolution, 1910–1940.* Albuquerque: University of New Mexico Press, 2002.

Gooch, Robert S., Paul A. Cherrington, and Yvonne Reinink. "Salt River Project Experience in Conversion from Agriculture to Urban Water Use." *Irrigation and Drainage Systems* 21, no. 2 (May 2007): 145–57. http://dx.doi.org/10.1007/s10795-007 -9026-2.

Goodstein, Phil H. *Denver from the Bottom Up: A People's History of Early Denver.* Denver: New Social Publications, 2004.

Granger, Byrd H. *Arizona's Names: X Marks the Place.* Tucson: Falconer, 1983.

Grimond, John. "A Special Report on Water." *The Economist* (May 22, 2010): 3–20.

Griswold del Castillo, Richard. *The Treaty of Guadalupe-Hidalgo: A Legacy of Conflict.* Norman: University of Oklahoma Press, 1990.

Gumerman, George J. *Exploring the Hohokam: Prehistoric Desert Peoples of the American Southwest.* Albuquerque: University of New Mexico Press, 1991.

Haddad, Brent M. *Rivers of Gold: Designing Markets to Allocate Water in California.* Washington, DC: Island Press, 2000.

Hadeed, S. J. "Several Water Resources Projects Survive Carter Cutback." *Water Pollution Federation* 49, no. 5 (May 1977): 727–30.

Hadjigeorgalis, Ereney. "A Place for Water Markets: Performance and Challenges." *Review of Agricultural Economics* 31, no. 1 (2008): 50–67.

Hanak, Ellen. "Finding Water for Growth: New Sources, New Tools, New Challenges." *Journal of the American Water Resources Association* 43, no. 4 (August 2007): 1024–35. http://dx.doi.org/10.1111/j.1752-1688.2007.00084.x.

Hanak, Ellen. "Stopping the Drain: Third-Party Responses to California's Water Market." *Contemporary Economic Policy* 23, no. 1 (January 2005): 59–77. http://dx.doi.org/10.1093 /cep/byi006.

Harkin, Michael E., and David Rich Lewis, eds. *Native Americans and the Environment: Perspectives on the Ecological Indian.* Lincoln: University of Nebraska Press, 2007.

Hart, John M. *Empire and Revolution: The Americans in Mexico since the Civil War.* Berkeley: University of California Press, 2002.

Hartman, L. M., and Don Seastone. *Water Transfers: Economic Efficiency and Alternative Institutions.* Baltimore: Johns Hopkins University Press, 1970.

Harvey, Mark W. T. "Echo Park, Glen Canyon, and the Postwar Wilderness Movement." *Pacific Historical Review* 60, no. 1 (February 1991): 43–67.

Harvey, Mark W. T. *A Symbol of Wilderness: Echo Park and the American Conservation Movement.* Albuquerque: University of New Mexico Press, 1994.

Hathaway, Pamala L., Paul W. Hirt, and Mary G. Wallace. *Institutional Response to a Changing Water Policy Environment.* Tucson: University of Arizona Water Resources Research Center, 1991.

Haury, Emil W. *The Hohokam, Desert Farmers & Craftsmen: Excavations at Snaketown, 1964–1965.* Tucson: University of Arizona Press, 1976.

Hayes, Douglas L. "The All-American Canal Lining Project: A Catalyst for Rational and Comprehensive Groundwater Management on the United States–Mexico Border." *Natural Resources Journal* 31 (Fall 1991): 803–27.

Hays, Samuel P. *Beauty, Health, and Permanence: Environmental Politics in the United States, 1955–1985.* New York: Cambridge University Press, 1987. http://dx.doi.org/10.1017/CBO9780511664106.

Hays, Samuel P. "From Conservation to Environment: Environmental Politics in the United States since World War II." *Environmental Review* 6, no. 2 (Fall 1982): 14–29. http://dx.doi.org/10.2307/3984153.

Hays, Samuel P. *A History of Environmental Politics since 1945.* Pittsburgh: University of Pittsburgh Press, 2000.

Hendricks, William Oral. Guillermo Andrade and Land Development of the Mexican Colorado River Delta, 1874–1905. PhD diss., University of Southern California, Los Angeles, 1967.

Hennessy, Megan. "Colorado River Water Rights: Property Rights in Transition." *University of Chicago Law Review* 71, no. 4 (Autumn 2004): 1661–87.

Hiltzik, Michael. *Colossus: Hoover Dam and the Making of the American Century.* New York: Free Press, 2010.

Hoffman, Abraham. "Origins of a Controversy: The U.S. Reclamation Service and the Owens Valley–Los Angeles Water Dispute." *Journal of the Southwest* 19, no. 4 (Winter 1977): 333–46.

Howe, Charles W. "Increasing Efficiency in Water Markets: Examples from the Western United States." In *Water Marketing: The Next Generation,* ed. Terry L. Anderson and Peter J. Hill. New York: Rowman & Littlefield Publishers, 1997.

Howe, Charles W., and Christopher Goemans. "Water Transfers and Their Impacts: Lessons from Three Colorado Water Markets." *Journal of the American Water Resources Association* 39, no. 5 (October 2003): 1055–65. http://dx.doi.org/10.1111/j.1752-1688.2003.tb03692.x.

Howitt, Richard E. "Empirical Analysis of Water Market Institutions: The 1991 California Water Market." *Resource and Energy Economics* 16, no. 4 (1994): 357–71.

Hulse, James W. *Nevada's Environmental Legacy: Progress or Plunder.* Reno: University of Nevada Press, 2009.

Hundley, Norris, Jr. "The Dark and Bloody Ground of Indian Water Rights: Confusion Elevated to Principle." *Western Historical Quarterly* 9, no. 4 (October 1978): 454–82. http://dx.doi.org/10.2307/967447.

Hundley, Norris, Jr. "Clio Nods: *Arizona v California* and the Boulder Canyon Act—A Reassessment." *Western Historical Quarterly* 3, no. 1 (January 1972): 17–57. http://dx.doi.org/10.2307/967706.

Hundley, Norris, Jr. *Dividing the Waters: A Century of Controversy between the United States and Mexico.* Berkeley: University of California Press, 1966.

Hundley, Norris, Jr. *The Great Thirst: Californians and Water—A History.* Berkeley: University of California Press, 2001 [1992].

Hundley, Norris, Jr. "The Politics of Water and Geography: California and the Mexican-American Treaty of 1944." *Pacific Historical Review* 36, no. 2 (May 1967): 209–26.

Hundley, Norris, Jr. *Water and the West: The Colorado Compact and the Politics of Water in the American West.* Berkeley: University of California Press, 1975.

Ingram, Helen M., Nancy K. Laney, and David M. Gillilan. *Divided Waters: Bridging the U.S.-Mexico Border.* Tucson: University of Arizona Press, 1995.

Ingram, Helen M., Lawrence A. Scaff, and Leslie Silko. "Replacing Confusion with Equity: Alternatives for Water Policy in the Colorado River Basin." In *New Courses for the Colorado River: Major Issues for the Next Century,* ed. Gary D. Weatherford and F. Lee Brown, 186–95. Albuquerque: University of New Mexico Press, 1986.

Iseman, Tom. "Banking on Colorado Water." *PERC Reports: The Property and Environment Research Center* 28, no. 1 (Spring 2010).

Israel, Morris, and Jay R. Lund. "Recent California Water Transfers: Implications for Water Management." *Natural Resources Journal* 35, no. 1 (1995): 1–32.

Jackson, Donald C. *Building the Ultimate Dam: John S. Eastwood and the Control of Water in the West.* Norman: University of Oklahoma Press, 1995.

Jackson, Donald C., and Norris Hundley, Jr. "Privilege and Responsibility: William Mulholland and the St. Francis Dam Disaster." *California History* 82, no. 3 (2004): 8–47. http://dx.doi.org/10.2307/25161743.

Jacobs, Jeffrey W., and James L. Wescoat Jr. "Managing River Resources: Lessons from Glen Canyon Dam." *Environment* 44, no. 2 (March 2002): 8–19. http://dx.doi.org/10.1080/00139150209605595.

James, George Wharton. *The Wonders of the Colorado Desert.* Boston: Little, Brown, 1911.

Jenkins, Garry. *The Wizard of Sun City: The Strange True Story of Charles Hatfield, the Rainmaker Who Drowned a City's Dreams.* New York: Basic Books, 2005.

Johnson, Dennis W. *The Laws That Shaped America: Fifteen Acts of Congress and Their Lasting Impact.* New York: Routledge, 2009.

Kahrl, William L. *Water and Power: The Conflict over Los Angeles Water Supply in the Owens Valley.* Berkeley: University of California Press, 1982.

Kalt, William D. *Tucson Was a Railroad Town: The Days of Steam in the Big Burg on the Main Line.* Mountlake Terrace, WA: VTD Rail Publications, 2007.

Keith, John. "Water Quality Management and Irrigated Agriculture: Potential Conflicts in the Colorado River Basin." *American Journal of Agricultural Economics* 59, no. 5 (December 1977): 948–54. http://dx.doi.org/10.2307/1239869.

Kershner, Frederick D., Jr. "George Chaffey and the Irrigation Frontier." *Agricultural History* 27, no. 4 (October 1953): 115–22.

Kirwan, Laura, and Daniel McCool. "Negotiated Water Settlements: Environmentalists and American Indians." In *Trusteeship in Change: Toward Tribal Autonomy in Resource Management,* ed. Richmond Clow and Imre Sutton. Boulder: University Press of Colorado, 2001.

Kluger, James R. *Turning on Water with a Shovel: The Career of Elwood Mead.* Albuquerque: University of New Mexico Press, 1992.

Koppes, Clayton R. "Public Water, Private Land: Origins of the Acreage Limitation Controversy, 1933–1953." *Pacific Historical Review* 47, no. 4 (November 1978): 607–636.

Krall, Lisi. "US Land Policy and the Commodification of Arid Land (1862–1920)." *Journal of Economic Issues* 35, no. 3 (September 2001): 657–74.

Kriska, Andrea M., Aramesh Saremi, Robert L. Hanson, Peter H. Bennett, Sayuko Kobes, Desmond E. Williams, and William C. Knowler. "Physical Activity, Obesity, and the Incidence of Type 2 Diabetes in a High-Risk Population." *American Journal of Epidemiology* 158, no. 7 (2003): 669–75.

Kroeber, Alfred Louis, and Clifton B. Kroeber. *A Mohave War Reminiscence, 1854–1880*. New York: Dover, 1994.

Kupel, Douglas E. *Fuel for Growth: Water and Arizona's Urban Environment*. Tucson: University of Arizona Press, 2003.

Laflin, Pat. *The Salton Sea: California's Overlooked Treasure*. Indio, CA: Coachella Valley Historical Society, 1999 [1995].

Land, Barbara, and Myrick Land. *A Short History of Las Vegas*. Reno: University of Nevada Press, 1999.

Langston, Nancy. *Where Land & Water Meet: A Western Landscape Transformed*. Seattle: University of Washington Press, 2003.

Larson, Kelli L., Annie Gustafson, and Paul Hirt. "Insatiable Thirst and a Finite Supply: An Assessment of Municipal Water-Conservation Policy in Greater Phoenix, Arizona, 1980–2007." *Journal of Policy History* 21, no. 2 (2009): 107–37. http://dx.doi.org/10.1017/S0898030609090058.

LeVeen, Phillip. "A Political Economic Analysis of the 1982 Reclamation Reform Act." *Western Journal of Agricultural Economics* 8, no. 2 (1983): 255–65.

Lewis, William M., Jr., ed. *Water and Climate in the United States*. Boulder: University Press of Colorado, 2003.

Leydet, François, and David Brower. *Time and the River Flowing: Grand Canyon*. San Francisco: Sierra Club, 1964.

Libecap, Gary D. " 'Chinatown': Owens Valley and Western Water Reallocation—Getting the Record Straight and What It Means for Water Markets." *Texas Law Review* 83, no. 2055 (2005): 2055–89.

Libecap, Gary D. *Owens Valley Revisited: A Reassessment of the West's First Great Water Transfer*. Stanford: Stanford Economics and Finance, 2007.

Lichtenfeld, Christine. "Indian Reserved Water Rights: An Argument for the Right to Export and Sell." *Land and Water Review* 131 (1989): 131–52.

Lindsay, Stephanie. "Counting Every Drop: Measuring Surface and Ground Water in Washington and the West." *Environmental Law* (Northwestern School of Law) 39, no. 193 (2009): 193–214.

Logan, Michael F. *Desert Cities: The Environmental History of Phoenix and Tucson*. Pittsburgh: University of Pittsburgh Press, 2006.

Luckingham, Bradford. *Phoenix: The History of a Southwestern Metropolis*. Tucson: University of Arizona Press, 1989.

Luckingham, Bradford. *The Urban Southwest: A Profile History of Albuquerque, El Paso, Phoenix, and Tucson*. El Paso: Texas Western Press, 1982.

Lukas, Jeff (Western Water Association), Connie Woodhouse (University of Arizona), and Henry Adams (University of Arizona). Colorado River Streamflow, a Paleo Perspective. http://treeflow.info/lees/gage/html (accessed January 24, 2012).

Martin, Russell. *A Story That Stands Like a Dam: Glen Canyon and the Struggle for the Soul of the West.* Salt Lake City: University of Utah Press, 1999 (original work published 1989).

Martínez, Oscar Jáquez. *Troublesome Border.* Tucson: University of Arizona Press, 2006.

Mays, Larry. *Integrated Urban Water Management: Arid and Semi-arid Regions.* New York: Taylor & Francis, 2009.

McCaffrey, Stephen C. "The Harmon Doctrine One Hundred Years Later: Buried, Not Praised." *Natural Resources Journal* 36 (Summer 1996): 549–90.

McCool, Daniel. *Native Waters: Contemporary Indian Water Settlements and the Second Treaty Era.* Tucson: University of Arizona Press, 2002.

McCool, Daniel. "Precedent for the Winters Doctrine: Seven Legal Principles." *Journal of the Southwest* 29, no. 2 (Summer 1987): 164–78.

McCool, Daniel. "Winters Comes Home to Roost." In *Fluid Arguments: Five Centuries of Western Water Conflict,* ed. Char Miller, 120–38. Tucson: University of Arizona Press, 2001.

McGuire, Thomas R. "The River, the Delta, and the Sea." *Journal of the Southwest* 45, no. 3 (Autumn 2003): 371–410.

McNally, Mary. "Water Marketing: The Case of Indian Reserved Rights." *Water Resources Bulletin* 30, no. 6 (December 1994): 963–70. http://dx.doi.org/10.1111/j.1752-1688.1994 .tb03344.x.

McNitt, Frank. *Navajo Wars.* Albuquerque: University of New Mexico Press, 1972.

McPhee, John. *Encounters with the Archdruid.* New York: Farrar, Straus and Giroux, 1971.

Megdal, Sharon. "Front-Row View of Federal Water Lawmaking Shows Process Works: US-Mexico Transboundary Aquifer Assessment Act Pondered, Passed, and Signed." *Arizona Water Resource* 15, no. 3 (January-February 2007): 1–2.

Megdal, Sharon. "Water Bills Included on Legislative Agenda." *Arizona Water Resource* (March/April 2003), 1–3.

Mehan, G. Tracy, III. "Energy, Climate Change, and Sustainable Water Management." *Environment Reporter* 38, no. 48 (2007): 1–7.

Meko, David M., Connie A. Woodhouse, Christopher A. Baisan, Troy Knight, Jeffrey J. Lukas, Malcolm K. Hughes, and Matthew W. Salzer. "Medieval Drought in the Upper Colorado River Basin." *Geophysical Research Letters* 34, no. 10 (May 24, 2007): L10705. http://dx.doi.org/10.1029/2007GL029988.

Metcalf, Barbara Ann. Oliver M. Wozencraft in California, 1849–1887. MA thesis, University of Southern California, 1963.

Meyer, Michael C. *Water in the Hispanic Southwest: A Social and Legal History, 1550–1850.* Tucson: University of Arizona Press, 1984.

Meyers, Charles J., and Richard A. Posner. *Market Transfers of Water Rights.* Washington, DC: National Water Commission, 1971.

Meyers, Charles J., and Richard L. Noble. "The Colorado River: The Treaty with Mexico." *Stanford Law Review* 19, no. 2 (January 1967): 367–419. http://dx.doi.org/10.2307 /1227558.

Miller, Frederic P., Agnes F. Vandome, and John McBrewster. *History of Denver.* Saarbrücken, Germany: VDM Publishing, 2009.

Miller, Jon R. "The Political Economy of Western Water Finance: Cost Allocation and the Bonneville Unit of the Central Utah Project." *American Journal of Agricultural Economics* 69, no. 2 (May 1987): 303–10. http://dx.doi.org/10.2307/1242280.

Miller, Norman. *Environmental Politics: Stakeholders, Interests, and Policymaking*, 2nd ed. New York: Routledge, 2009.

Mize, Ronald L., and Alicia C.S. Swords. *Consuming Mexican Labor: From the Bracero Program to NAFTA*. Toronto: University of Toronto Press, 2011.

Moehring, Eugene P., and Michael S. Green. *Las Vegas: A Centennial History*. Reno: University of Nevada Press, 2005.

Moore, Charles V. "Impact of Federal Acreage Limitation Policy on Western Irrigated Agriculture." *Western Journal of Agricultural Economics* 7 (December 1982): 301–16.

Moore, Mik. "Coalition Building between Native American and Environmental Organizations in Opposition to Development." *Organization & Environment* 11, no. 3 (1998): 287–313. http://dx.doi.org/10.1177/0921810698113002.

Moss, Joyce. *Spanish and Portuguese Literatures and Their Times*. Detroit: Gale Group, 2002.

Muehlmann, Shaylih. Where the River Ends: Environmental Conflict and Contested Identities in the Colorado Delta. PhD diss., University of Toronto, 2008.

Mulholland, Catherine. *William Mulholland and the Rise of Los Angeles*. Berkeley: University of California Press, 2000.

Murphy, James J., Ariel Dinar, Richard E. Howitt, Steven J. Rassenti, and Vernon L. Smith. "The Design of 'Smart' Water Market Institutions Using Laboratory Experiments." *Environmental and Resource Economics* 17, no. 4 (December 2000): 375–94. http://dx.doi.org/10.1023/A:1026598014870.

Musoke, Moses S., and Alan L. Olmstead. "The Rise of the Cotton Industry in California: A Comparative Perspective." *Journal of Economic History* 42, no. 2 (June 1982): 385–412. http://dx.doi.org/10.1017/S0022050700027480.

Nash, George H., and Kendrick A. Clements. *The Life of Herbert Hoover*. New York: W. W. Norton, 1983.

Nash, Roderick Frazier. *Wilderness and the American Mind*. New Haven: Yale University Press, 2001 (original work published 1967).

National Research Council of the National Academies. *Colorado River Basin Water Management: Evaluating and Adjusting to Hydroclimate Variability*. Washington, DC: National Academies Press, 2007.

Natural Resources Law Center, Summer Program. *Water and Growth in the West*. Boulder: Natural Resources Law Center, University of Colorado School of Law, and the William and Flora Hewlett Foundation, 2000.

Nelson, Sarah M. *Denver: An Archeological History*. Boulder: University Press of Colorado, 2008.

Newcom, S. Joshua. "Deciding about the Colorado River Delta: Rejuvenated Wetlands Raise New Issues about Where Flood Flow Should Go." *River Report: Water Education Foundation* (Spring 1999): 1–11.

Newell, Alan S. "First in Time: Tribal Reserved Water Rights and General Adjudications in New Mexico." In *Fluid Arguments: Five Centuries of Western Water Conflict*, ed. Char Miller. Tucson: University of Arizona Press, 2001.

O'Brien, Kevin M., and Robert R. Gunning. "Water Marketing in California Revisited: The Legacy of the 1987–92 Drought." *Pacific Law Journal* (Sacramento, CA) 25, no. 3 (1993–1994): 1053–85.

Officer, James E. *Hispanic Arizona, 1536–1856*. Tucson: University of Arizona Press, 1987.

Orr, David W. *Earth in Mind: On Education, Environment, and the Human Prospect*. Washington, DC: Island, 2004.

Orsi, Richard J. *Sunset Limited: The Southern Pacific Railroad and the Development of the American West, 1850–1930*. Berkeley: University of California Press, 2005.

Palmer, Gary B. "Water Development Strategies in the Colorado River Basin: Expansion versus Involution." *Anthropological Quarterly* 51, no. 2 (April 1978): 99–117. http://dx.doi .org/10.2307/3317767.

Paulson, Larry J., and John R. Baker. "Nutrient Interactions among Reservoirs on the Colorado River." *Publications: Water Resources,* Paper 58, 1980. http://digitalscholarship .unlv.edu/water_pubs/58 (accessed March 26, 2011).

Pearson, Byron E. "We Have Almost Forgotten How to Hope: The Hualapai, the Navajo, and the Fight for the Central Arizona Project, 1944–1968." *Western Historical Quarterly* 31, no. 3 (Autumn 2000): 297–316. http://dx.doi.org/10.2307/969962.

Peña, Guillermo, De La. "Rural Mobilizations in Latin America since 1920." In *Latin America: Politics and Society since 1930*, ed. Leslie Bethell. New York: Cambridge University Press, 1998. http://dx.doi.org/10.1017/CBO9780511626081.008.

Phelps, Charles E., Nancy Y. Moore, and Morlie Hammard Graubard. Efficient Water Use in California: Water Rights, Water Districts, and Water Transfers. R-2386-CSA/RF RAND Corporation, 1978.

Phillips, Sarah T. *This Land, This Nation: Conservation, Rural America, and the New Deal*. New York: Cambridge University Press, 2007. http://dx.doi.org/10.1017/CBO9780511618703.

Pisani, Donald J. "The Dilemmas of Indian Water Policy, 1887–1928." In *Fluid Arguments: Five Centuries of Western Water Conflict*, ed. Char Miller. Tucson: University of Arizona Press, 2001.

Pisani, Donald J. *To Reclaim a Divided West: Water, Law, and Public Policy, 1848–1902*. Albuquerque: University of New Mexico Press, 1992.

Pisani, Donald J. *Water and American Government: The Reclamation Bureau, National Water Policy, and the West, 1902–1935*. Berkeley: University of California Press, 2002.

Pisani, Donald J. *Water, Land, and Law in the West: The Limits of Public Policy, 1850–1920*. Lawrence: University Press of Kansas, 1996.

Plog, Stephen. *Ancient Peoples of the American Southwest*. New York: Thames and Hudson, 1997.

Pontius, Dale, principal investigator. *Colorado River Basin Study for the Western States Water Policy Review Advisory Commission*. Tucson: SWCA, Inc., Environmental Consultants, March 1997. http://wwa.colorado.edu/colorado_river/docs/pontius%20colorado.pdf (accessed February 28, 2012).

Powell, James Lawrence. *Dead Pool: Lake Powell, Global Warming, and the Future of Water in the West*. Berkeley: University of California Press, 2008.

Prelog, Andrew J. Water Scarcity and Rapid Complex Change in Colorado: An Evaluation of Historical Patterns of Water Appropriation and Socio-Demographic Growth. MS thesis, Colorado State University, 2007.

Price, Monroe E., and Gary D. Weatherford. "Indian Water Rights in Theory and Practice: Navajo Experience in the Colorado River Basin." *Law and Contemporary Problems* 40, no. 1 (Winter 1976): 97–131. http://dx.doi.org/10.2307/1191333.

Proffitt, Thurber Dennis. *Tijuana: The History of a Mexican Metropolis*. San Diego: San Diego State University Press, 1994.

Radosevich, George. *Evolution and Administration of Colorado Water Law, 1876–1976*. Fort Collins, CO: Water Resources Publications, 1976.

Ramírez, David Piñera. *Los Orígenes de las Poblaciones de Baja California: Factores Externos, Nacionales y Locales*. Mexicali, Baja California México: Universidad Autónoma de Baja California, 2006.

Ramírez, David Piñera y Jesús Ortiz Figeroa, eds. *Historia de Tijuana*, 2nd ed. Tijuana: Universidad Autónoma de Baja California, 1989.

Reisner, Marc. *Cadillac Desert: The American West and Its Disappearing Water*. New York: Penguin Books, 1993 (original work published 1986).

Richards, Leonard L. *The California Gold Rush and the Coming of the Civil War*. New York: Alfred A. Knopf, 2007.

Righter, Robert W. *The Battle over Hetch Hetchy: America's Most Controversial Dam and the Birth of Modern Environmentalism*. New York: Oxford University Press, 2005.

Rogers, Jedediah. "Robert B. Griffith Water Project (Formerly Southern Nevada Water Program)." Denver: Bureau of Reclamation, 2001. http://www.usbr.gov/projects /ImageServer?imgName=Doc_1305577385454.pdf (accessed September 20, 2012).

Rogers, Peter P., M. Ramon Llamas, and Luis Martinez-Cortina, eds. *Water Crisis: Myth or Reality?* Marcelino Botin Water Forum 2004. New York: Taylor & Francis Group, 2006.

Rothman, Hal K., and Mike Davis, eds. *The Grit beneath the Glitter: Tales from the Real Las Vegas*. Berkeley: University of California Press, 2002.

Rowley, William D. *The Bureau of Reclamation: Origins and Growth to 1945*, vol. 1. Denver: Bureau of Reclamation, 2006.

Saffell, Cameron Lee. "Common Roots of a New Industry: The Introduction and Expansion of Cotton Farming in the American West." PhD diss., Iowa State University, Ames, 2007.

Samaniego López, Marco Antonio, coordinador. *Breve Historia de Baja California*. Mexicali, Baja California México: Universidad Autónoma de Baja California, 2006.

Sanchez, Roberto A. "Water Quality Problems in Nogales, Sonora." *Environmental Health Perspectives* 103, Suppl. 1 (February 1995): 93–97. Medline:7621811.

Sauder, Robert A. *The Lost Frontier: Water Diversion in the Growth and Destruction of Owens Valley Agriculture*. Tucson: University of Arizona Press, 1994.

Sauder, Robert A. *The Yuma Reclamation Project: Irrigation, Indian Allotment, and Settlement along the Lower Colorado River*. Reno: University of Nevada Press, 2009.

Schackley, M. Steven, and Steven Lucas-Pfingst. *The Early Ethnography of the Kumeyaay*. Berkeley: University of California Press, 2007.

Schlosberg, David. *Defining Environmental Justice: Theories, Movements, and Nature*. New York: Oxford University Press, 2007. http://dx.doi.org/10.1093/acprof:oso/9780199286294.001.0001

Schulte, Steven C. *Wayne Aspinall and the Shaping of the American West*. Boulder: University Press of Colorado, 2002.

Seasholes, Kenneth, Barbara Tellman, Gary Woodard, and Joe Gelt. *Water in the Tucson Area: Seeking Sustainability*. Water Resources Research Center, no. 20, July 31, 1999. http://wrrc.arizona.edu/publications/water-tucson-area-seeking-sustainability (accessed September 18, 2012).

Seldin, Chris. "Interstate Marketing of Indian Water Rights: The Impact of the Commerce Clause." *California Law Review* 87, no. 6 (December 1999): 1545–80. http://dx.doi.org/10.2307/3481051.

Sementelli, Arthur Jay. "Naming Water: Understanding How Nomenclature Influences Rights and Policy Choices." *Public Works Management & Policy* 13, no. 1 (July 2008): 4–11. http://dx.doi.org/10.1177/1087724X08321165.

Service, Robert F. "Water Resources: As the West Goes Dry." *Science* 303, no. 5661 (February 20, 2004): 1124–27. http://dx.doi.org/10.1126/science.303.5661.1124. Medline:14976291.

Sheridan, Thomas E. "The Big Canal: The Political Ecology of the Central Arizona Project." In *Water, Culture, and Power: Local Struggles in a Global Context*, ed. John M. Donahue and Barbara Rose Johnston, 163–86. Washington, DC: Island Press, 1998.

Shiva, Vandana. *Water Wars: Privatization, Pollution and Profit*. Cambridge, MA: South End Press, 2002.

Sholders, Mike. "Water Supply Development in San Diego and a Review of Related Outstanding Projects." *Journal of San Diego History* 48, no. 1 (Winter 2002): 61–70.

Shurts, John. *Indian Reserved Water Rights: The Winters Doctrine in Its Social and Legal Context, 1880s–1930s*. Norman: University of Oklahoma Press, 2000.

Sirgo, Henry. *Establishment of Environmentalism on the U.S. Political Agenda in the Second Half of the Twentieth Century: The Brothers Udall*. Lewiston, NY: Edwin Mellen, 2004.

Smith, Duane A. *The Trail of Gold and Silver: Mining in Colorado, 1859–2009*. Boulder: University Press of Colorado, 2009.

Smith, Thomas Gary. *John Saylor and the Preservation of America's Wilderness*. Pittsburgh: University of Pittsburgh Press, 2006.

Smythe, William Ellsworth. *History of San Diego, 1542–1908: The Modern City*, vol. II. San Diego: The History Company, 1908.

Soden, Dennis L. *The Environmental Presidency*. Albany: SUNY Press, 1999.

Sokolow, Alvin, Sonja Varea Hammond, Maxwell V. Norton, and Evan E. Schmidt. "California Communities Deal with Conflict and Adjustment at the Urban-Agricultural Edge." *California Agriculture* 64, no. 3 (July-September 2010): 121–28. http://dx.doi.org/10.3733/ca.v064n03p121.

Song, Lisa. "Thirsty Cities: Water Management in a Changing Environment." *Earth: The Science behind the Headlines* (December 31, 2009). http://www.earthmagazine.org.

Sonnenfeld, David A. "Mexico's 'Green Revolution,' 1940–1980: Towards an Environmental History." *Environmental History Review* 16, no. 4 (Winter 1992): 28–52. http://dx.doi.org /10.2307/3984948.

Sonnichsen, C. L. *Tucson: The Life and Times of an American City.* Norman: University of Oklahoma Press, 1987 (original work published 1982).

Spangenberg, Joachim H., Stefanie Pfahl, and Kerstin Deller. "Towards Indicators for Institutional Sustainability: Lessons from an Analysis of Agenda 21." *Ecological Indicators* 2, nos. 1–2 (November 2002): 61–77. http://dx.doi.org/10.1016/S1470-160X(02)00050-X.

Sprouse, Terry W. "Water Issues on the Arizona-Mexico Border: The Santa Cruz, San Pedro and Colorado Rivers, An Issue Paper." Water Resources Research Center, College of Agriculture and Life Sciences, University of Arizona, February 2005. http://ag.arizona .edu/azwater/files/terrypaper1.pdf (accessed July 7, 2010).

Stegner, Wallace. *Beyond the Hundredth Meridian: John Wesley Powell and the Second Opening of the West.* Boston: Houghton Mifflin, 1954.

Stegner, Wallace, ed. *This Is Dinosaur: Echo Park Country and Its Magic Rivers.* New York: Alfred A. Knopf, 1955.

Stene, Eric A. *Yuma Project and Yuma Auxiliary Project.* Denver: Historic Reclamation Projects, Bureau of Reclamation, 1996. http://www.usbr/gov/projects//ImageServer?img Name=Doc_1271086556202.pdf (accessed August 8, 2010).

Stevens, Joseph E. *Hoover Dam: An American Adventure.* Norman: University of Oklahoma Press, 1988.

Stewart, Kenneth M. "The Aboriginal Territory of the Mohave Indians." *Ethnohistory* (Columbus, OH) 16, no. 3 (Summer 1969): 257–76. http://dx.doi.org/10.2307/481587.

Storey, Lee Herold. "Leasing Indian Water off the Reservation: A Use Consistent with the Reservation's Purpose." *California Law Review* 76, no. 1 (January 1988): 179–220. http:// dx.doi.org/10.2307/3480500.

Stout, Joseph Allen. *Schemers & Dreamers: Filibustering in Mexico, 1848–1921.* Fort Worth: Texas Christian University Press, 2002.

Strathman, Theodore Andrew. "Land, Water, and Real Estate: Ed Fletcher and the Cuyamaca Water Company, 1910–1926." *San Diego History* 50, no. 3 (Summer/Fall 2004): 124–44.

Stuart, Paul. "United States Indian Policy: From the Dawes Act to the American Indian Policy Review Commission." *Social Service Review* 51, no. 3 (September 1977): 451–63. http://dx.doi.org/10.1086/643524.

Sturgeon, Stephen C. *The Politics of Western Water: The Congressional Career of Wayne Aspinall.* Tucson: University of Arizona Press, 2002.

Taylor, Lawrence D. "The Mining Boom in Baja California from 1850 to 1890 and the Emergence of Tijuana as a Border Community." *Journal of the Southwest* 43, no. 4 (Winter 2001): 463–92.

Thomashow, Mitchell. *Bringing the Biosphere Home: Learning to Perceive Global Environmental Change.* Cambridge, MA: MIT Press, 2002.

Turner, Frederick Jackson. "The Significance of the Frontier in American History." Essay presented to the American Historical Association, 1893. Ann Arbor: University Microfilms, 1966 [1893].

Tyler, Daniel. *The Last Water Hole in the West: The Colorado–Big Thompson Project and the Northern Colorado Water Conservancy District*. Niwot: University Press of Colorado, 1992.

Tyler, Daniel. *Silver Fox of the Rockies: Delphus E. Carpenter and Western Water Compacts*. Norman: University of Oklahoma Press, 2003.

Tyus, Harold M., and James F. Saunders III. "Nonnative Fish Control and Endangered Fish Recovery: Lessons from the Colorado River." *Fisheries* (Bethesda, MD) 25, no. 9 (2000): 17–24. http://dx.doi.org/10.1577/1548-8446(2000)025<0017:NFCAEF>2.0.CO;2.

Udall, Stewart L. *The Quiet Crisis*. New York: Avon Books, 1964.

Utley, Robert. *A Life Wild and Perilous: Mountain Men and the Paths to the Pacific*. New York: Henry Holt, 1997.

Vanderwood, Paul J. *Satan's Playground: Mobsters and Movie Stars at America's Greatest Gaming Resort*. Durham, NC: Duke University Press, 2010.

Vaux, J. J., Jr., and R. E. Howitt. "Managing Water Scarcity: An Evaluation of Interregional Transfers." *Water Resources Research* 20, no. 7 (July 1984): 785–92. http://dx.doi.org/10.1029/WR020i007p00785.

Vollman, William. *Imperial*. New York: Penguin, 2010.

Walsh, Casey. "'To Come of Age in a Dry Place': Infrastructures of Irrigated Agriculture in the Mexico-U.S. Borderlands." *Southern Rural Sociology* 24, no. 1 (2009): 21–43.

Walton, Brett. "Water Rights: Arizona Senators Jon Kyl and John McCain Meet with Navajo Nation Leaders." *Circle of Blue*, April 5, 2012. http://www.circleofblue.org/waternews/2012/world/water-rights-arizona-senators-john-kyle-and-john-mccain-meet-with-navajo-nation-leaders/ (accessed October 22, 2012).

Walton, John. *Western Times and Water Wars: State, Culture, and Rebellion in California*. Berkeley: University of California Press, 1992.

Ward, Evan R. *Border Oasis: Water and the Political Ecology of the Colorado River Delta, 1940–1975*. Tucson: University of Arizona Press, 2003.

Ward, Evan R. "Crossroads on the Periphery: Yuma County Water Relations, 1922–1928." MA thesis, University of Georgia, Athens, 1997.

Ward, Evan R. "'The Politics of Place': Domestic and Diplomatic Priorities of the Colorado River Salinity Control Act (1974)." *Journal of Political Ecology* 6 (1999): 31–56.

Ward, Evan R. "The Twentieth-Century Ghosts of William Walker: Conquest of Land and Water as Central Themes in the History of the Colorado River Delta." *Pacific Historical Review* 70, no. 3 (August 2001): 359–85. http://dx.doi.org/10.1525/phr.2001.70.3.359.

Waters, Frank. *Colorado*. New York: Rinehart, 1946.

"Wayne N. Aspinall: 'I Fitted the Epoc.'" *Shale Country* (December 1982): 16–17.

Weatherford, Gary D., and F. Lee Brown. *New Courses for the Colorado River: Major Issues for the Next Century*. Albuquerque: University of New Mexico Press, 1986.

Weatherford, Gary, Mary Wallace, and Lee Herold Storey. *Leasing Indian Water: Choices in the Colorado River Basin*. Washington, DC: The Conservation Foundation and the John Muir Institute, 1988.

Weatherford, Gary, Mary Wallace, Lee Herold Storey, and F. Lee Brown, eds. *New Courses for the Colorado River: Major Issues for the Next Century*. Albuquerque: University of New Mexico Press, 1985.

Wehr, Kevin. *America's Fight over Water: The Environmental and Political Effects of Large-Scale Water Systems*. New York: Routledge, 2004.

Weinberg, Marca. "Assessing a Policy Grab Bag: Federal Water Policy Reform." *American Journal of Agricultural Economics* 84, no. 3 (August 2002): 541–56. http://dx.doi.org /10.1111/1467-8276.00318.

Weldon, Elizabeth. "Practically Irrigable Acreage Standard: A Poor Partner for the West's Water Future." *William and Mary Environmental Law and Policy Review* 25, no. 1 (2000): 203–31.

Wescoat, James L., Jr. "Impacts of Federal Salinity Control on Water Rights Allocation Patterns in the Colorado River Basin." *Annals of the Association of American Geographers* 76, no. 2 (June 1986): 157–74. http://dx.doi.org/10.1111/j.1467-8306.1986.tb00110.x.

West, Elliot. *The Contested Plains: Indians, Goldseekers, & the Rush to Colorado*. Lawrence: University Press of Kansas, 1998.

Wheeler, William Bruce, and Michael J. McDonald. *TVA and the Tellico Dam, 1936–1979*. Knoxville: University of Tennessee Press, 1986.

White, Richard. *"It's Your Misfortune and None of My Own": A New History of the American West*. Norman: University of Oklahoma Press, 1991.

White, Richard. *The Organic Machine: The Remaking of the Columbia River*. New York: Hill and Wang, 1995.

Whiteman, David C. *Mountain Meteorology: Fundamentals and Applications*. New York: Oxford University Press, 2000.

Wiegner, Kathleen K. "The Water Crisis: It's Almost Here." *Forbes* (August 20, 1979): 56–63.

Wild, Peter. *The Opal Desert: Explorations of Fantasy and Reality in the American Southwest*. Austin: University of Texas Press, 1999.

Wilhite, Donald A. *Drought and Water Crisis: Science, Technology, and Management Issues*. Boca Raton, FL: CRC Press, Taylor & Francis, 2005.

Wilkinson, Charles F. *Crossing the Next Meridian: Land, Water, and the Future of the West*. Washington, DC: Island Press, 1992.

Wilson, Frank. "A Fish out of Water: A Proposal for International Instream Flow Rights in the Lower Colorado River." *Colorado Journal of International Law and Policy* 5 (1994): 249–72.

Wohl, Ellen. *Disconnected Rivers: Linking Rivers to Landscapes*. New Haven: Yale University Press, 2004.

Wohl, Ellen. *Virtual Rivers: Lessons from the Mountain Rivers of the Colorado Front Range*. New Haven, CT: Yale University Press, 2001.

Worster, Donald. *A River Running West: The Life of John Wesley Powell*. New York: Oxford University Press, 2000.

Worster, Donald. *Rivers of Empire: Water, Aridity, and the Growth of the American West*. New York: Oxford University Press, 1985.

Worster, Donald. *Under Western Skies: Nature and History in the American West*. New York: Oxford University Press, 1992.

Young, McGee. "From Conservation to Environment: The Sierra Club and the Organizational Politics of Change." *Studies in American Political Development* 22, no. 2 (2008): 183–203.

Zetland, Jason David. Conflict and Cooperation within an Organization: A Case Study of the Metropolitan Water District of Southern California. PhD diss., University of California, Davis, 2008.

Index

94–95, 210; Colorado River allocation, 17–19, 20, 21, 22, 51, 179, 240; and Grand Canyon dams, 157–58; hydroelectric power, ix, 45–46; irrigated agriculture in, 32, 33, 34, *36*, 56(n40); Navajo water rights, 164–66; water rights, 22, 37, 38, 42, 43–45, 119; water transfers, 212–13; water use, *36*, 40–41, 97, 236–37

Arizona Groundwater Management Act, 160

Arizona Interstate Stream Commission, 47

Arizona National Guard, and Praker dam, 22

Arizona Power Authority, 47

Arizona State University, School of Sustainability, 137

Arizona v. California, 41, 48–51, 162, 166, 210

Arizona Water Banking Authority, 212, 222

Arizona Water Rights Settlement Act, 161

Armijo, Antonio, 132

Army Corps of Engineers, Tres Rios Constructed Wetlands, 126

arsenic, in groundwater, 164

Ashurst, Henry, 184, 200(n29)

Aspinall, Wayne, 23, 75; and Central Arizona Project, 69, 70, 73; Upper Colorado River Basin Storage Project, 52–54, 67

Aspinall Unit, 62

Audubon Society, 65

Australia, water marketing, 208

Aztec (N.M.), 164

Azurix Corporation, 220

Babbitt, Bruce, on water policy, 98–99, 100

Baggley, George F., 157

Baja California, 6, 132, 180, 182, 184, 185, 199(n7); Colorado River water, 18, 179; farmers' activism in, 188–89; mining in, 127–29. *See also* Mexicali; Tijuana

Barr, Burton W., on CAP, 94–95

Bartlett Dam, 126

Basic Magnesium, Incorporated (BMI), 133

beach habitat, in Grand Canyon, 99

Bee River, 181

biosphere preserves, 193

Black Canyon, 20, 62

Blake, William P., 11–12

Bloomfield, 164

Blue Mesa Dam, 62

Blue Ridge Dam, 126

Blythe, Thomas Henry, 12

BMI. *See* Basic Magnesium, Incorporated

Bonneville Unit, 91

bonytail, 63, 103

Border Sanitation and Wastewater Quality Summit, 195

Border XXI Program, 194

Border 2012, 194–95

Border 2020, 195

Bosque Redondo, 150

Boulder Canyon, 62; dam planned for, 17, 20–21

Boulder Canyon Bill, 19–20

Boulder Canyon Project Act, 20, 42, 49, 50, 131

Bracero Program, 186

Brady, Brian J., 221

Bridge Canyon, proposed dam at, 47, 69, 70, 71–72, 157

Brock Reservoir, Warren H., 196–97, 217–18

Brower, David, 65, 66, 67, 68, 71, 72

Brundtland, Gro Harlem, 88

Brundtland Report, 88

Burch, Leonard, 169

Bureau of Indian Affairs, 103, 151, 154

Bureau of Reclamation, xiii, 40, 47, 52, 73, 75, 87, 89, 99, 100, 119, 134, 162, 194, 242; basin-wide planning, 238, 239; canal projects, 115, 191, 217; changes in, 92, 93, 96–97; dam projects, 15–17, 22–23, 66; environmental impact studies, 239–40; and Indian nations, 157, 163–64; irrigation projects, 24, 38; Hopi-Navajo water rights settlement, 240–41; project costs, 44–45; selenium contamination program, 102–3; water conservation policies, 196–97; water sources, 210, 211; Wellton-Mohawk Project, 76–77, 187–88

Cadillac Desert (Reisner), xi, 88–89

Cadiz Water Project, 220

Cahuilla, Lake, 11–12

Calexico, 130

California, 7, 41, 43, 46, 98, 119; *Arizona v. California,* 49–51; canal lining project, 131–32, 191; and Central Arizona Project, 47–51, 74, 156; Colorado River allocation, 17–19, 20, 21–22, 179, 240; and Colorado River Storage Project, 67–69; drought, 89, 96; hydroelectric power, ix , 16–17; irrigated agriculture in, 32,

39, 56(n40); water marketing, 210–14, 219–21; water rights, 38, 100, 196; water shortages, 93–94; water use, *33*, 37, 45, 97, 235–36

California Aqueduct, 22, 211

California, Gulf of, 6, 62

California Department of Water Resources, 212

California Gold Rush, 7, 117

California State Water Project, 211, 219

California Superior Court, 214

California Supreme Court, 100

Calloway, O. P., 12

canals, 12, 17, 20, 21, 46, 115, 123; drainage, 76–77, 188; lining of, 131–32, 191, 214, 217. *See also* All-American Canal; Central Arizona Project

Canyonlands National Park, 62

Cárdenas, Cuauhtémoc, 188, 189

Cárdenas, García López de, 6

Cárdenas, Lázaro, 130–31, 185

Carey, Joseph Maull, 11

Carey Act (Federal Desert Land Act), 11

Carnegie Corporation, 160

Carpenter, Delph, 17

Carter, Jimmy, 87; on water policy, 90–92

CAWCD. *See* Central Arizona Water Conservation District

CDC. *See* Colorado Development Company

Central Arizona Project (CAP), 40, 41, 75, 89, 90, 100, 126, 156, 190, 213, 236; and California water sources, 93–94, 210, 211–12; Carter on, 91–92; and Grand Canyon dams, 71–72, 157–58; Indian nations and, 159, 160–62, 172(n43); origins of, 46–51; Tucson and, 113–14, 121; Stewart Udall on, 69–71; supporters of, 94–95; US Congress debate over, 73–74

Central Arizona Water Conservation District (CAWCD), 95

Central Utah Project (CUP), 38, 75–76, 78, 91

Central Valley Project, 119

Chaffey, George, 12

Chandler, Harry, 20, 117, 130, 180, 181, 182, 184

Chandler (Ariz.), 236

Chapman, Oscar, 66

Cheesman, Walter S., 123

Cheesman Dam, 123–24

Chemehuevi, 156, 160

Cherry Creek, 122

China, 208

Chinese, on Mexicali Valley farms, 181, 182

chlorofluorocarbons (CFCs), eliminating use of, 96

chub, humpback, 63, 103–4

Churchill, Edward, 44–45

Citizens for All-American Canal Safety, 191

Citizens United for Resources and Environment, 191

Clean Colorado River Alliance, 103

Clean Water Act, 77, 89, 95, 168

climate change, x, 5–6, 223, 236

Clinton, Bill, 98

cloud seeding, 74

Coachella Canal, 21, 131, 191, 217

Coachella Valley, 21, 38

Coachella Valley Water District, 21, 100, 213, 214

coal-fired power plants, 74, 159,.166

coal mining, 74, 166

Cocopah, 150, 156, 160, 197

Cold War, 48, 78, 124, 188–89

Colorado, 48, 62, 74, 155, 222, 224; Animas–La Plata Project, 161, 168–69; Bureau of Reclamation projects in, 16, 53; Colorado River allocation, 17–19; irrigated agriculture in, 33, *34*; oil shale, 77–78; selenium contamination, 102–3; water projects for, 73, 91; water use, 215, 237–38

Colorado–Big Thompson Project, 23, 124

Colorado Cattleman's Association, 40

Colorado Desert, irrigation for, 11–12

Colorado Development Company (CDC), 12–13

Colorado River Aqueduct, 115, 119

Colorado River Association of California, 47

Colorado River Delta Water Trust, 197, 217

Colorado River Indian Reservation, 151, 156, 160

Colorado River Basin Project Act, 74

Colorado River Basin Salinity Control Act, 77, 89, 190, 234

Colorado River Basin Management Study, 100–101

Colorado River Commission, 133, 134

Colorado River Compact, 18–19, 20, 37, 42, 78, 134, 210, 234, 237; Indian rights and, 154, 163; Mexico and, 183, 184

Colorado River Development Commission, 133

Colorado River Governance Initiative, 197

Colorado River Land Company (CRLC), 130, 180, 181–82, 185

Treaty of Guadalupe Hidalgo, 7, 117, 179
Tres Rios Constructed Wetlands Demonstration
    Project, 126
trout, and humpback chub recovery, 103
Tucson, x, 47, 95, 215, 237; Central Arizona
    Project, 113–14; water sources, 119–21, 136, 236
Tucson Water and Sewage Department, 120
Tucson Water Company, 120
Turner, Frederick Jackson, 10
TVA. *See* Tennessee Valley Authority
typhoid, 123

Udall, Morris, 77; on Central Arizona Project, 69,
    70, 72, 158; and Sierra Club, 72–73
Udall, Stewart, 76, 156, 189; and Central Arizona
    Project, 70–71, 159, 172(n43); *The Quiet Crisis*,
    69
Ulloa, Francisco de, 6
Uncompaghre River Project, 16
Union Pacific Railroad, 134
United Nations, 193; environmental issues, 88,
    95–96
United Nations Conference on Environment and
    Development (Earth Summit), 88, 96
US Air Force Academy, 124
US Congress, 10, 77, 157; Boulder Canyon Bill,
    19–20; and Central Arizona Project, 48, 71,
    73–74; Navajo-Gallup Water Supply Project,
    163–64
US Department of Energy, 79
US Department of the Interior, 15, 66, 239;
    irrigated agriculture, 38–39; selenium studies,
    101–2; water policies, 98–99; on water short-
    ages, 92–93; water transfers, 213–14
US Fish and Wildlife Service, 99, 101–2, 103,
    193–94
US Geological Survey, 15, 103, 152; on selenium
    pollution, 101–2
US-Mexico border: cooperation, 194–98; delta
    preservation, 193–94; water salinity issues,
    187–90; water use, 179–80, 190–91
United States–Mexico Transboundary Aquifer
    Assessment Act, 196
US-Mexico Water Treaty, 96
US State Department, 183
United States Reclamation Service, 15. *See also*
    Bureau of Reclamation

US Supreme Court: *Arizona v. California*, 41,
    42, 46, 48–51; Colorado River water, 22, 156;
    Winters case, 153–54
University of Arizona, Institute of the
    Environment, 164
University of Nevada, Las Vegas, water use survey,
    138–39
upper basin: Colorado River Compact, 18, 19, 215,
    224; Colorado River Storage Project, 66–68;
    water projects, 51–54; water rights, 155–56. *See
    also various projects; states*
Upper Colorado River Basin Compact, 51–52, 156
Upper Colorado River Basin Storage Act, 44
Upper Colorado River Basin Storage Project,
    52–53
Upper Colorado River Endangered Fish
    Recovery Program, 103
uranium, 78, 164
Uranium Reduction Company, 78–79
urban areas, 113–14, 141(n27), 242; water market-
    ing, 215, 216; water rights transfers, 100, 127;
    water use, 98, 136–39. *See also various cities*
Utah, 7, 48, 53, 74, 164, 238; Central Utah Project,
    75–76, 91; Colorado River allocation, 17–19,
    51; energy development, 77, 78; irrigated
    agriculture in, 32, 34–35, 38
Ute Mountain Utes, 160
Ute Nations: Animas–La Plata Project, 159, 160,
    161, 168–69
utility companies, 208

Valley Forward, 137
Vanderhoof, John D., 77–78
vegetable production, x, 131
Verde River, 126, 161
Volcano Lake, 181

Waddell Dam, 126
Walker, William, 180
Walter B. Griffith Project, 134
Warren, Francis E., 11
wastewater treatment, 131, 234; recycling, 209,
    235–36, 237; Tijuana, 128–30, 195
water banks, 209, 222, 224, 237
water delivery projects, Navajo and Hopi reserva-
    tions, 165, 166
water management plans, basin-wide, 196